An Investigation into the Detection and Mitigation of Denial of Service (DoS) Attacks

S.V. Raghavan and E. Dawson
Editors

An Investigation into the Detection and Mitigation of Denial of Service (DoS) Attacks

Critical Information Infrastructure Protection

Springer

Editors
S.V. Raghavan
Department of Computer Science
and Engineering
Indian Institute of Technology Madras
Chennai
India
svr@cs.iitm.ernet.in

E. Dawson
Information Security Institute
Queensland University of Technology
Brisbane
Australia
e.dawson@qut.edu.au

ISBN 978-81-322-0276-9 e-ISBN 978-81-322-0277-6
DOI 10.1007/978-81-322-0277-6
Springer New Delhi Dordrecht Heidelberg London New York

Library of Congress Control Number: 2011937969

Printed on acid-free paper

Springer is part of Springer Science+Business Media (www.springer.com)

Preface

Over the last decade, Denial of Service (DoS) and Distributed Denial of Service (DDoS) attacks have emerged as one of the greatest threats to communications networks such as the Internet. The scale and sophistication of DoS attacks extends to critical infrastructure such as water and electricity supply which increasingly use the Internet as a communications channel.

This publication reports on a collaborative research project into DoS and DDoS attacks undertaken jointly by a consortium comprising the Society for Electronic Transactions and Security (SETS) (India), the Indian Institute of Technology (IIT) (Madras), Anna University (Chennai), the National Institute of Technology (NIT) (Tiruchirapalli), Thiagarajar College of Engineering (TCEC) (Madurai), PSG College of Technology (PSG) (Coimbatore) and Queensland University of Technology (QUT) (Australia).

Various technical aspects related to the detection and mitigation of DoS and DDoS attacks are presented. Possible vulnerabilities in emerging technologies such as service-oriented architectures and IPv6 to DoS attacks are highlighted. In addition, policy and legal dimensions of DoS and DDoS attacks are discussed.

We acknowledge the contributions to this research from the main investigators listed as authors of chapters.

Chennai, India
Brisbane, Australia

S.V. Raghavan
E. Dawson

List of Authors
(in alphabetical order)

Anna University, Chennai, India

- C. Chellapan (drcc@annauniv.edu)
- H. Narasimhan (nhari88@gmail.com)
- V. Varadarajan (venk1989@gmail.com)

Indian Institute of Technology Madras, Chennai, India

- S.P. Meenakshi (spmeena@cse.iitm.ac.in)
- C. Pandu Rangan (prangan@iitm.ac.in)
- S.V. Raghavan (svr@cs.iitm.ernet.in)
- B. Ravindran (ravib@iitm.ac.in)

PSG College of Technology, Coimbatore, India

- A. Nadarajan (anitha_nadarajan@mca.psgtech.ac.in)

Queensland University of Technology, Brisbane, Australia

- E. Ahmed (e.ahmed@qut.edu.au)
- S. Bhatia (s.bhatia@qut.edu.au)
- C. Boyd (c.boyd@qut.edu.au)
- W.J. Caelli (w.caelli@qut.edu.au)
- A. Clark (a.clark@qut.edu.au)
- E. Dawson (e.dawson@qut.edu.au)
- J. Georgiades (jenny.georgiades@qut.edu.au)
- J. Gonzalez-Nieto (j.gonzaleznieto@qut.edu.au)
- H. Liu (hua.liu@connect.qut.edu.au)
- G. Mohay (g.mohay@qut.edu.au)
- S. Panichprecha (s.panichprecha@ist.qut.edu.au)
- D. Schmidt (schmidda@qut.edu.au)
- J. Smith (j4.smith@qut.edu.au)

- D. Stebila (stebila@qut.edu.au)
- S. Suriadi (s.suriadi@qut.edu.au)
- A.B. Tickle (ab.tickle@qut.edu.au)

Society for Electronic Transactions and Security, Chennai, India

- S.M. Bhaskar (smb@nic.in)
- L. Kuppusamy (l.kuppusamy@qut.edu.au)[1]
- J. Rangasamy (j.rangasamy@qut.edu.au)[1]
- R. Vijayasarathy (vijayasarathy@setsindia.net)[2]

Thiagarajar College of Engineering (TCE), Madurai, India

- S.M. Shalinie (shalinie@tce.edu)

[1]Currently enrolled as a PhD student at the Queensland University of Technology, Brisbane, Australia.

[2]Currently enrolled as a PhD student at the Indian Institute of Technology Madras (IITM), Chennai, India.

Contents

List of Acronyms

ACL	access control list
ACMA	Australian Communications and Media Authority
ACS	Australian Computer Society
ADDP	Australian Defence Doctrine Publication
ADF	Australian Defence Force
AFP	Australian Federal Police
AGD	Australian Government Attorney-General's Department
AGIMO	Australian Government Information Management Office
AHTCC	Australian High Tech Crime Centre
AISRF	Australia-India Strategic Research Fund
ANFIS	Adaptive Neuro-Fuzzy Inference System
ANN	artificial neural network
APEC	Asia-Pacific Economic Cooperation
ARP	address resolution protocol
AusCERT	Australian Computer Emergency Response Team
AWC	abnormal window count
BS	base station
BTS	base transceiver station
CANFIS	Cumulative-sum-based Adaptive Neuro-Fuzzy Interface System
CC	Common Criteria
CCRA	Common Criteria Recognition Arrangement (International)
CERT	Computer Emergency Response Team
CERT-Australia	CERT Australia
CERT-In	CERT India
CIAC	Critical Infrastructure Advisory Council
CIP	Critical Infrastructure Protection
CIR	Critical Infrastructure Resilience
CNI	Critical National Infrastructure
COI	Community of Interest (Australia)
CORBA	common object request broker architecture
CPA	Change Point Analysis

CSOC	Computer Security Operations Centre
CUSUM	cumulative sum
DAD	duplicate address detection
DARPA	Defense Advanced Research Projects Agency
DBCDE	Department of Broadband, Communications and the Digital Economy (Australia)
DDoS	distributed denial of service
DEC	Digital Equipment Corporation (USA)
DFAT	Department of Foreign Affairs and Trade (Australia)
DHS	Department of Homeland Security (USA)
DIA	Defence Intelligence Agency (India)
DII	Defence Information Infrastructure
DIT	Department of Information Technology (India)
DMM	DoS Mitigation Module
DNS	Domain Name System
DNSSEC	Domain Name System – Security Extensions
DOM	document object model
DoS	Denial of Service
DoT	Department of Telecommunications (India)
DPMC	Department of Prime Minister and Cabinet (Australia)
DRDO	Defence Research and Development Organisation (India)
DSD	Defence Signals Directorate (Australia)
DSTO	Defence Science and Technology Organisation (Australia)
DTS	Department of Telecommunication Services (India)
EC	European Community
EMP	Electromagnetic Pulse
ENISA	European Network and Information Security Agency
EU	European Union
EW	electronic warfare
FDI	Foreign Direct Investment (India)
FIPB	Foreign Investment Planning Board (India)
FIRST	Forum for Incident Response and Security Teams (International)
FIS	Fuzzy Inference System
FPGA	field-programmable gate array
FRED	Fair Random Early Detection
GII	Global Information Infrastructure
GoI	Government of India
GOSIP	Government Open Systems Interconnection Profile
GovCERT.au	Australian Government Computer Emergency Readiness Team (now absorbed into CERT-Australia)
GSM	Global System for Mobile Communications
HF	high frequency
HMM	hidden Markov model
HTTP	hyper-text transfer protocol

IA	Information Assurance
IANA	Internet Assigned Numbers Authority (International)
IB	Central Intelligence Bureau (India)
IBM	International Business Machines Inc. (USA)
ICANN	Internet Corporation for Assigned Names and Numbers
ICERT	Indian Computer Emergency Response Team (India)
ICMP	Internet control message protocol
ICMPv6	Internet control message protocol version 6
ICS	industrial control system
ICT	Information and Communication Technology
IDL	interface definition language
IDS	intrusion detection system
IETF	Internet engineering task force
IFIP	International Federation for Information Processing
IIA	Internet Industry Association (Australia)
IIT	Indian Institute of Technology (Madras)
IKE	Internet Key Exchange
INFOSEC	Information Security
IO	Information Operations
IP	Internet protocol
IPS	intrusion prevention system
IPSA	Ionospheric Prediction Service Australia
IPsec	Internet protocol security
IPv4	Internet protocol version 4
IPv6	Internet protocol version 6
ISAKMP	Internet Security Association and Key Management Protocol
ISP	Internet Service Provider
IT	Information Technology
ITSEAG	Information Technology Security Expert Advisory Group (Australia)
ITU-T	International Telecommunications Union
JCB	Joint Cipher Bureau (India)
JFK	Just Fast Keying
LOIC	Low Orbit Ion Cannon
MALT	Mutual Legal Assistance Agreement
MIP	mobile Internet protocol
MTU	maximum transmission unit
NA	neighbour advertisement
NAT	network address translation
NATO	North Atlantic Treaty Organization
NB	Naive Bayesian
NBN	National Broadband Network (Australia)
NCI	National Critical Infrastructure
ND	neighbour discovery
NIC	National Informatics Centre (India)

NII	National Information Infrastructure
NIIP	National Information Infrastructure Protection
NIPRNet	Non-classified Internet Protocol Router Network (USA)
NIST	National Institute of Standards and Technology
NIT	National Institute of Technology (Tiruchirapalli)
NS	neighbour solicitation
NSA	National Security Agency (USA)
NSF	National Science Foundation (USA)
NSP	Not Seen Previously
NSST	National Security Science and Technology (Australia)
NTP	National Telecom Policy (India)
NUD	neighbour unreachability detection
OASIS	Organization for the Advancement of Structured Information Standards
OECD	Organisation for Economic Cooperation and Development
OEEC	Organisation for European Economic Co-operation
OSI	Open Systems Interconnection
PC	Personal Computer
PCA	Principal Component Analysis
PCO	Public Call Office (India)
PLC	programmable logic controller
PMC	Department of the Prime Minister and Cabinet (Australia)
PPM	Probabilistic Packet Marking
PSG	PSG College of Technology (Coimbatore)
QoS	quality of service
QUT	Queensland University of Technology (Australia)
RA	router advertisement
RAAF	Royal Australian Air Force
RAN	Royal Australian Navy
RBF	Radial Basis Function
RED	random early detection
RFC	request for comment
RMI	remote method invocation
ROC	Receiver Operating Characteristic
RPC	remote procedure call
RS	router solicitation
RST	request security token
RSTR	request security token response
RTU	remote terminal unit
SCADA	supervisory control and data acquisition
SCADA CoI	SCADA community of interest
SETS	Society for Electronic Transactions and Security (India)
SOA	service-oriented architecture
SOAP	simple object access protocol
SSAC	Security and Stability Advisory Committee

SSE	Smart and Secure Environment
SSH	secure shell
SSL	Secure Socket Layer
SVM	support vector machine
TCE	Thiagarajar College of Engineering (Madurai)
TCP	transmission control protocol
TCP/IP	transmission control protocol/Internet protocol
TCSEC	Trusted Computer System Evaluation Criteria (USA)
TDSAT	Telecommunications Dispute Settlement and Appellate
TISN	Trusted Information Sharing Network
TRA	traffic rate analyser
TRAI	Telecom Regulatory Authority of India (India)
TS	token service
TTL	time to live
UDDI	universal description, discovery and integration
UDP	User Datagram Protocol
UNCITRAL	United Nations Commission on Trade Related Laws
UNCLOS	United Nations Convention on the Law of the Sea
UNESCO	United Nations Educational, Scientific and Cultural Organization
URL	Uniform Resource Locator
W3C	World Wide Web Consortium
WAT	window arrival time
WCF	Windows Communication Foundation
WSDL	web service description language
WSI	Web Services Interoperability organization
WSS	Web services SOAP security
WWW	World Wide Web
XML	eXtensible Markup Language

Chapter 1
Introduction

S.V. Raghavan and E. Dawson

Contemporary society has grown increasingly reliant on information and the systems used to store, process, and communicate that information. Consequently very few aspects of modern-day life would continue to operate smoothly in the absence of functioning information and communications systems. This increasing societal dependence on information and communications technologies in general and communications networks in particular is most obvious when the delivery of services via these systems and networks is disrupted even for relatively short periods. Such situations in which access to networked services by legitimate customers or clients is deliberately disrupted are collectively categorised as 'denial of service' or DoS attacks. These are the subject of the ensuing discussion.

Whilst the Morris worm severely disrupted the nascent Internet as it existed in 1988, the historical record places the first reported large-scale and deliberate Denial of Service (DoS) attack via the public Internet as occurring back in August 1999 at the University of Minnesota [2,7]. Very soon after that incident, in February 2000, a group of popular e-commerce websites, i.e. Yahoo.com, cnn.com and eBay.com, also suffered from a denial of service attack. This was the first example of a deliberate large-scale attack directed at rendering commercial sites unreachable from the Internet [4]. It was also, arguably, the first example of a DoS attack causing significant and direct financial loss and was an omen for what was to follow.

In the decade since those initial incidents, the scale, sophistication and frequency of denial of service attacks have increased dramatically. Currently, one particularly potent form of attack is to harness hundreds of thousands (if not millions) of

S.V. Raghavan (✉)
Department of Computer Science and Engineering, Indian Institute of Technology Madras, Chennai, India
e-mail: svr@cs.iitm.ernet.in

E. Dawson
Information Security Institute, Queensland University of Technology, Brisbane, Australia
e-mail: e.dawson@qut.edu.au

S.V. Raghavan and E. Dawson (eds.), *An Investigation into the Detection and Mitigation of Denial of Service (DoS) Attacks: Critical Information Infrastructure Protection*,
DOI 10.1007/978-81-322-0277-6_1, © Springer India Pvt. Ltd. 2011

attacker-controlled computers (also known as bots). Under the direction of a botnet controller and using sophisticated command-and-control structures, such bots are orchestrated so that collectively they direct tens of gigabits of Internet traffic against a selected target. This type of attack is commonly known as a distributed denial of service (DDoS) attack, and withstanding such attacks necessitates a dramatic increase in the scale and complexity of resources deployed by the target.

The problem of denial of service is now so pervasive that virtually every computer connected to a network has the potential to be affected in some way. This even extends to computer and network systems that underpin basic infrastructure services including water, sewage, electricity, train operations,[1] safety monitoring systems at nuclear power plants[2] and the operation of bank ATMs.[3] As things stand, DoS and DDoS attacks present significant challenges to the continued use of the Internet for critical communications.

Since the very first DoS attack, analysts have expended considerable effort in endeavouring to fathom the motivation for such attacks. As expected there is no single or even dominant motivator. Over time, DoS attacks have been used to extract revenge, gain prestige, as well as for monetary ends [1]. A recent trend is, ostensibly, the use of DoS attacks for political ends [5] with the goal of conveying political messages or disrupting national critical infrastructures. Attackers have also extended DoS attacks to cellular telecommunication networks. Several new kinds of DoS attacks, such as user de-registration request spoofing, location update request spoofing and fake base transceiver station (BTS)-based DoS attacks are becoming prevalent.

At the outset, the security community did not acknowledge that DoS was an explicit security concern in its own right. Instead it was believed DoS attacks resulted from a failure to adequately address integrity concerns [3]. However, this perception soon changed once the threat posed by DoS and DDoS attacks to the orderly functioning of computer networks and to the services they deliver was more clearly appreciated. Nowadays the increasing frequency and scale of DoS and DDoS attacks over computer networks and the associated strain on the supporting communications infrastructure leaves no doubt that DoS is a serious and legitimate concern that must be addressed.

Given the pernicious nature of the threat posed by DoS attacks and the direct impact of such attacks on the ability of organisations to deliver services to their clients and customers, an increasing number of commercial and governmental organisations are now more willing to invest considerable resources to addressing the problem. This can take a variety of forms including having access to significant reserves of telecommunications bandwidth and server processing capacity. In fact a number of commercial organisations provide DoS and DDoS detection and mitigation as a service. Their business model is built on ensuring continuity of

---

[1] http://catless.ncl.ac.uk/Risks/23.35.html#subj10.1

[2] http://catless.ncl.ac.uk/Risks/22.88.html#subj5

[3] http://catless.ncl.ac.uk/Risks/22.52.html#subj7.1

service between client and legitimate customers. The recent success of Amazon in fending off a DoS/DDoS attack highlights the direct benefit from the (considerable) investments in upgrading their infrastructure to make it more resilient to attack.[4]

A decade of experience in dealing with DoS and DDoS attacks suggests that such attacks come in many guises, are continuously evolving and that the problem is far from being conclusively solved. Consequently, investigation into DoS and DDoS attacks continues to be an important and active area of research.

The aim of this publication is to report on a collaborative research project into DoS and DDoS attacks undertaken jointly by a consortium comprising Society for Electronic Transactions and Security (SETS) (India), Indian Institute of Technology (IIT) (Madras), Anna University (Chennai), National Institute of Technology (NIT) (Tiruchirapalli), Thiagarajar College of Engineering (TCE) (Madurai), PSG College of Technology (PSG) (Coimbatore) and Queensland University of Technology (QUT) (Australia). This publication offers insight into the complexity of the problem to be solved as well as the breadth of the research being conducted into the various facets of the DoS/DDoS problem.

Chapter 2 sets the context for the subsequent discussion by undertaking a detailed technical review of DoS and DDoS attacks. An appreciation of the number of different vectors that have been exploited to launch such attacks can be gleaned from the number and size of the DoS/DDoS taxonomies that have appeared in recent years. The chapter also examines the architecture of Botnets, which, as mentioned previously, are a particularly pernicious manifestation of DDoS attack scenarios. The remainder of Chap. 2 addresses the task of responding to a DoS/DDoS attack. Unlike confidentiality and integrity which are both supported by a wide range of formal tools, models and techniques, approaches for addressing availability requirements (in the presence of a malicious and intelligent adversary) are still embryonic. To date, a substantial part of the research effort in the DoS/DDoS arena has been directed towards the basic tasks of reliably detecting the onset of an attack and then mitigating its effect. Chapter 2 provides an overview of that research. Chapter 2 also reviews a set of pertinent case studies which have been selected because they exemplify a specific facet of the problem.

As indicated previously, the set of perpetrators of deliberate DoS/DDoS attacks is potentially very large, international and technically capable. Also their motives range across the full gamut of vices: revenge, prestige, politics, and money. However, and notwithstanding the work that has been done as outlined in Chap. 2, there are limits to what can be achieved solely as a technical response. Over time it has become increasingly clear that, even if armed with knowledge of the identities and motives of potential perpetrators, there is very limited scope for an individual organisation to mitigate the impact of the specific case of a high-rate flooding attack through anticipation or through pre-emption at its primary source. In addition, other forms of deliberate attack are increasingly occurring, e.g. prevention of access to the Internet's 'Domain Name System (DNS)', etc.

---

[4] http://money.cnn.com/2010/12/09/technology/amazon_wikileaks_attack/index.htm?source=cnn_bin&hpt=Sbin

Chapter 3 broadens the discussion of the DoS/DDoS problem by examining it through the dimensions of policy, both public and private, and law. A key focus of the discussion is on the obligations imposed on telecommunications carriers to provide secure delivery of telecommunications services to the end user as well as defining the role of ISP's in the security of the Internet and the 'digital economy'. This review extends to identifying standard, minimum requirements for suppliers and manufacturers of ICT products, systems and services. The question is posed as to whether or not by shipping products, such as software that has known and inherent vulnerabilities, makers of such products become liable for attacks that occur via those products. Chapter 3 also discusses the relationship between corporate governance and ICT and the existence of a duty of care owed by directors in ensuring secure IT practices as well as the obligations imposed on users of the infrastructure in relation to DoS and protection of their infrastructure. Chapter 3 also delves into the entirely problematic question as to how DoS attacks should be investigated, by whom and under what legal directives, legislation or like instruments. The chapter also reviews the legal mechanisms for the enforcement of recovery from any breaches, particularly where such breaches may propagate within an economy. It also discusses the vexed problem of determining who is responsible for any liability or loss that occurs. The chapter also reviews relevant topics in the area of so-called information or cyber warfare where DoS takes on a national security concern. Relevant legislation is assessed to check if they offer guidance on how to approach the issue of vulnerability and DoS/DDoS attack at different levels as well as discussing the lessons from the seminal Maroochy Shire, 'Queen v Boden' case in Queensland, Australia. In India, the role and function of the 'IT Act, 2000', as recently amended, is also examined as a major input into the discussion as well as an analysis of the situation governing 'Internet Service Providers (ISP)' in India.

An integral part of any research into the DoS/DDoS problem is the capability to conduct experiments in order to demonstrate proof-of-concept of ideas and strategies. Chapter 4 discusses the design, development and implementation of the testbeds that were developed to underpin the technical dimension of the research. It begins with a review of existing testbed architectures and associated design strategies, and discusses the challenges in emulating the complexity and scale of even small subsets of the Internet environment under experimental laboratory conditions. Although the Indian and Australian testbeds were developed separately to emulate different aspects of network structure, they still share many software components, such as the generation of emulated DDoS traffic using many discrete IP-addresses from a small number of physical machines, and the measurement of the effects of an attack on individual services. The setting up and execution of individual experiments, however, differs on the two platforms, because the larger Indian testbed was conceived as a resource to be shared by many institutions in Tamil Nadu, whereas the Australian testbed is a dedicated local resource.

Chapter 5 is the first in a set of chapters that cover the key technical aspects of the research into the detection and mitigation of DoS and DDoS attacks and in particular high-rate flooding attacks. It begins with a detailed review of techniques for the real-time detection of anomalous traffic that is characteristic of a DDoS event. It highlights that investigations into detection techniques have drawn upon disciplines as diverse as machine learning, probability theory, statistics, pattern recognition, data mining and adaptive control. The chapter emphasises that, to date, no single classification technique has emerged that consistently outperforms all others under the highly variable traffic conditions in the Internet. One promising strategy is to operate with an ensemble of classifiers each tuned to a particular nuance that could signify the presence of anomalous traffic and hence a DDoS event. In keeping with this, members of the research consortium have investigated different but complementary kinds of anomaly-detection technique. For example, the QUT (Australia) research team has developed a technique based on their previous research work on Change Point Analysis (CPA) that had proven successful in identifying anomalous traffic events. This technique uses the Not Seen Previously (NSP) algorithm to detect the onset of a DDoS attack, exploiting the fact that a specific characteristic of DDoS attacks is that such attacks are closely associated with the appearance of a large number of new users (i.e. IP addresses) at the target site. On the other hand, the IIT-Madras (India) research team has developed an innovative and efficient machine learning approach, one that is suitable for hardware implementation, for classifying TCP and UDP flood attacks. Chapter 5 presents the results obtained in each case and discusses future prospects.

Central to the strategy of mitigating the impact of a DoS attack is restricting connections only to authorised or legitimate users. This requires mechanisms that can differentiate legitimate users from others who may be participating in the DoS attack. Authentication can help identify legitimate users and easily drop connections from non-legitimate users, potentially reducing denial of service threats. However, most conventional forms of cryptographic authentication, such as digital signatures, involve computationally expensive verification operations, causing the authentication protocol to itself become a denial of service vulnerability. Client puzzles are a form of lightweight authentication that typically have low cost verification. Chapter 6 provides an in-depth discussion of cryptographic techniques for lightweight authentication using client puzzles and other fast verification techniques. These authentication techniques can be combined to achieve the principle of *gradual authentication*, allowing servers to commit only limited resources before obtaining some level of assurance regarding a client's identity.

In Chap. 7, the focus of discussion shifts to the protection of service-oriented architectures (SOAs) (implemented using web services) from DoS and DDoS attacks. Basically, SOAs, implemented using *web services*, seek to use open and interoperable standards to facilitate easier enterprise application integration, to provide application flexibility and to facilitate the dynamic composition of applications from disparate services. Despite their role in delivering sophisticated web applications, as discussed in Chap. 7, web services are often poorly protected from DoS attacks [6, Table 2-2, Sect. 2.6, pp. 2-17]. Chapter 7 discusses various points of DoS vulnerabilities in web services technologies. This is followed by a detailed description of a set of experiments to validate the identified points of DoS vulnerabilities and a proposal of a general architecture that can be used to mitigate DoS attacks in web services. Finally, the chapter details a specific proposal which applies cryptographic client-puzzle techniques (discussed in Chap. 6) to mitigate DoS attacks in a web services environment by requiring clients to expend some resources to solve a puzzle first before access to the web service is granted. In particular, we detail our work in integrating client puzzles into a web services application built upon the Microsoft .NET Windows Communication Foundation framework and a set of experiments to analyse the effectiveness of client puzzles in mitigating DoS attacks under various scenarios.

ICT technologies and particularly data communications technologies evolve both to address limitations in the existing technology as well as to provide additional functions that can underpin enhanced services to the online community. Whilst such developments in technology may also eliminate certain known vulnerabilities that can be exploited in a DoS/DDoS attack, it is common for any new technology to come with its own inherent set of new vulnerabilities. These can, in turn, be exploited to create an entirely new set of DoS/DDoS attacks. Chapter 8 focuses on one such new technology namely the new version of the IP protocol (IPv6) designed to replace the existing version IPv4. The purpose of Chap. 8 is to explore the vulnerabilities of IPv6 in more detail and, in particular, those vulnerabilities that could be exploited in a DoS attack. The rationale for selecting IPv6 goes beyond its role as the designated successor for IPv4 in the Internet. The ubiquity of the TCP/IP suite of protocols (of which IPv4 is an integral part) and their widespread availability in commercial-off-the-shelf (COTS) products means that the impact of the exploitation of vulnerabilities in the Internet protocol version 6 (IPv6) has the potential to extend well beyond those of the Internet. One such domain, and one that is the subject of discussion in Chap. 8, involves computer and network-based systems that are used to monitor and control critical infrastructure particularly in utilities such as energy and water. The final section of the chapter discusses this issue and examines the efficacy of tools to model and review risks emanating from the propagation of identifiable vulnerabilities in IPv6 deployment in such monitoring and control systems.

Overall, this work offers a timely and important contribution to the problem of DoS and DDoS attacks and highlights the merits in developing solutions collaboratively across a consortium of research institutions.

References

1. Carl, G., G. Kesidis, R.R. Brooks, and S. Rai. 2006. Denial-of-service attack – Detection techniques. *IEEE Internet Computing* 10(1): 82–89.
2. Garber, L. 2000. Denial-of-service attacks rip the internet. *Computer* 33(4): 12–17.
3. Gligor, V.D., M. Blaze, and J. Ioannidis. 2000. Denial of service – Panel discussion. In *Security Protocols Workshop, Lecture notes in computer science*, vol. 2133, 194–203. Springer, Berlin.
4. Grice, C. 2000. How a basic attack crippled yahoo. http://news.cnet.com/2100-1023-236621.html. Accessed 16 Feb 2011.

5. Nazario, J. 2011. Political DDoS: Estonia and beyond (invited talk). In *17th USENIX Security Symposium*, San Jose, July 2008. http://streaming.linux-magazin.de/events/usec08/tech/archive/jnazario/. Accessed 16 Feb 2011.
6. Singhal, A., T. Winograd, and K. Scarfone. 2007. *Guide to Secure Web Services – Recommendations of the National Institute of Standards and Technology*. National Institute of Standards and Technology, Gaithersburg.
7. Zuckerman, M.J. 2000. How the government failed to stop the world's worst internet attack. *USA Today*, p. 1A.

Chapter 2
Background

A.B. Tickle, E. Ahmed, S.M. Bhaskar, G. Mohay, S. Panichprecha, S.V. Raghavan, B. Ravindran, D. Schmidt, and S. Suriadi

In Chap. 1, we have introduced the serious consequences that a Denial of Service (DoS) attack could pose on our society which is increasingly reliant on information and the systems used to store, process, and communicate that information. However, the DoS problem has various dimensions and definitions.

In this chapter, we introduce the concept of Denial of Service (DoS) and how it relates to the problem of distributed denial of service (DDoS). The aims of this chapter are to allow readers to appreciate the complexity of the DoS/DDoS problem and to understand why this problem affects virtually every system connected to a network. Furthermore, this chapter also attempts to give readers an overview of the research that has been conducted in both detecting and mitigating DoS attacks.

In Sect. 2.1, an overview of the concept of DoS is provided, followed by an explanation of the various types of DDoS attacks in Sect. 2.2 to demonstrate the multi-dimensional nature of DDoS and the prevalence of DoS problem in various layers of computer and network systems. An explanation of botnet (which is one of the key enablers in launching a DDoS attack) and its architecture is provided in Sect. 2.3.

In Sect. 2.4, we provide a review of the various DoS/DDoS detection techniques that have been developed. The primary focus is on those techniques that are capable

G. Mohay (✉) • E. Ahmed • S. Panichprecha • D. Schmidt • S. Suriadi • A.B. Tickle
Information Security Institute, Queensland University of Technology, Brisbane, Australia
e-mail: g.mohay@qut.edu.au; e.ahmed@qut.edu.au; s.panichprecha@isi.qut.edu.au; schmidda@qut.edu.au; s.suriadi@qut.edu.au; ab.tickle@qut.edu.au

S.M. Bhaskar
Society for Electronic Transactions and Security, Chennai, India
e-mail: smb@nic.in

S.V. Raghavan • B. Ravindran
Department of Computer Science and Engineering, Indian Institute of Technology Madras, Chennai, India
e-mail: svr@cs.iitm.ernet.in; ravib@iitm.ac.in

S.V. Raghavan and E. Dawson (eds.), *An Investigation into the Detection and Mitigation of Denial of Service (DoS) Attacks: Critical Information Infrastructure Protection*, DOI 10.1007/978-81-322-0277-6_2, © Springer India Pvt. Ltd. 2011

of identifying anomalies in traffic volume in real time. As we will see, a number of powerful and sophisticated techniques drawn from the realm of time-series analysis are currently being used in this area.

Section 2.5 details the various mitigation techniques that have been proposed. Importantly, this section illustrates the ongoing difficulty and challenges in mounting an effective defence to denial of service attacks.

Finally, Sect. 2.6 provides some case studies of denial of service attacks, including how those attacks were detected and the mitigation strategies applied.

The discussion forwarded in this chapter is limited to the technical aspects of the DoS problem. A detailed discussion on the legal and policy issues related to the problem of denial of service is provided in Chap. 3.

2.1 Denial of Service (DoS) and Distributed Denial of Service (DDoS)

Denial of service has many definitions and dimensions. One of the most direct definitions of denial of service, provided by the International Telecommunications Union (ITU-T) recommendation X.800 [27] is as follows:

> *denial of service*: 'The prevention of authorized access to resources or the delaying of time-critical operations'.

The Committee on National Security Systems (CNSS) information assurance glossary definition [65] is more general and identifies denial of service as:

> Any action or series of actions that prevents any part of an [information system] from functioning.

Denial or degradation of service may result from malicious or benign actions. These actions may originate locally or remotely from the service, or user, experiencing denial or degradation of service. The communications bandwidth, memory buffers, computational resources, or the network protocol or application processing logic of the victim, or any systems on which the victim depends for delivering service (the domain name system or credit card payment service for example), may be targeted. The ultimate goal of a DoS attack is to compromise the availability of services.

A distributed denial of service (DDoS) attack is a variant of the more generic DoS attack. The key feature of a DDoS attack is the large number of hosts used to launch such an attack. It is common to see up to hundreds of thousands (if not millions) of hosts being used to launch a DDoS attack. Often, these hosts are nothing more than a set of computers which have been corrupted without the knowledge of their owners (through a Trojan or a backdoor program) such that they can be remotely controlled by a 'master' entity. These hosts are known as 'bots' or zombies, and a collection of such bots under the control of a 'master' entity is known as a 'botnet'. Typically, a bot master launches a DDoS attack by commanding the bots to send some traffic to a specified victim to overwhelm, exhaust, or corrupt the victim's resources. A detailed explanation of DDoS attacks and botnet architecture is provided in Sects. 2.2 and 2.3 respectively.

2.2 Distributed Denial of Service Taxonomy

To combat DDoS attacks, we need to understand the multi-dimensional nature of DDoS attacks. Such an understanding helps us to design the most effective mitigation strategies based on the types of expected DDoS attacks and the types of resources we want to protect in a given environment.

To give readers a flavour of the complexity of the DDoS problem, in this section, we present two 'dimensions' from which the DDoS problem can be viewed: mode of attack and target of attack.[1] By understanding the various modes of DDoS attacks, we can appreciate the need to employ multiple mitigation techniques as there is no 'one solution fits all' option in combating DDoS. Furthermore, by understanding the various targets of DDoS attack, we can appreciate not only the prevalence of the DDoS problem in virtually any network and computer systems, but also the scale of impacts that a DDoS attack can produce depending on which resources it targets.

2.2.1 *Mode of Attack*

The highly interconnected and increasingly critical nature of systems attached to the Internet presents adversaries with a rich source of targets to achieve their aims. An attacker mounting a resource exhaustion–style attack can deny service at a victim system, or at a system on which the victim depends, either via high-rate flooding (brute-force) attacks or via semantic (non-flooding) attacks.

2.2.1.1 High-Rate Flooding Attack

High-rate flooding (or brute-force) attacks aim to pre-emptively consume a critical resource of the victim that will prevent the target from communicating with other network clients. They are typically based on flooding the victim with spurious network packets, but may also target the electromagnetic radio spectrum thereby jamming communications.

Brute force attacks require the attacker to have access to sufficient resources (network bandwidth for example) to overwhelm the victim, or the systems on which

---

[1] Readers who are interested in the details of the taxonomy of DDoS should refer to the paper by Mirkovic and Reiher [44].

it depends. This resource symmetry simplifies detection, making the attacker more conspicuous to network monitors. In the past, the cost of entry, in terms of resources required, to mount such attacks is also high and the attacker must amass the required resources before being able to mount an attack. However, while this may have been true in the past, recently, the proliferation of botnets suggests that amassing such resources is not overly difficult. Once the resources have been acquired, however, attacks are simple to execute.

The resources available to an attacker or attackers (via vulnerable, misconfigured or compromised systems) are likely, for the foreseeable future, to far outstrip the resources available to any legitimate subnetwork or server system seeking to provide services to the Internet community. The creativity and willingness of adversaries to adapt their tactics should not be underestimated and require consideration of more intelligent attack strategies that exploit the semantic aspects of protocols and applications.

A classic example of a DoS flooding attack is the 'TCP SYN flooding attack' [14]. This attack exploits flaws in the three-way-handshake TCP protocol. In a normal TCP connection, a client initiates a connection by sending a TCP SYN packet to the server. The server responds to the SYN packet with a SYN/ACK packet. Then, the client responds with an ACK packet to complete the three-way handshake. In a TCP SYN flooding attack, an attacker sends a large number of half-open TCP connections to the victim. In particular, the attacker sends only SYN packets to the victim host and does not send the ACK (last step in the three-way handshake) to the victim host. The victim keeps the half-open connections open which eventually exhausts the resources on the victim host.

Such high-rate flooding attacks are conceptually very simple and can be easily launched in a small environment (e.g. in a LAN). However, in the case of large, commercial services, servers usually have much more resources compared to a desktop or workstation class computer. Hence, this attack is less likely to succeed. In order to launch the same attack, attackers must build a network of computers whose resource is much larger than a single server. Thus, botnet-based distributed denial-of-service attacks were developed. The details of distributed denial-of-service attacks are discussed below.

A DDoS attacks can be further subdivided into two categories based on how the distributed attack is coordinated: manually, through human coordination, or automated.

Human Coordination of High-Rate Flooding Attack

This type of DDoS attack requires human coordination [61] and a large number of participants to succeed. Examples of such attacks are F5 attack [64], Slashdot effect, and the Estonia incident [61]. In order for the attack to work, the attacker has to orchestrate a large number of users to simultaneously (or continuously for the F5 attack) send service requests to the victim host. Due to the large volume of data targeted at the victim host, the bandwidth, memory, or CPU of the victim host will eventually be exhausted. An F5 attack consists a large group of people simultaneously and continuously holding down the F5 key (page reload key) when all of them open a certain web page on their web browser. An example of the F5 attack incident occurred in 2006 when a group of students launched the attack against their school website [64]. A student created a web page which was hosted on the school server. The student then asked other students to open the page using a web browser and reload the page by holding down the F5 key (reload/refresh). By holding down the F5 key, the web browser repeatedly sends HTTP requests to the web server. Thus, the accumulation of all HTTP requests generates a large volume of data. Another example of a human coordination high-rate flooding DDoS attack is the Slashdot effect [63]. Slashdot (http://www.slashdot.org) is a website which hosts links to science and information technology news headlines. The Slashdot effect occurs when the link to a (victim) website with limited resources (bandwidth and hardware) is posted on the Slashdot website and the link is loaded by Slashdot visitors. Due to the popularity of the Slashdot website, the number of visitors to Slashdot per day is large and thus the request from the visitors could cause the victim web server to go offline. Although the intention of the visitors is legitimate, posting a link in such a way could lead to a DDoS attack.

Automated or Semi-automated High-Rate Flooding Attack

Automated attacks or semi-automated attacks are launched by an attacker who either exploits the weakness in network design or has control over a large number of hosts. Three categories of these types of attacks are amplification, reflection, and botnet. The details of these categories are as follows.

Amplification attacks:

An amplification attack comprises an attacker, an amplifier network and a victim host. Basically, for the purposes of mounting this type of attack, any network of host computers that permits broadcast messages constitutes an amplifier network. The amplification occurs because a single (broadcast) packet can trigger a response packet from each host in the network. The number of response packets can be very large and, since the destination IP address of all these response packets is now that of the intended victim, the total volume of response packets floods the bandwidth of the network to which the victim host is connected. Figure 2.1 illustrates the amplification attack mechanism. To launch the attack, the attacker sends one or more packets with the source IP address of the packets forged to be the IP address of the victim host and the destination IP address set equal to the broadcast IP address for the network on which the victim resides.

Examples of amplification attacks are the Smurf attack [13] and the Fraggle attack [37].

The *Smurf attack* [13] is based on the ICMP echo request (ping command). To launch the attack, the attacker sends an ICMP echo request packet with the

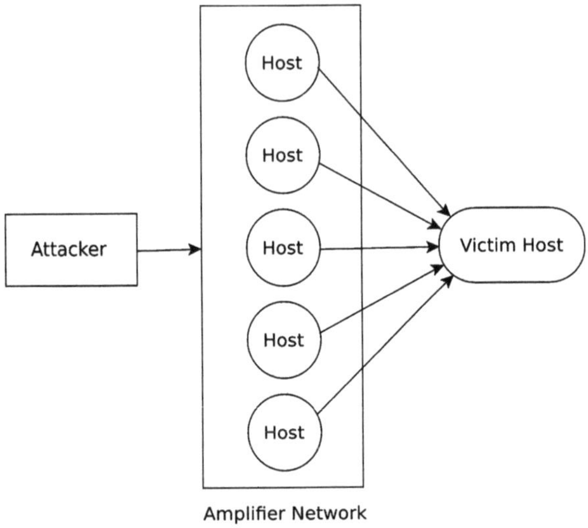

Fig. 2.1 Amplification attack mechanism

destination IP address set to the broadcast address of the (amplifier) network and the source IP address set equal to that of the victim host. Once the ICMP echo request is sent to the broadcast address, all machines in the broadcast domain will send an ICMP echo response packet to the victim host. For instance, in a C class network 192.168.0.0/24, the broadcast address is 192.168.0.255 and there are up to 254 machines in the network. Hence, if the attacker sends an ICMP echo request to the broadcast address, up to 254 machines could respond to the request.

The *Fraggle attack* (or oscillation attack) [37] is similar to the Smurf attack except it uses the UDP protocol instead of the ICMP protocol. The Fraggle attack utilises the UDP echo (port 7) and UDP character generator (port 19) services. The UDP echo service will respond to a packet with the same payload. The UDP character generator service will respond to a packet with a string of characters. To launch the attack, the attacker sends a UDP packet, with a spoofed source IP address of the victim host and source port 7, to the broadcast address where the destination port is port 19. Once the packet arrives at the broadcast address, all hosts will respond with a string of characters to UDP echo on the victim host. Hence, a large volume of data will be bounced between all hosts in the broadcast network and the victim host.

Reflection attacks:

A reflection attack comprises an attacker, a set of reflectors, and a victim host. For the purposes of mounting this type of attack, a reflector is any host which responds to an incoming packet by returning a packet to the source IP address of the packet it receives [51]. Examples of reflectors are web servers, mail servers, and DNS servers.

Figure 2.2 illustrates the reflection attack mechanism. To launch a reflection attack, an attacker first creates a list of IP addresses or DNS names of reflectors,

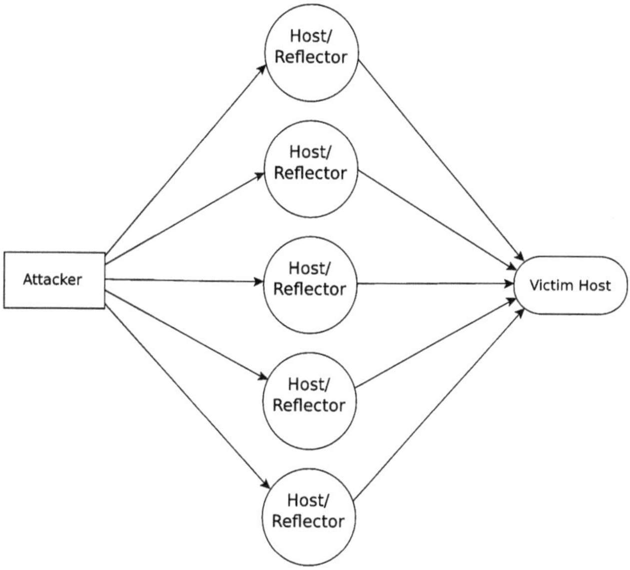

Fig. 2.2 Reflection attack mechanism

for example, any publicly accessible web servers. The attacker then sends packets to the reflectors where the packets have spoofed the source IP address of the victim. Since the reflector will respond to any packet, the reflectors will send responding packets to the victim. Hence, the bandwidth of the victim is flooded through multiple channels from the Internet.

The Reflection attack differs from the Amplification attack discussed previously because in this case the attacker must send an individual packet to each (reflector) host in order to initiate a response. Hence there is no corresponding amplification effect.

Generally, reflection attacks are used to disguise the attacker and to add complexity (packets originated from multiple sources) to other forms of attacks, for example, botnet-based attacks. In some cases, reflectors are used in conjunction with the amplification technique where the attacker sends a packet to the broadcast address of the reflectors, in order to increase the effectiveness of the attack. Such attacks are hard to detect and mitigate.

Botnet-based attacks:

A bot (or zombie) is a computer which has been compromised and is remotely controlled by the computer of an attacker. Botnets are networks of interconnected bots. Botnets are used as a tool by an attacker or a group of attackers to launch high-rate flooding DDoS attacks. In addition, botnets are also used to spread spam or phishing e-mails, distribute malware, and for sniffing network traffic [3].

To launch a botnet-based attack, the attacker may employ one or a combination of any attack techniques that we have discussed above, that is DoS attack, amplification attacks, and reflection attacks. The amplification technique is used to make the DDoS attack even more powerful. The reflection attack technique is used to hide the origin of the address of the attacker or rather the agents. When they occur in combination, such techniques will readily exhaust available network bandwidth so as the attack is much more difficult to mitigate. Consequently, due to the distributed nature of botnets and particularly that the attacks originate from such a large number of individual sources, botnet-based attacks are very hard to prevent and mitigate.

In Sect. 2.3, the details of various botnet architectures are discussed.

2.2.1.2 Semantic Attack

Semantic attacks, or attacks that target the logic or resource allocation strategies of the target, are more difficult to execute than brute force attacks as they require detailed knowledge of the protocol, operating system, or application being targeted. Typically, the resources required by an attacker (in terms of storage, bandwidth, or processing capability) are asymmetric, allowing a relatively weak adversary to deny or degrade service at a much more powerful server. Stated differently, semantic attacks allow the resources available to an adversary to be effectively utilised in disrupting services, that is, a modestly resourced attacker may impact a large number of systems.

A few examples of classic semantic attacks include the ping of death, the land attack, and the teardrop attack. The ping of death [30] is an attack based on the ICMP protocol. The attack involves an attacker sending malformed ping packets to a victim machine. In particular, the malformed ping packets are oversized packets (65,535 bytes) whereas a normal ping packet is 64 bytes. Hence, upon receiving the oversized packets, the victim machine will crash. The *ping of death* attack affected most operating systems. This problem has been fixed in most operating systems released after 1997.

The *land attack* [40] is a remote DoS attack that targets flaws in the implementation of TCP/IP stack in the operating system. The attacker sends a spoofed TCP packet to the victim machine where the values of source and destination IP addresses are the same. When the victim machine receives such a packet, it communicates to itself continuously until the machine crashes.

The *teardrop attack* [12] takes advantage of the vulnerabilities in the software module that is responsible for reassembling fragmented TCP/IP packets. The attacker launches the attack by sending malformed (overlapping or oversized) fragmented TCP/IP packets to the victim machine. The victim machine will crash when it tries to reassemble the fragmented packets, forcing the machine to reboot.

Recent technologies, including web services (discussed in Chap. 7) and IPv6 (discussed in Chap. 8), are not immune against semantic DoS attacks. There have been many semantic attacks against these recent technologies, including the oversized cryptography attack [29], coercive XML parsing attack [29], neighbour advertisement spoofing attack [29], and many others.

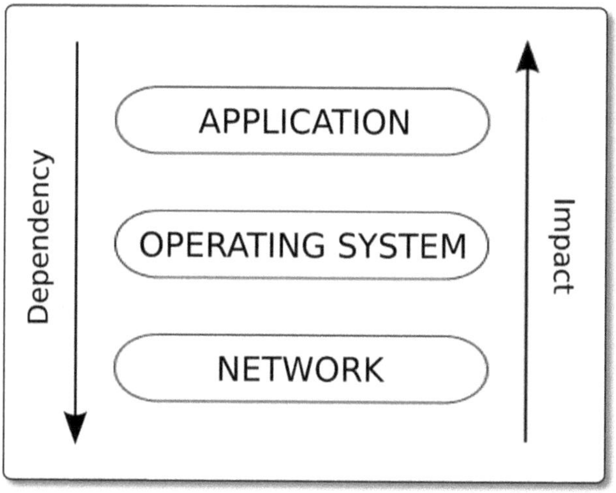

Fig. 2.3 DoS dependency and propagation

2.2.2 *Target of Attack*

Knowing the resources being targeted by a DDoS attack is useful in determining the best allocation of resources to mitigate such an attack. Furthermore, it also helps us appreciate the prevalence of the DDoS problem: that it affects virtually any systems that are connected to a network.

DDoS attacks may target the network, client, server, or router operating systems, or specific client or server applications [48]. These different targets can be viewed as a hierarchy in which attacks at one layer will impact all higher layers (see Fig. 2.3). For example, a DoS (or DDoS) attack targeting network resources may cause a client to be unable to connect to a particular network to access services. Similarly, a DoS attack targeting vulnerability in a victim's operating system could result in legitimate users being unable to access applications that rely on the operating system's services.

2.2.2.1 Attacks Targeting Networks

Current Internet protocols and their implementation present ample opportunity for the maligned to act. In particular, given the reliance of many modern applications on the availability of network services (such as those applications delivered through the web services and cloud computing technologies), an attack on the availability of network services could result in legitimate clients being unable to access those applications.

The reliance on network services in today's computer systems only highlights the wide reach of security threats from the outside world. Most of today's computer systems need some form of network service to function properly, such as receiving program updates, security patches, and activating some programs. The interconnections of many corporate systems to the Internet also mean that those corporate systems are subjected to DoS and DDoS attacks by malicious entities from all over the world. Even the traditionally isolated critical infrastructure systems, such as power plants and water distribution systems, are now increasingly connected to other networks (such as corporate networks), which may eventually expose these systems to threats from the Internet as well. These scenarios show that (1) a denial of network services to legitimate clients can have serious implications and (2) there are not many systems which are immune to denial of service threats due to the highly interconnected system architecture we have nowadays.

In the remainder of this section, we give few examples of the types of DoS/DDoS attacks that target network services.

One of the most commonly used transport-layer protocols is the transmission control protocol (TCP). In TCP, reset (RST) messages can be fabricated and cause the premature termination of a TCP session between two hosts. While this feature of TCP has been long understood, it was only recently that the feasibility of crafting RST packets with the correct values (source and destination IP addresses, source and destination TCP port numbers, and sequence number values) was fully understood [69].

The TCP congestion control algorithms can be abused in a number of ways by malicious and protocol non-compliant peers. For example, as Savage et al. [54] note, a greedy user can modify the TCP protocol implementation to optimistically acknowledge (ACK) segments of data it has not yet received. This has the affect of causing the sender of packets to increase the rate at which it injects traffic into the network, allowing the receiver to gain a greater and unfair access to available bandwidth. Sherwood et al. [55] further investigate the impact that optimistic acknowledgements can have on the network, finding that the attack presents a serious threat to network stability and may even be capable of causing Internet-wide congestion collapse. Such dramatic impacts can be achieved by an attacker with only modest resources owing to potential amplification effects. Based on typical operating systems and TCP parameters, they conclude that an attacker with access to a 56 kilobit per second modem can generate approximately 8.9 megabits per second of traffic. Further, when this technique is directed at distributed targets, it is unlikely to be easily detectable. Highly asymmetric and undetectable attacks must be considered a significant threat to ongoing network stability.

Kuzmanovic and Knightly [31] consider an attacker who exploits the retransmission time-out (RTO) characteristics of TCP flows and discover that a low rate denial of service attack is able to dramatically reduce throughput for a given TCP flow. Attackers using this technique can degrade TCP flows from remote locations, only need to possess limited bandwidth themselves, and will not generate traffic that will be detected by current monitoring systems. Such attack optimisation strategies may be viewed as an effective technique for converting what are essentially brute force attacks, into semantic attacks, highlighting the need for active research into solutions for both attack classes.

At the network layer, source address spoofing, for example, is trivially performed and widely used in denial of service attack packets to impede attack detection and response efforts. Various types of network-layer attack have been shown to be able to bring down large Internet provider sites, such as *eBay, E-Trade*, and *Yahoo*! [20] through the use of ping flooding attack. The *ping of death* attack explained earlier in Sect. 2.2.1.2 is also an example of attack targeting the network layer.

Despite widespread deployment for over 15 years, critical vulnerabilities are still being discovered in TCP [15, 35, 71]. This suggests that the process used to develop and evaluate Internet protocols makes no attempts to systematically identify or remove such vulnerabilities. By inference, the set of known vulnerabilities in emerging protocols such as IPv6 (which have been developed using the process of rough consensus and working code) will increase over time, hence, the necessity to investigate the DoS/DDoS vulnerabilities in IPv6 technologies.

As part of this monograph, we pay a particular attention to the DoS/DDoS attacks that may arise from the use of the emerging IPv6 network. It is envisaged that eventually, most networks will make the transition from the currently widely used IPv4 standard to the new IPv6 standard. However, as with most new technologies, there are always new vulnerabilities. In Chapter 8 of this monograph, we provide a detailed discussion on our research into the IPv6 DoS vulnerabilities, including the gap analysis to highlight which types of IPv6 DoS vulnerabilities that still cannot be effectively mitigated. We also provide an analysis of how IPv6 technologies may affect the security of existing control systems.

2.2.2.2 Attacks Targeting Cellular Telecommunication Network

DoS attacks can happen not only on IP networks as described previously, but also on cellular telecommunication networks. Often, these attacks exploit the lack of authentication systems in cellular networks. Several DoS attacks exploiting this known vulnerability have been described in detail in [1]. These attacks include the location update request spoofing, camping on false base station (BS), user deregistration request spoofing, legitimate roaming partner identity spoofing, push-service initiator identity spoofing, and Internet router identity spoofing.

For example, the user deregistration request spoofing attack works by sending a spoofed user deregistration request to the network. Due to the lack of authentication of such a request message, the user (the DoS victim in this case) is subsequently deregistered from the network, leaving him/her unreachable.

Alternatively, an attacker can also send a spoofed location update request message to the network. As a result, the network registers the user in this new location, while in reality, the user is not. Hence, any communication to this user will be directed to this new location in which the user does not exist.

2.2.2.3 Attacks Targeting Operating Systems

In addition to targeting vulnerabilities in network protocols, attackers may target the data structures and algorithms used by host or router operating systems in the processing of network datagrams to affect service availability. When a computer's operating system is attacked and becomes unavailable, it follows that a legitimate client who needs access to applications which rely on the availability of services of the computer's operating system (which virtually include all applications) will be denied access too.

In 1996, the CERT/CC of Carnegie Mellon University released an advisory detailing a denial of service attack that targeted a widespread vulnerability in the fragment reassembly code of a number of operating systems [10]. The attack crafted an oversized, and fragmented, Internet control message protocol (ICMP) datagram (referred to as a *ping-of-death* – explained earlier in Sect. 2.2.1.2), that when reassembled on vulnerable systems would cause a buffer overflow and consequently the system would crash, freeze, or reboot.

A high rate of TCP connection attempts can be used to exhaust available data structures for storing pending TCP connections via TCP SYN flooding attacks [11] and lead to system memory exhaustion or reboots.

2.2.2.4 Attacks Targeting Applications (Layer 7 Attacks)

Recently, the DDoS attacks using botnets have shifted towards targeting the Layer 7 or the application-layer resources [8]. Many DDoS/DoS attacks have been launched and new ones are still being discovered [4, 24].

The prevalence of common network technologies (such as Web/HTTP, FTP, SMTP, and others) has resulted in the delivery of many applications and services through these well-known interfaces. One main advantage of using common network technologies, especially HTTP, to launch a DDoS attack is that often these channels are not blocked by enterprise firewalls. This allows an easy access to the Web services applications by remote clients. However, this also means that we have now opened a direct channel from the outside world to our backend servers!

One common method of launching a Layer 7 attack is by overwhelming a web server by sending millions of seemingly valid requests. One good example of a technology that is susceptible to such an attack is the *web services* technology. *Web services* applications enable the delivery of various types of applications using (commonly) HTTP as the transport mechanism, and application messages are exchanged using the XML messaging structure. The exposure of these services on the network allows remote attackers to directly exploit various DoS vulnerability of XML-processing logic (such as coercive XML parsing, oversize XML payload, and others [29, 29]) through the *web services* interface.

In Chap. 7 of this monograph, we provide a detailed overview of the DoS/DDoS vulnerability of Web services technologies, as well as the results of our research in validating and mitigating DoS vulnerability in Web services applications using various techniques.

2.3 Botnet Architecture

Typically, a botnet-based attack comprises an attacker, a victim, and a set of bots (or zombies). The victim is the host being targeted for the DDoS attack. The bots are compromised computer systems which are used by the attacker to launch a DDoS attack. Hence the first step is for the attacker to release a worm or other form of malware to subvert a set of unprotected computer systems. The set of bots is collectively referred to as a botnet. The attacker also uses a separate command-and-control system to manage the botnet remotely. By using a large botnet, the attacker can launch a powerful and effective DDoS attack easily and make the attack harder for a network forensic investigator to trace back to the attacker and hence prevent.

In some cases, either amplification attacks or reflection attacks or both are employed by the attacker. Allowing each bot to initiate either or both forms of attack enhances the effectiveness of botnet-based DDoS attacks. The amplifier network enables the attacker to increase the volume of traffic to be used during the DDoS attack. The reflectors let the attacker hide the origin of the bots. Hence, the reflectors help hide the traces to the attacker.

There are four categories of botnet architecture: agent-handler model, IRC-based architecture, peer-to-peer architecture, and the advanced hybrid architecture.

2.3.1 Agent-Handler Model

In the agent-handler model, the bots (i.e. the compromised computer systems) are referred to as agents. The attacker also uses a separate layer of computer systems – referred to as handlers – as the command-and-control system to manage the agents (i.e. bots). Figure 2.4 shows the architecture of the agent-handler model. The attacker communicates with the handlers to establish the command-and-control (aka C&C or C2) system. Typically a handler is a powerful server with plenty of resources (bandwidth, memory, and processing power). In addition to receiving commands from the attacker, the handler is responsible for keeping track of agents and sending commands including configurations and updates to the agents. The owner of the compromised computer systems typically has no knowledge that an agent program has been installed in their machine or that they are part of a botnet. The agent software generally comes in the form of a Trojan or malware. The attacker uses agents as stepping stones to launch attacks against the victim.

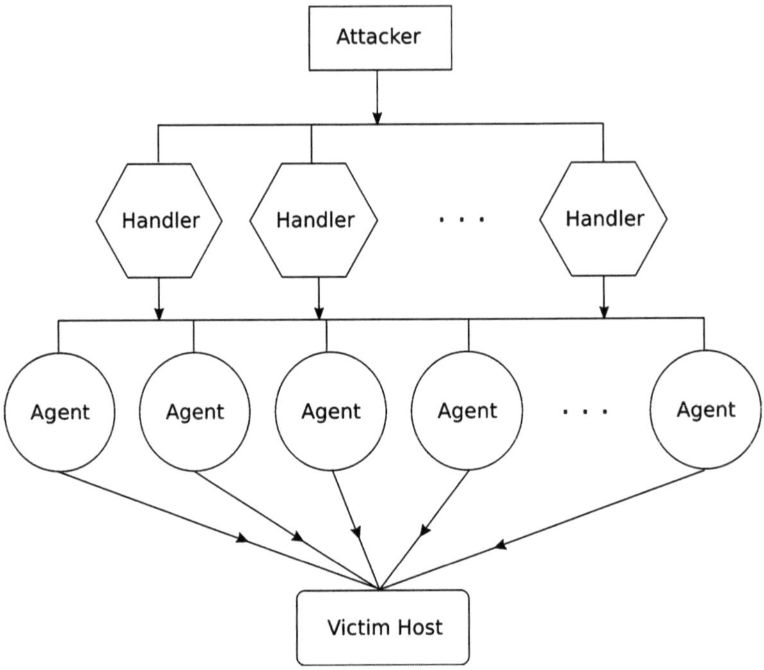

Fig. 2.4 Agent-handler model botnet architecture

It is possible for the attacker to employ amplification or reflection attacks or both. As discussed previously, allowing each agent to initiate either or both forms of attack enhances the effectiveness of the attacks. It also adds a layer of complexity that makes it difficult to trace the originators of the attack. In the early versions of some of the agent-handler botnets, such as Trin00 [16] and Tribe Flood Network (TFN) [18], communications between the attacker and handlers were sent in the clear. Newer botnets, such as Stacheldraht [17] and the newer version of TFN called TFN2K [6], use encrypted communication channels. Encrypted communications make the analysis and detection of the botnets more difficult.

The agent-handler architecture has a major limitation in that attackers must be able to communicate with handlers and handlers must be able to contact agents. If the attackers lose the communication, possibly by the implementation of new network filtering rules, the attackers lose control of the agents and thus cannot orchestrate the agents to attack a new target. Hence, this architecture has become less popular. The evidence of the decline in usage is in the fact that botnets using agent-handler architecture were the ones during the year 1999–2000, and newer botnets use either the IRC-based architecture or the peer-to-peer architecture.

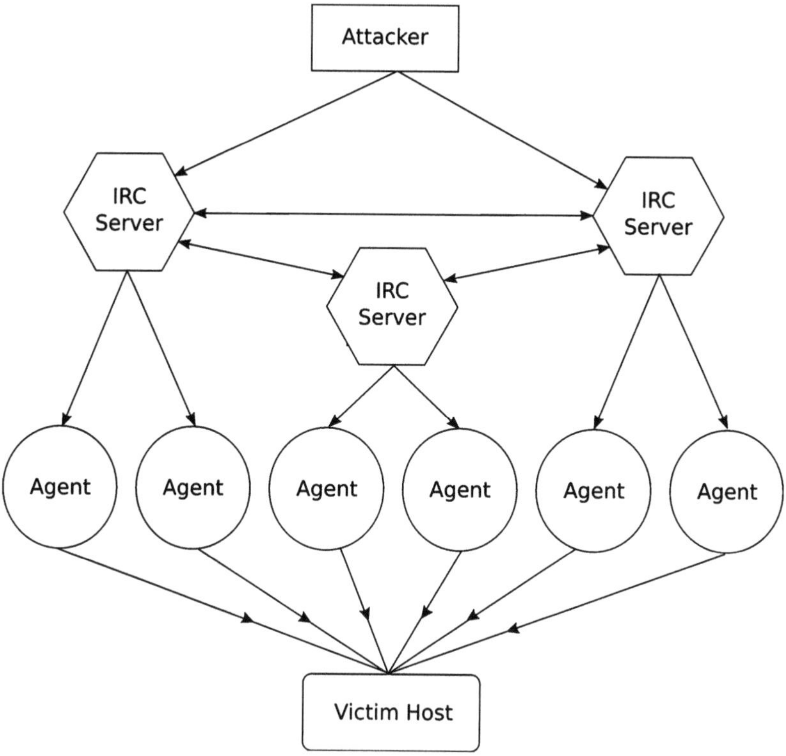

Fig. 2.5 IRC-based botnet architecture

2.3.2 *IRC-Based Botnet Architecture*

The Internet Relay Chat (IRC)-based botnet architecture addresses the limitation in the agent-handler architecture discussed previously. The IRC-based botnet replaces the handler with public IRC servers. Figure 2.5 illustrates the architecture of IRC-based botnets. When agents have been deployed, each agent connects to an IRC server and waits for commands. Attackers issue commands to agents through IRC channels using the IRC protocol. It is possible for the attacker to employ amplification or reflection attacks or both. As discussed previously, allowing each agent to initiate either or both forms of attack enhances the effectiveness of the attacks. It also adds a layer of complexity in order to hide the traces of the attackers. The communications between the attacker and IRC servers may be encrypted.

The two main differences between the IRC-based architecture and the agent-handler architecture are control structure and communications [19]. In the IRC-based architecture, each agent connects to one IRC server whereas in the agent-handler,

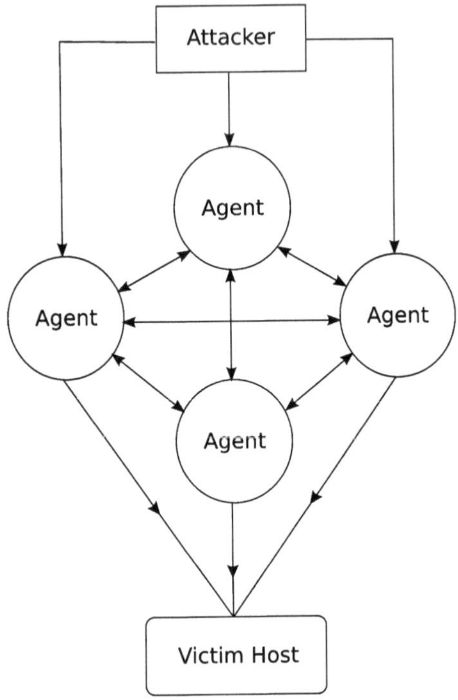

Fig. 2.6 P2P-based botnet architecture

each agent may connect to more than one handler. In terms of communications, the attacker uses the IRC protocol to communicate with its agents, while in the agent-handler architecture the attacker uses a proprietary protocol.

Examples of IRC-based botnets are Agobot [41], SDBot [62], Spybot [38], and Nugache [60].

2.3.3 Peer-to-Peer Based Botnet Architecture

Peer-to-Peer-based (or P2P-based) botnet architectures are more distributed than either the agent-handler model or the IRC-based architecture discussed previously. Figure 2.6 shows the architecture of the P2P-based botnet architecture. Unlike the agent-handler model and the IRC-based architecture, the P2P-based botnet architecture has no separate handlers. Commands are sent to agents (i.e. bots) through the P2P protocol. In this case each agent/bot is not only responsible for actually delivering the final attack but is also part of the command-and-control structure for managing other agents. Thus, the P2P-based botnet architecture is difficult to shutdown because of its highly distributed nature.

In addition to distributing commands, the P2P communication channel is used to distribute new versions of the bot software and to download new attack tools and a list of new targets. In order to make the attack or the communications much more difficult to detect and analyse, the communications may be encrypted. Examples of P2P-based botnet architecture are Storm [59], Rambot [25], Slapper [2], and Sinit [58].

2.3.4 Advanced Hybrid Peer-to-Peer Botnet Architecture

Hybrid P2P botnets behave both as clients and servers in a traditional P2P file sharing system. An attacker can inject his/her commands into any hosts of this botnet. Each host periodically connects to its neighbours to retrieve orders issued by their commander. As long as a new command shows up, the host will forward this command to all nearby servant bots immediately. Such architecture combines the following features [68]: (1) it requires no bootstrap procedure; (2) a bot only has a limited list of its neighbour bots; thus, even if this bot is captured, the captor can only obtain the limited list of neighbour bots, not the complete list of bots; (3) an attacker can easily manage the entire botnet by issuing a single command.

2.4 Detection Techniques

2.4.1 Detecting High-Rate Flooding Attacks

The management of high-rate flooding attacks involves two separate but inter-related phases. The first management phase is attack detection and in this section of the report we will examine two aspects of detection in more detail. One area of focus is on detecting the establishment of a network of (compromised) agent computers (aka a botnet). Establishing a botnet of some form is a common precursor to unleashing some of the most virulent forms of high-rate flooding attacks. Hence detecting the formation of a botnet community may provide an advance warning of an impending attack [21]. However, the main aspect of the detection process that we will cover in this section is that of recognising that an attack is underway.

One of the characteristic features of a high-rate flooding attack is the appearance of a large volume of traffic in the target computer's network. The key, however, is in isolating those situations where any (significant) changes in the composition and volume of the network traffic represent actual anomalous (and hence potentially malicious) behaviour. This topic is the key focus of Chap. 5. It is worth noting here that efficient and timely detection are only the first phase of the management strategy for dealing with a high-rate flooding attack.

The second management phase is attack-mitigation and this too is discussed in detail in Chap. 5. Mitigating the impact of a high-rate flooding attack by reducing network traffic load is built upon techniques for detecting as accurately as possible, errant traffic within the total traffic flow.

2.4.1.1 Botnet-Based Attack Detection

In the case of a botnet-based attack, mitigation or attack reaction involves examining ways to disrupt the associated command-and-control structure of the botnet. Necessarily this requires that we have in place techniques for detecting botnet activity. We address this immediately below.

As was discussed previously, botnets constitute some of the most virulent forms of high-rate flooding attacks. Detecting the formation of a botnet community may therefore provide an advance warning of an impending attack [21]. Also, as will be discussed in the following section, early detection of a botnet community affords the potential to mitigate the impact of a botnet attack by disrupting or limiting the functioning of the command-and-control structure required for managing the botnet herd [60].

Consequently a considerable amount of effort has been directed to detecting botnet communities (see, e.g. [47]). As with the preceding discussion on detecting traffic anomalies, detecting botnet communities relies essentially on the same three elements: (1) recording and/or measurement of certain parameters of interest; (2) data analysis; and (3) decision-making.

For example an early indicator of an impending botnet attack is the appearance of a 'new' worm that rapidly compromises a set of (unprotected) computer systems that will become the botnet herd [21]. Typically such worms manifest themselves as newly installed program code and a number of organisations and individuals regularly monitor computer systems looking for evidence that new code has been installed. (One approach to collecting evidence of this aspect of botnet activity is via 'honeypots' or 'honeynets' discussed previously [34].) Once such a code is detected it is then subject to rigorous analysis to ascertain its characteristics and there is now in place a comprehensive alert network for disseminating information about newly discovered worms rapidly [21, 23].

Another approach is to search for evidence of botnet command-and-control traffic or so-called C2 traffic. Because such C2 traffic may, for example, utilise one or more different communications protocols (e.g. HTTP, IRC, P2P) [36] or advanced cryptographic processes [25], accurate detection of botnet command-and-control traffic is a very complex activity. Hence this is currently an area of active research. As is to be expected there is a considerable degree of overlap between the techniques discussed previously for detecting traffic anomalies and techniques for detecting botnet command-and-control traffic. For example, Strayer et al. [61] examine what they term 'flow characteristics' such as bandwidth, packet timing, and burst duration looking for evidence of botnet C2 activity. Similarly Lu et al. [36] attempt an even finer level of granulation in identifying botnet command-and-control traffic by examining payload signatures and flow characteristics.

2.4.2 *Detecting Semantic Attacks*

Though not always the case, semantic-based DoS attacks do often come with some characteristics which can be distinguished from normal packets.

Semantic attacks often target a known vulnerability of a platform. Once the vulnerability of the platform is discovered, a specially crafted message is then sent to the victim. For example, attacks targeting a known vulnerability in IBM Data Power appliances[2] which are commonly used to handle web services messages rely on the sending of a specifically crafted ICMP packet. This ICMP packet thus has a certain characteristic which upon inspection will reveal its malicious intention.

Another example of semantic attack is the oversized XML document attack [29] which attempts to exhaust a server's resources by sending a large XML payload. Similar to the previous example, this type of attack can be easily detected by examining the characteristics of the XML payload first (such as the number of elements, the level of deepest nesting, and others) before parsing it.

Semantic attack can also target known vulnerabilities in network protocols. The *ping of death*, *land* attack, and *teardrop* attack are classic examples of semantic attacks targeting vulnerabilities in network protocols.

The attacks mentioned in this section have the commonality of being able to be detected by examining the content of the packet to identify if they contain any structure or data that will cause DoS on the recipient of those messages. In other words, semantic attacks can often be detected by using signature-based intrusion detection system (IDS) and common application-layer network firewalls.

2.5 Mitigation Techniques

2.5.1 *Mitigating High-Rate Flooding Attacks*

As discussed previously, a high-rate flooding attack involves directing a large volume of spurious network traffic towards a target host (and its associated network) with the intent to disrupt its ability to service its legitimate clients. As was discussed previously, at a technical level such attacks involve two key phases: (a) set-up (typically by exploiting vulnerabilities and establishing botnets), and (b) traffic generation. In the preceding chapter we examined ways of detecting such attacks. In this chapter we examine ways of mitigating the impact of such attacks.

[2] http://secunia.com/advisories/38256

Mitigating the impact of a high-rate flooding attack essentially relies on three main strategies. The first is deterrence. The aim is either to reduce the threat significantly or, in an ideal situation, to eliminate the threat entirely. This includes, for example, identifying and disabling botnet communities before they are able to wreck havoc. A second strategy, and one that is more localised, is the adoption and consistent and rigorous implementation of good security practices. The aim of this second strategy is to reduce vulnerability by making every element or component within a network more resilient. In traditional risk-management terms this is so-called 'hardening'. Unfortunately the 'public good' of hardening as a collective strategy against network flooding attacks is greatly diminished if only a comparatively small proportion of computer users make the effort to protect their individual systems. The third strategy, and again one that is essentially localised in its implementation, comprises the set of techniques for managing an actual high-rate flooding attack in real time.

In concert, a combination of these three strategies appears to offer both potential and actual victims of high-rate flooding attacks some degree of optimism that the impact of such attacks can be mitigated. We address the first two topics immediately below, and the third in Chap. 5.

2.5.1.1 Deterrence

Clearly, the most direct means of reducing the severity of an attack of this type is to dissuade people from mounting such an attack in the first place. The underlying premise is that a high likelihood of detection and successful prosecution coupled with penalties (particularly involving a custodial sentence) will act as a suitable deterrent. This line of reasoning underpins our judicial system and is considered applicable to almost every criminal situation. However judging by the number and frequency of high-rate flooding attacks that still occur [55], and notwithstanding the concerted efforts being made by law enforcement agencies in a number of countries [28], so far this approach appears to have met with only limited success. The first problem is one of attribution. Finding evidence that directly links an individual or group of individuals to an actual attack is a painstaking and resource-intensive exercise. The second problem is one of jurisdiction. In particular the current situation is that perpetrators still seem to be able to operate with impunity in certain countries [23, 28, 60]. Also, as shown in two recent reports [55, 70], a large number of incidents simply appear to go unreported because authorities are perceived to have only limited resources available to investigate each incident.

A second deterrence strategy is targeted at identifying and disabling or disrupting botnet communities as they form and before they are able to unleash their attack. As discussed in the previous chapter, the starting point is detection. However, given the resilience and adaptability of the botnet controllers, the overall prospects for successfully mitigating the impact of a high-rate flooding DDoS attack by targeting the botnet command-and-control infrastructure appear to be somewhat limited [25]. In particular, one of the key problems is that perpetrators have an appreciation of the technical limitations of the detection techniques discussed in the previous section and also the capability to utilise increasingly sophisticated techniques (e.g. strong encryption) to mask their activities particularly in the command-and-control phase [25, 28].

2.5.1.2 Adoption and Rigorous Implementation of Good Security Practices

The second strategy for mitigating the impact of a high-rate flooding attack is to make every element or component in the network more resilient to attack. As we have discussed previously, a high-rate flooding attack requires that the attacker generate a network traffic stream that is large enough to overwhelm the resources of the intended target/victim. Given that the capacity of these resources (e.g. network bandwidth, CPU processing power of the servers) tends to increase over time, a successful attack using, for example, a brute-force strategy relies on the attacker being able to marshal a commensurate increase in resources under their control (e.g. a larger botnet herd). Establishing and maintaining this 'resource symmetry' [56] between attacker and defender associated with a brute force attack makes the attacker more conspicuous to detection and offers a potential mitigation strategy by denying an attacker access to the resources needed to mount an attack.

This mitigation strategy is the adoption and rigorous implementation of good security practices. Whilst, unfortunately, such a strategy is seen as almost bordering on the prosaic, in reality it is simply about following the basics of information security. It involves, for example, installing malware detection systems [60] including in particular, implementing procedures for identifying and blocking so-called spam e-mails since these are one of the primary vehicles for distributing malware [28]. It also involves educating users not to open unknown or suspicious e-mail attachments or links since these remain as one of the primary means by which worms such as the Storm worm (discussed previously) propagate. It involves having a comprehensive and actively enforced security policy that mandates, for example, the systematic application of security updates and 'patches'. When applied rigorously at the level of an individual user computer system, the cumulative effect of this strategy would be to make it increasingly difficult for the attacker to achieve the level of resource symmetry required to mount a successful brute force attack that is preventing as many individual computer systems as possible from being compromised and then being used in an attack, reduces the size of the botnet herd and hence restricts the potential volume of traffic that could be unleashed in a flooding attack. However, and notwithstanding continuous exhortations about the importance of adopting good information security practices, there is still a yawning chasm between the desirable and the reality. This was amply demonstrated recently by the extremely high rate of spread of the Conficker (Downadup) worm through individual computer systems that had not been updated with the most recent security patches even though these patches had been released several months previously [21].

The adoption and rigorous implementation of good security practices extends beyond the computer systems of individual users. For example an extremely potent form of amplification attack discussed previously is to exploit vulnerabilities in the operation of DNS servers [46] and for which corrective code has not been applied. Good security practices also involve, for example, implementing rules in (particularly border) routers and firewalls that inter alia [53]:

1. Block all inbound traffic where the source address is from your internal networks (aka ingress filtering)
2. Block all outbound traffic where the source address is not from your internal networks (aka egress filtering)
3. Block all inbound and outbound traffic where the source or destination addresses are from the private address ranges
4. Block all source-routed packets
5. Block all broadcast packets, including directed broadcasts
6. Block all packet fragments

These types of security measures, for example, reduce the opportunity for an attacker to inject spurious traffic onto a network. Other security practices that could, if deployed properly, significantly mitigate the impact of a high-rate flooding attack involve the implementation of explicit security protocols such as IPsec and DNSsec, and, ultimately, IPv6 (which mandates the use of IPsec). Central to the functioning of the IPsec protocol, for example, is the (mutual) authentication of the communicating parties and the protection (i.e. confidentiality and integrity) of the traffic between them by using cryptographic techniques. However, in situations where one of the parties is a compromised computer system such as we have discussed previously (i.e. a ‘bot’), then the validity of the authentication process becomes somewhat problematic. In addition, the authentication process potentially consumes significant resources itself and, in a high-rate flooding attack, could conceivably therefore become the point of failure.

Of themselves these actions may ultimately only indirectly mitigate the impact of a high-rate flooding attack. However, they will at least go a substantial way to assuring the integrity of all the components within a given network perimeter which will assist in the defence process once an attack is detected. Moreover making individual components in a network more resilient to compromise reduces the risk of a direct semantic attack [56] in which a relatively weak attacker may still compromise a more powerful target without first creating a set of resources to generate a flood of spurious traffic (i.e. ‘resource asymmetry’ cf. ‘resource symmetry’ discussed previously [56]).

2.5.2 *Mitigating Semantic Attacks*

As discussed in Sect. 2.4.2, semantic attacks can be detected using a variety of IDS and application-level firewalls. The mitigation of semantic attacks is therefore straightforward: one needs to configure the IDS and/or firewalls to drop (or at least quarantine) those packets which match malicious packet signatures. The adoption and rigorous implementation of good security policies (detailed in Sect. 2.5.1.2) and user education may also alleviate the risk of semantic-based attacks.

2.6 Case Studies

2.6.1 Case Study 1: Storm Botnet

The 'Storm' botnet (aka Storm Worm botnet, Peacomm, Win32, Nuwar, Trojan.Peacomm, W32.DAM) was first reported in January 2007 [59] when a number of anti-spam websites were the target of denial-of-service attacks. (For the purpose of this discussion, the worm used to create the individual bots will be referred to as the Storm bot, and the network of bots will be referred to as the Storm botnet.)

During the year 2007, the Storm botnet launched DDoS attacks against several websites which included anti-spam sites and competitive spammer sites. Researchers who tried to analyse the Storm bot, for example, attach a debugger to the bot or run DNS query on IP address that could possibly be the attacker, were themselves subjected to DDoS attacks (high-rate flooding, TCP SYN flooding, ICMP flooding) [22]. In particular, and illustrative of the perniciousness of this type of threat that was discussed previously in Sect. 2.1, the company Blue Security was forced into bankruptcy through its efforts in combating the threat. From [66], 'Somebody wrote a [botnet], and Blue Security did a really good job of fighting, so [the attackers] did a DDoS and took it off the Net for awhile. Blue Security went to the best anti-DDoS technology on earth. The next onslaught came and [Blue Security's defences] worked. So the botnet herder stole two other people's botnets. With three botnets, [the attack] worked, to the point where the ISP said, I am not going to let you take down my entire ISP to protect you, you're on own. And Blue Security is now out of business'. It has been estimated that by September 2007, the Storm botnet may have compromised up to 50 million computers [57]. Compared to other bots that appeared at the same period of time, the Storm bot incorporated a number of novel botnet techniques such as code obfuscation, code encryption, virtual-machine detection, and peer-to-peer (P2P) communications. The Storm bot continues to evolve and still poses a highly potent threat.

The Storm bot targets machines running the Microsoft Windows operating systems (except Windows 2003 Server) [57]. The bots propagate through Trojan attachment in spam e-mail. The Storm bot scans the compromised machine for e-mail addresses and then sends spam e-mails to those addresses with a Trojan attached to the spam. Once a machine is infected, it becomes part of the Storm botnet. The infected machines perform several operations, such as sending spam e-mails, downloading new attack tools, and launching DDoS attacks. Details of the operations and the analysis of the Storm bot are discussed in the next section.

2.6.1.1 Analysis of the Storm Bot

As indicated above, when it was released in 2007, the Storm bot was one of the most advanced bots of its time and remains active. The Storm bot implements several protective mechanisms against detection and analysis. Firstly, the body of the Storm bot is encrypted using three techniques: XOR, TEA, and TIBS [60]. The Storm bot is decrypted when a user (victim) executes the Trojan (Storm bot) that comes with a spam e-mail where the spam is sent from another infected computer system (i.e. a bot). Once the code is decrypted, the Storm bot attaches itself to the 'tcpip.sys'. This results in the Storm bot being started every time the machine is turned on. In addition, the Storm bot installs rootkit on to the machine. The rootkit provides functions to the Storm bot to terminate any process that may harm the bot such as firewalls and antivirus programs (see [52] for the full list of such programs). Some versions of the Storm bot are equipped with virtual machine detection function. If the Storm bot detects that it is running under a virtual machine, which is normally used by researchers to analyse bots, the bot will behave differently such as stop running.

The Storm bot comprises four modules: spamming, DDoS attack, download and update, and P2P communication. The details of each module are as follows.

Spamming Module

The Storm bot uses the spamming module to propagate [52]. The Storm bot downloads the spam e-mail template from other infected machines. The Storm bot scans the disk of the infected machine to gather information about hosts, users, mailing lists with a set of specific patterns such as 'yahoo.com', 'gmail.com', 'f-secur', root@', and 'bugs@'. Once the spamming module has collected sufficient information, it sends spam e-mail containing the Storm bot to the set of addresses it has gathered. By early 2008, the spamming module had evolved to include phishing attacks. These phishing e-mails were targeted at bank customers particularly in Europe [67].

DDoS Attack Module

The Storm bot employs several high-rate flooding attack techniques such as TCP SYN flooding and ICMP ping flooding [59]. The Storm herder orchestrates all of its infected machines (agents) to send a large volume of network traffic simultaneously to the target machine. Such an attack succeeds due to the large number of infected machines.

Download and Update Module

This is module is in fact a part of the P2P communication module. However, the ability to download and update itself is one of the important features of the

Storm bot. The Storm bot can download new version of the Storm bot, additional functionality, spam e-mail templates, updated anti-forensic techniques, and attack tools [60].

P2P Communication

The Storm bot is one of the first widespread bots that replaces IRC-based C&C communication with peer-to-peer communication. The Storm bot uses existing and popular P2P protocols namely eDonkey and Overnet. The communications between peers are encrypted. The list of active peers, machines that are still infected and reachable, on each infected machine is updated frequently.

2.6.1.2 Detection

Because the spamming module in the Storm bot generates a large volume of network traffic [52], one method to detect the Storm bot is to analyse network traffic volume. In particular, a key indicator is the presence of a large volume of outbound traffic on TCP port 25 (SMTP) originating from the infected machine.

Another detection solution is to locate specific files, for example, spooldr.ini, on the suspicious machine [60]. The spooldr.ini contains a list of peers, that is other infected machines.

2.6.1.3 Mitigation

Since the Storm botnet implements the P2P architecture for its command-and-control structure, mitigating the impact of an attack by disrupting the command-and-control communication is difficult. However, there are a few mechanisms that can mitigate the spreading of Storm. The primary mechanism is by educating users. Users should be aware that e-mail attachments may contain malware or Trojan, in this case Storm, and should not execute them. The only propagation mechanism implemented in Storm is through attaching executable Trojan with spam e-mails. Hence if users do not execute the Trojan, the Storm bot will stop spreading.

2.6.2 Case Study 2: Agobot Botnet

Agobot (aka Gaobot, Polybot, and Phatbot) was first discovered in late 2002 [41]. The first version of Agobot was developed by Axel Gembe [9]. The Agobot is an IRC-based botnet. Agobot targets all versions of Microsoft Windows. The Agobot is one of the most sophisticated bots [26].

The most interesting characteristic of the Agobot is the fact that it probably has the most variants [26]. This is due to the availability of the graphical user interface (GUI) tool which allows anyone to easily create variants of the Agobot. The GUI tool provides detailed documentation of the functionalities of the bot that users can choose to incorporate into the new variation of Agobot. Hence, even a novice user can create new bots with little or no knowledge about Agobot's technical details.

2.6.2.1 Analysis of the Agobot

The main functionalities of the Agobot and its variants are as follows.

IRC-Based Command and Control

Agobot communicates with the attacker through the IRC protocol. Once a machine is infected with Agobot, the bot connects to an IRC server and waits for further commands from the attacker. Amongst other commands (see [5] and [26] for a list of available commands), the attacker can issue commands to orchestrate the Agobot botnet to launch a DDoS attack a specific target.

In the later variants of Agobot, in particular, Phatbot, the IRC-based command and control communication has been replaced with peer-to-peer (P2P) communication. The P2P communication adds complexity to Phatbot, and thus makes it harder to shut the botnet down.

Propagation

Agobot has two mechanisms to propagate itself: (a) using a peer-to-peer (P2P) network file-sharing application and (b) using buffer overflow exploits. In the P2P method, Agobot propagates itself using the P2P file-sharing application Kazaa (aka KaZaa). When Agobot first infects a machine, it creates a copy of itself in Kazaa folders. These folders are automatically read by the Kazaa P2P client program, and thus Agobot is shared over the P2P network.

In the buffer overflow exploit method, Agobot utilises the large collection of buffer overflow exploits against Microsoft Windows, such as MS03-007 WebDAV [43] and MS03-026 RPC DCOM [42], to propagate itself. The bot propagates using this method only when it receives command from the attacker. When the bot receives the command, it first scans a set of machines (based on a set of IP addresses provided by the attacker) for the vulnerabilities that are known by the bot and copies itself to vulnerable machines.

DDoS Attack Module

Agobot contains the following DDoS attacks [5]: UDP flood, SYN flood, HTTP flood, ICMP flood, Targa flood (random IP protocol, fragmentation and fragment offset values, and spoofed source address), and wonk flood (one SYN packet followed by 1,023 ACK packets). These attacks are launched when the botnet receives an order and the target from the attacker.

Information Gathering and Scanning

This module scans the drive of the infected machine for Paypal passwords, and AOL keys. Also, this module is capable of sniffing network traffic, key logging, and searching Windows registry.

Self-defence

This module defends Agobot on infected machine from being removed. This is achieved though several mechanisms, for example, disabling access to antivirus sites and stopping firewall and antivirus processes.

In addition, Agobot detects if it is running under debuggers (such as OllyDebug, SOftIce, and procdump) or VMWare. If the debugger or VMWare is detected, Agobot will behave differently, that is, stop running malicious activities.

Polymorphism

Polymorphism refers to the mechanisms to modify code but retain the algorithm. This process can also be referred to as mutation. The purpose of polymorphism is to defeat signature-based detection systems, for example, intrusion detection system and antivirus program. Agobot supports four types of polymorphic encoding strategies: XOR, rotate left, rotate right, and swap consecutive bytes.

2.6.2.2 Detection

One of the two approaches to Agobot propagation utilises buffer overflow exploits. Hence monitoring such exploits will help detect the bot [26].

Another detection solution that may work in a limited environment is to monitor IRC activities. Typically a corporate network does not allow IRC activities. Therefore, if IRC activities originating from a host in the network are detected, that host may be infected.

2.6.2.3 Mitigation

Agobot uses the IRC protocol to communicate with the attacker. Hence, by blocking outgoing IRC protocol messages in a network the bot will be prevented from causing any damage. However, this may be applicable only in a corporate network where there is a policy that prohibits the use of IRC. infectionvectors.com [26] suggests that installing application proxies, which provide a similar result, will stop Agobot from contacting the attacker.

Another mitigation solution is to keep the software up to date. As discussed above, Agobot uses a set of buffer overflow exploits as stepping stones to propagate. Hence, if machines are fully patched, the propagation will be stopped.

2.6.3 *Case Study 3: Operation Payback*

Operation Payback is the name given to an overall campaign conducted by a group known as *Anonymous* which was formed to launch a series of retaliatory DDoS attacks against anti-piracy organisations [7, 32]. This group was responsible for launching a series of large-scale DDoS attacks against those organisation (including PayPal, Visa, Mastercard, Amazon EC2, and others) who withdrew services to the whistle-blower website WikiLeaks. This group has a set of DDoS attack mechanisms which we will discuss in this case study.

2.6.3.1 Analysis of Operation Payback

Operation Payback uses a tool called Low Orbit Ion Cannon (LOIC) [32] to launch the DDoS attacks. This tool consists of several features which are discussed below.

Command-and-Control Structure

At the simplest level, the user downloads the LOIC tool and then enters the URL or IP address of the victim(s). The tool then launches DDoS attack against the victim(s). Alternatively, users can use the *Hivemind* feature of this tool which allows them to connect to a botnet where the users' machine can be controlled remotely via IRC channel or a Twitter account.

The bot controller then directs the bots to open TCP connections to the victim (port 80) and start overloading the victim with a series of both valid and meaningless HTTP requests. Furthermore, these bots can also flood the victim with UDP packets.

Voluntary Participant

The LOIC tool can be downloaded by participants who voluntarily join the network.

Multi-vector Attack

A characteristic difference between LOIC tool and those discussed in the previous case studies is that LOIC tool is a collection of different attack strategies such as high-rate SYN and UDP flooding, malformed UDP and HTTP packets, low bandwidth attack (e.g. slowloris[3]), and many others.

2.6.3.2 Detection

Due to the multi-vector attack nature of the LOIC tool, the detection of such attacks requires a combination of various DoS detection techniques, including signature-based detection, efficient hardware-based packet filtering devices, as well as the more advanced techniques, such as high-rate real-time packet analysis and machine learning algorithms.

2.6.3.3 Mitigation

In the case of Operation Payback against Amazon EC2 system, Amazon has managed to mitigate this attack quite successfully. This is mainly due to the fact that Amazon EC2 system has a significant spare server capacity which allows it to handle a massive traffic increase without failing. In other words, the resource imbalance often associated with DoS attack has been inverted: it seems that in the case of DDoS attack against the Amazon EC2, the 'victim' has more resources than the attackers.

References

1. 3rd Generation Partnership Project. 2000. DoS attacks to 3G networks and users. ftp://www.3gpp.org/tsg_sa/WG3_Security/TSGS3_15_Washington/Docs/PDF/S3-000571.pdf. Accessed 17 Feb 2011.
2. Arce, I., and E. Levy. 2003. An analysis of the slapper worm. *IEEE Security & Privacy* 1(1): 82–87.
3. Bächer, P., T. Holz, M. Kötter, and G. Wicherski. 2008. Honeynet project: Know your enemy: Tracking botnets. http://www.honeynet.org/papers/bots. Accessed 8 Feb 2009.
4. Banks, Z. 2009. Slowloris HTTP denial of service. http://hackaday.com/2009/06/17/slowloris-http-denial-of-service/. Accessed 27 Jan 2011.
5. Barford, P., and V. Yegneswaran. 2006. An inside look at botnets. *Malware detection*, eds. In M. Christodorescu, S. Jha, D. Maughan, D. Song, and C. Wang, Advances in Information Security, 171–191. Berlin: Springer Science+Business Media, LLC.

[3] http://ha.ckers.org/slowloris/

6. Barlow, J. 2000. Axent releases a full tfn2k analysis. http://www.securiteam.com/securitynews/5YP0G000FS.html. Accessed 10 Feb 2009.
7. Bradley, T. 2010. Operation payback: Wikileaks avenged by hacktivists. http://www.pcworld.com/businesscenter/article/212701/operation_payback_wikileaks_avenged_by_hacktivists.html. Accessed 28 Jan 2011.
8. Brenner, B. 2010. Layer 7 increasingly under DDoS gun. http://www.csoonline.com/article/526263/report-layer-7-increasingly-under-ddos-gun. Accessed 27 Jan 2011.
9. Bryan-Low, C. 2007. How legal codes can hinder hacker cases. *The Wall Street Journal*. Jan 17, p. A8.
10. Computer Emergency Respone Team (CERT). 1996. Denial-of-service attack via ping. http://www.cert.org/advisories/CA-1996-26.html. Accessed Aug 2004.
11. Computer Emergency Respone Team (CERT). 1996. SYN flooding attack. Available: http://www.cert.org/advisories/CA-1996-21.html. Accessed Aug 2004.
12. CERT/CC. 1997. Cert advisory ca-1997-28 ip denial-of-service attacks. http://www.cert.org/advisories/CA-1997-28.html. Accessed 11 Feb 2009.
13. CERT/CC. 1998. Cert advisory ca-1998-01 smurf ip denial-of-service attacks. http://www.cert.org/advisories/CA-1998-01.html. Accessed 12 Feb 2009.
14. Cheswick, W.R., and S.M. Bellovin. 1994. *Firewalls and Internet Security: Repelling the Wily Hacker*. Addison-Wesley Professional.
15. CPNI. 2008. Advice on the sockstress vulnerabilities (FICORA 193744). Advisory, Centre for the Protection of National Infrastructure.
16. Dittrich, D. 1999. The DoS Project's "trinoo" distributed denial of service attack tool. http://staff.washington.edu/dittrich/misc/trinoo.analysis. Accessed 16 Feb 2011.
17. Dittrich, D. 1999. The "stacheldraht" distributed denial of service attack tool. http://staff.washington.edu/dittrich/misc/stacheldraht.analysis. Accessed 16 Feb 2011.
18. Dittrich, D. 1999. The "tribe flood network" distributed denial of service attack tool. http://staff.washington.edu/dittrich/misc/tfn.analysis. Accessed 16 Feb 2011.
19. Dittrich, D., and S. Dietrich. 2007. Command and control structures in malware. *The USENIX Magazine* 32(6). http://www.usenix.org/publications/login/2007-12/openpdfs/dittrich.pdf. Accessed 16 Feb 2011.
20. Garber, L. 2000. Denial-of-service attacks rip the internet. *Computer* 33(4): 12–17.
21. Goodin, D. 2009. Superworm seizes 9m PCs, 'stunned' researchers say. http://www.theregister.co.uk/2009/01/16/9m_downadup_infections/. Accessed 16 Feb 2011.
22. Greene, T. 2007. Storm worm strikes back at security pros. http://www.networkworld.com/news/2007/102407-storm-worm-security.html. Accessed 11 Feb 2009.
23. Grimes, R.A. 2009. Fighting malware: An interview with Paul Ferguson. http://www.infoworld.com/d/security-central/fighting-malware-interview-paul-ferguson-447. Accessed 16 Feb 2011.
24. Higgins, K.J. 2010. Researchers to demonstrate new attack that exploits HTTP. http://www.darkreading.com/vulnerability-management/167901026/security/attacks-breaches/228000532/index.html. Accessed 27 Jan 2011.
25. Hund, R., M. Hamann, and T. Holz. 2008. Towards next-generation botnets. In *European Conference on Computer Network Defense. EC2ND 2008*, 33–40.
26. infectionvectors.com. 2004. Agobot and the "kitchen sink", Dublin, Ireland http://www.infectionvectors.com/vectors/Agobot_&_the_Kit-chen_Sink.pdf. Accessed 28 Jan 2011.
27. International Telecommunication Union. 1991. Data communication networks: Open systems interconnection (OSI); security, structure and applications–security architecture for open systems interconnection for CCIT applications. Recommendation X.800, Telecommunication Standardization Sector of ITU, Geneva, Switzerland.
28. (International Telecommunication Union) ITU. 2008. Itu botnet mitigation toolkit: Background information. Technical report.
29. Jensen, M., N. Gruschka, and R. Herkenhöner. 2009. A survey of attacks on web services. *Computer Science – R&D* 24(4): 185–197.

30. Kenny, M. 1997. Ping of death. http://insecure.org/sploits/ping-o-death.htm. Accessed 11 Feb 2009.
31. Kuzmanonvic, A., and E.W. Knightly. 2006. Low-rate TCP-targeted denial of service attacks and counter strategies. *IEEE/ACM Transactions on Networking* 14(4): 683–696.
32. Labovitz, C. 2010. The internet goes to war. http://asert.arbornetworks.com/2010/12/the-internet-goes-to-war/. Accessed 28 Jan 2011.
33. Lee, K., J. Kim, K.H. Kwon, Y. Han, and S. Kim. 2008. DDoS attack detection method using cluster analysis. *Expert Systems with Applications* 34(3): 1659–1665.
34. Li, Z., A. Goyal, and Y. Chen. 2008. Honeynet-based botnet scan traffic analysis. In *Botnet detection: Countering the largest security threat*, eds. W. Lee, C. Wang, and D. Dagon, 25–44. Berlin: Springer.
35. Louis, J.C., and R.E. Lee. 2011. Introduction to sockstress. http://insecure.org/stf/tcpdos/outpost24-sect-sockstress.pdf. Accessed 16 Feb 2011.
36. Lu, W., M. Tavallaee, and A.A. Ghorbani. 2009. Automatic discovery of botnet communities on large-scale communication networks. In *ACM Symposium on InformAtion, Computer and Communications Security (ASIACCS'09)*, 1–10, Sydney, Australia.
37. Martin, J. 2004. Denial of service (dos) attacks. http://www.securitydocs.com/library/2616. Accessed 1 Feb 2011.
38. McAfee. 2003. W32/spybot worm gen. http://vil.nai.com/vil/content/v_100282.htm. Accessed 13 Feb 2009.
39. McPherson, D., C. Labovitz, M. Hollyman, J. Nazario, and G.R. Malan. 2008. Worldwide infrastructure security report. Technical report, Arbor Networks.
40. meltman@lagged.net. 1997. The LAND attack (IP DOS). http://insecure.org/sploits/land.ip.DOS.html. Accessed 11 Feb 2009.
41. Trend Micro. 2002. Worm_agobot.a. http://www.trendmicro.com/VINFO/VIRUSENCYCLO/default5.asp?VName=WORM_AGOBOT.A. Accessed 13 Feb 2009.
42. Microsoft. 2003. Buffer overrun in RPC interface could allow code execution. Technical report MS03-026. http://www.microsoft.com/technet/security/bulletin/MS03-026.mspx. Accessed 28 Jan 2011.
43. Microsoft. 2003. Unchecked buffer in windows component could cause server compromise. Technical report MS03-007. http://www.microsoft.com/technet/security/bulletin/MS03-007.mspx. Accessed 28 Jan 2011.
44. Mirkovic J., and P. Reiher. 2004. A taxonomy of ddos attack and ddos defense mechanisms. *ACM SIGCOMM Computer Communication Review* 34(2): 39–53. http://www.cis.udel.edu/sunshine/publications/ccr.pdfarticlesteven.
45. Mölsä, J. 2005. Mitigating denial of service attacks: A tutorial. *Journal of Computer Security* 13(6): 807–837.
46. Moscaritolo, A. 2009. New style of DNS amplification can yield powerful DDoS attacks. http://www.scmagazineus.com/new-style-of-dns-amplification-can-yield-powerful-ddos-attacks/article/126839/. Accessed 16 Feb 2011.
47. Nazario, J. 2008. Political DDoS: Estonia and beyond. In *USENIX Security '08*. USENIX. http://streaming.linux-magazin.de/events/usec08/tech/archive/jnazario/.
48. Needham, R.M. 1993. Denial of service. In *The 1st ACM Conference on Computer and Communications Security*, 151–153, Fairfax.
49. Nikander, P., J. Kempf, and E. Nordmark. 2007. IPv6 neighbor discovery (ND) trust models and threats. http://www.ietf.org/rfc/rfc3756.txt. Accessed 10 Feb 2011.
50. Padmanabhuni, S., V. Singh, K.M.S. Kumar, and A. Chatterjee. 2006. Preventing service oriented denial of service (PreSODoS): A proposed approach. In *ICWS '06: Proceedings of the IEEE International Conference on Web Services*, 577–584, Washington, IEEE Computer Society.
51. Paxson, V. 2001. An analysis of using reflectors for distributed denial-of-service attacks. *ACM SIGCOMM Computer Communication Review* 31(3): 38–47. http://www.icir.org/vern/papers/reflectors.CCR.01.pdf.

52. Porras, P., H. Saidi, and V. Yegneswaran. 2007. A multi-perspective analysis of the storm (Peacomm) worm. http://www.cyber-ta.org/pubs/StormWorm/report. Accessed 16 Feb 2011.
53. Riley, S. 2006. Configure your router to block DOS attempts. http://blogs.technet.com/steriley/archive/2006/07/10/Configure-your-router-to-block-DOS-attempts.aspx.
54. Savage, S., N. Cardwell, D. Wetherall, and T. Anderson. 1999. TCP congestion control with a misbehaving receiver. *SIGCOMM Computer Communication Review* 29(5): 71–78.
55. Sherwood, R., B. Bhattacharjee, and R. Braud 2005. Misbehaving TCP receivers can cause internet-wide congestion collapse. In *CCS '05: Proceedings of the 12th ACM Conference on Computer and Communications Security*, 383–392, New York, ACM Press.
56. Smith, J. 2007. *Denial of service: Prevention, modelling and detection*. Ph.D. thesis, Information Security Institute, Queensland University of Technology, Brisbane, Australia.
57. Spiess, K. 2007. Worm 'Storm' gathers strength. http://www.neoseeker.com/news/7103-worm-storm-gathers-strength/. Accessed 12 Feb 2009.
58. Stewart, J. 2003. Sinit P2P trojan analysis. http://www.secureworks.com/research/threats/sinit/. Accessed 13 Feb 2009.
59. Stewart, J. 2007. Storm worm DDoS attack. http://www.secureworks.com/research/threats/storm-worm/?threat=storm-worm. Accessed 11 Feb 2009.
60. Stover, S., D. Dittrich, J. Hernandez, and S. Dietrich. 2007. Analysis of the storm and nugache trojans: P2P is here. *The USENIX Magazine* 32.
61. Strayer, W.T., D. Lapsely, R. Walsh, and C. Livadas. 2008. Botnet detection based on network behavior. In *Botnet detection: Countering the largest security threat*, eds. vol. 36 Advances in information security, 1–24. Berlin: Springer.
62. Symantec. 2002. Backdoor.sdbot. http://www.symantec.com/security_response/writeup.jsp?docid=2002-051312-3628-99&tabid=1. Accessed 13 Feb 2009.
63. Terdiman, D. 2004. Solution for slashdot effect? http://www.wired.com/science/discoveries/news/2004/10/65165. Accessed 1 Feb 2011.
64. UNIONTOWN. 2006. Student accused of trying to crash school's computer system. http://www.wkyc.com/news/news_article.aspx?ref=RSS&storyid=45721. Accessed 16 Feb 2011.
65. US Committee on National Security Systems. 2006. National information assurance (IA) glossary. Instruction 4009, CNSS.
66. Vaas, L. 2007. Storm worm botnet lobotomizing anti-virus programs. http://www.eweek.com/c/a/Security/Storm-Worm-Botnet-Lobotomizing-AntiVirus-Programs/. Accessed 12 Feb 2009.
67. Vamosi, R. 2008. Phishers now leasing the storm worm botnet. http://news.cnet.com/8301-10789_3-9847276-57.html. Accessed 11 Feb 2009.
68. Wang, P., S. Sparks, and C.C. Zou. 2010. An advanced hybrid peer-to-peer botnet. *IEEE Transactions on Dependable and Secure Computing* 7(2): 113–127. ftp://www.3gpp.org/tsg_sa/WG3_Security/TSGS3_15_Washington/Docs/PDF/S3-000571.pdf. Accessed 16 Feb 2011.
69. Watson, P.A. 2004. Slipping in the window: TCP reset attacks. Technical whitepaper, CanSecWest. http://cansecwest.com/core04/cansecwest04.iso.
70. Wilson, C. 2008. Botnets, cybercrime, and cyberterrorism: Vulnerabilities and policy issues for congress. Technical report, US Dept of State. http://www.fas.org/sgp/crs/terror/RL32114.pdf. Accessed 16 Feb 2011.
71. Yamaguchi, F. 2008. TCP denial of service vulnerabilities. http://ftp.ccc.de/congress/25c3/video_h264_720x576/25c3-2909-en-tcp_denial_of_service_vulnerabilities.mp4. Accessed 16 Feb 2011.

Chapter 3
Policy and Law: Denial of Service Threat

W.J. Caelli, S.V. Raghavan, S.M. Bhaskar, and J. Georgiades

3.1 Decades of Indecision: 'Weapons of Mass Disruption'

3.1.1 Introduction

A set of relevant quotes could 'set the scene' for research into and discussion of the policy and law aspects of DoS/DDoS against global, national and defence information infrastructures (GII, NII, DII), national critical infrastructure (CNI) and the nation state itself (Information Warfare, Cyber-warfare, Electronic Warfare).

> Conficker is still out there, and that's 28 terabits/second. If that thing was pointed at any U.S. national interest or any national interest, it would go down in a heartbeat.... My take is that every CIO should be shivering in a state of panic. *Mr Simon Crosby, CTO, Citrix Systems Inc., Florida, USA, 27 December 2010.* [120]

> 'If you spend more on coffee than on IT security, then you will be hacked,' Clarke said during his keynote address. 'What's more, you deserve to be hacked'. *Richard A. Clarke, 19 February 2002.* [109]

W.J. Caelli (✉)
Faculty of Science and Technology, Queensland University of Technology, Brisbane, Australia
e-mail: w.caelli@qut.edu.au

S.V. Raghavan
Department of Computer Science and Engineering, Indian Institute of Technology Madras, Chennai, India
e-mail: svr@cs.iitm.ernet.in

S.M. Bhaskar
Society for Electronic Transactions and Security, Chennai, India
e-mail: smb@nic.in

J. Georgiades
Faculty of Law, Queensland University of Technology, Brisbane, Australia
e-mail: jenny.georgiades@qut.edu.au

S.V. Raghavan and E. Dawson (eds.), *An Investigation into the Detection and Mitigation of Denial of Service (DoS) Attacks: Critical Information Infrastructure Protection*,
DOI 10.1007/978-81-322-0277-6_3, © Springer India Pvt. Ltd. 2011

There are some who believe we are going to have an electronic Pearl Harbor, so to speak, before we really make computer security the kind of priority that many of us believe it deserves to be. *Former US Senator Sam Nunn, 1998.* [84]

With the Republican ascension in Congress, the conservative abhorrence to regulation is intensifying. Barring a virtual 9/11, getting IT security regulation enacted will prove tough. *Eric Chabrow, Gov Info Security, 7 January 2011.* [80]

The pioneer in systems development in the online world has been the aviation sector, especially nimble low-cost carriers unburdened by legacy systems. In any given period at Qantas's budget offshoot Jetstar, at least *three out of four fares* are sold online through the group's website Jetstar.com. *The Australian, 8 January 2011.* [106]

...cyber security issues are also closely related to homeland security of India. On the front of cyber security as well India has to cover a long distance. India must develop cyber security capabilities as soon as possible. *Indian blog site, 8 January 2011.* [55]

We need to build greater resilience into Queensland's infrastructure. *Maj. Gen. M. Slater, DSC, AM, CSC – Leader, Queensland's Flood Recovery Taskforce, ABC Local Radio 612, 10 January 2011.*

3.1.2 Background: Dimensions in Policy and Law Research in Security and Resilience of National and Global Information Infrastructures

The research reported and discussed in this chapter comes about as a result of scientific and technological research cooperation between Australia and India. This program, the AISRF, was described in January 2011 by the Hon Kevin Rudd MP, Australian Minister for Foreign Affairs and Shri S.M Krishna, Indian External Affairs Minister, in the following terms following joint Ministerial talks in Australia (Fig. 3.1) [125]:

The Ministers welcomed the contribution, which the Australia-India Strategic Research Fund (AISRF) is making to the strong growth in cooperation in science and research, one of the most dynamic parts of the bilateral relationship. They noted that since its launch in 2006, the AISRF had supported over seventy innovative joint projects between Indian and Australian researchers and leading Australian and Indian institutions in strategically selected areas of scientific endeavours, including renewable energy, nanotechnology, agricultural research and biotechnology. Mr Rudd highlighted that Australia's commitment of $65 million over the current life of the Fund, which is matched by the Government of India, made the AISRF Australia's largest fund dedicated to bilateral research with any country. Mr. S.M. Krishna noted that the Fund was one of India's largest sources of support for international science.

This statement indicates the importance that both countries have given to matters of joint scientific and technology research, including matters related to the multidisciplinary research area of security and resilience in national information infrastructure (NII) and in the basic science, engineering and technology that underlies the products, systems and services created and delivered by the ICT industry worldwide.

Fig. 3.1 Photo includes Left to right: Mr Raghavendra Shastry, Adviser to the Minister of External Affairs, Her Excellency Mrs Sujatha Singh, Indian High Commissioner, The Hon. S.M. Krishna, Indian Minister of External Affairs, The Hon. Kevin Rudd MP, Australian Minister for Foreign Affairs, Ms Vijaya Latha Reddy, Secretary (East) Ministry of External Affairs. Melbourne, 20 January 2011

On 1 January 2011, Australia's national newspaper, *The Australian*, published an editorial [61] that looked back some 30 years, with the release of Australian Federal Government Cabinet documents, and forwards to the second decade of the twenty-first century, in the following editorial:

> Three decades ago, our policymakers joined the dots and recognised that globalisation would shake Australia to the core unless the country embarked on fundamental structural reform... Now as we enter the second decade of the 20th century we are embarking on a new round of globalisation that offers as many challenges to our politicians and business leaders as did the global changes of the 1980s. One challenge is the digital revolution of *high-speed broadband and ubiquitous connectivity* that is changing the way we live and work. (Emphasis by the authors)

From national and international business, government and diplomacy to social networks that rival the size of any large nation state or even region[1], the world has

[1]On 1 January 2011, the social network 'Facebook', based on server computers and disk storage devices, connected via the global Internet claimed a population of over 500 million 'active members' with over 50% making use of the network each day. Moreover, each one of those members is claimed to have an average of 130 'friends'. Any denial of service to this structure is now seen by many members as being of major significance, if not, catastrophic. Even small periods

entered a decade of rapid and largely unpredictable change but one in which the digital revolution will play a major and even definitive role. Indeed, the decade of the 'digital economy' and even the 'Internet Nation' has arrived, as long predicted. The impact of this digital economy was heralded recently in the following reported statement from a Google Inc. executive:

> There's really two economies running in parallel in the world today. There's the regular economy – everything you read in your newspaper every day. . . . Then there's the second economy that runs completely parallel to it, which is unencumbered from any of that. It's called the digital economy. In the digital economy right now, everything is going gangbusters. [65]

Meanwhile, that term 'Internet Nation' was referred to in an address entitled 'The New Internet Nation State' delivered in December 2010 by Mr R. Beckstrom, President and Chief Executive Officer of Internet Corporation for Assigned Names and Numbers (ICANN). He summarised the concept as follows:

> Once again we must reinvent the nation state. It is now time to design the *New Internet Nation State*. What will it look like? It must be built on openness, transparency and trust, with greater collaboration. Not theoretical collaboration, but the real kind that connects people. [72]

However, he did not comment at all on the problem of defence of that new 'state' against those who would attack it and did only marginally on its governance. As early as 1997/1998 Esther Dyson commented on the social effects of the Internet and the effects that this would have on governments. She stated as follows:

> Now, with the advent of the Net, we are privatising government in a new way – not only in the traditional sense of selling things off to the private sector, but by allowing organizations independent of traditional governments to take on certain 'government' regulatory roles. [89]

This movement for national governments to withdraw from policy, regulation and even enforcement, particularly during the time of the information technology 'revolution', has also been seen as government actually becoming more ineffective, particularly as the critical infrastructures of the nation are privatised. For the USA,

---

of downtime, as reported on 30 December 2010 by *The Australian* newspaper, may cause concern and frustration for its users. The question could easily be asked as to whether or not the 'citizens' of that 'Internet Nation', e.g. 'Facebook', and other 'social networks', could see DoS to the NII as simply an act of denying access by a citizen to his or her 'home country', their 'homeland'. Essentially these 'Internet' nations are operating largely outside the constraints of the geographic nation states of the eighteenth and nineteenth centuries. They are looking for 'governance' from a new, private sector 'government', often a privately or publicly held company based in one of the traditional nation states, e.g. 'Facebook'. The 'citizens' of this new 'State' in cyberspace could even be seen as maintaining human relationships on levels similar to the more traditional situation, e.g. maintaining a hierarchy of associations with other 'citizens' of the 'State'. [16].

Fareed Zakaria [139], the Mumbai, Indian-born, prominent journalist expressed this sentiment, referring likewise to the last 30 years or so, in the following way:

> ... it has developed a highly dysfunctional politics. An antiquated and overly rigid political system to begin with – about 225 years old – has been captured by money, special interests, a sensationalist media, and ideological attack groups.... the political process has been far more partisan and ineffective over the last three decades.

As discussed later in this chapter, Dyson also pointed out the distinction between the telecommunications/data network systems that provide the connection service and the 'end-of-line' equipment, namely the computers, that are attached and which provide the information services of interest. From the end-user viewpoint, the former is largely transparent. She again stated:

> Ironically, most 'Net' security risks don't actually happen on the Internet, they happen *at either end of an Internet connection,* where an intruder gets in or some information gets out.

Denial of service (DoS) attacks can easily disrupt services provided by an information system. Indeed, DoS attack has become a weapon which has been regularly used for gaining business edge over competitor companies by a maligned company, or even in organising 'cyber war' by an enemy country. Therefore, to protect Critical Information Infrastructure, there should be proper legal framework and policies against such Denial of Service attacks. Since cyber space has no boundaries there is a strong need for harmonisation of policy, legal and regulatory frameworks for critical information infrastructure protection.

The problem of DoS is ever-growing. A survey carried out by VeriSign in 2009 from a base of 400 DDoS professionals showed the future of DDoS threat in the coming years, as follows, given as a percentage of respondents who answered against a set of statements [51]:

It will increase substantially	25%
It will increase slightly	51%
It will neither increase or decrease	10%
Certain aspects (e.g. network level only) might decrease but others will increase (e.g. application-level attacks)	5%
It will decrease slightly	7%
It will decrease substantially	1%
Don't know/no opinion	3%

From an historical perspective, DoS (Denial of Service) attacks used to be less popular in the late 1980s, as performing them required more technical knowledge than was then available as compared to other more conventional attacks. They were often performed by individuals, using the physical resources of one, or a limited number of personal computers. With the start of 1990s, several websites were reported attacked by DoS attacks that rendered those sites largely inaccessible. By the year 2000, countless websites, including those of major businesses such as ebay, Amazon, Microsoft, and others were subjected to DoS attacks. The next generation of DoS attacks demonstrated that attackers were gaining more control. Rather than just flooding packets and trusting infected computers to perform as desired, attackers developed sophisticated malware, such as operating systems 'backdoors' or 'bots', that allowed them to issue commands to all the infected computers in a captured network of computers, coordinating and controlling the DoS attack. From an Indian national perspective, by the first decade of the twenty-first century, the 'IT Act 2000' came into the picture to provide legal support and redress against DoS attacks in India.

DoS attacks became more prevalent when a sufficient number of organisations and individuals became connected to the Internet to furnish a viable target. As detailed later, 'players' in the DoS/DDoS situation can come from a number of sources with a similar number of motivations, summarised as follows [69].

3.1.2.1 Politically Motivated Denial of Service Attacks

DoS attacks may be part of a declared or undeclared 'war' in the cyber world. The reported DoS attacks against the country of Estonia serve as an often quoted example. In 2007, a large and sustained distributed denial-of-service attack on several Estonian websites was reported, including those of government ministries and the prime minister's Reform Party. These attacks were allegedly launched from the Russian soil. It is said that it was done in a protest against relocation of the 'Bronze Soldier,' a Soviet-era war memorial commemorating an unknown Russian who died fighting the Germans in World War II.

3.1.2.2 Monetary Gain

Attackers may use DoS as an extortion tool, particularly against online gambling services, where lost downtime equates directly to lost revenue. Threats of repeat attacks are then accompanied by demands for payment. Alternatively, DoS may be used to disrupt the service of a competitor, or by attacks and false rumour depress a company's shares, which can be sold later at a profit [82].

3.1.2.3 Personal Achievement

Attackers may be motivated by a desire to impress their peers, or to exact revenge against a perceived enemy. The personal satisfaction is in proportion to the profile of the victim. This type of attack is prevalent among young hackers and may arise from simple boredom.

3.1.2.4 Information Warfare

Some attacks (such as the Estonian attacks) may be carried out for political reasons, including terrorism, typically against government entities. This term, however, is not reliably defined against acknowledged international definitions of warfare, e.g. by the United Nations.

As detailed later, India has taken a major step in legal response to these threats.

To provide legal and regulatory support to India's exponentially growing computer software industry, E-commerce and BPOs, as detailed later in this chapter, the 'IT Act 2000' was introduced and implemented. The growth of both the telecommunications and computer software and services sector as a part of India's gross domestic product (GDP) provides a vital background for the legal and policy decisions that have flowed over the last few years. For example, Section 43 of the IT Act 2000 defines Denial of Service (DoS) attacks. It defines DoS attacks as follows:

> Flooding a computer resource with more requests than it can handle. This causes the resource to crash thereby denying access of service to authorized users.

Examples of DoS attacks include attempts to 'flood' a network, to disrupt connections between two machines, to prevent a particular individual from accessing a service and to disrupt service to a specific system or person. The DoS/DDoS and resilience problem covers both the communications paths and the end computer systems but vulnerability mainly still lies with the computers at each end, as detailed further in this chapter. At the same time the second decade of the twenty-first century could be marked by a major change in how people use computers and telecommunications networks. The situation is that mobile, wireless connected systems are rapidly becoming the 'mainstream' for people to interact with information systems and to perform necessary and useful/entertaining transactions. In both Australia and India, this has become a major technical and business trend that is actually accelerating.

For example, mobile phones, essentially now equivalent to the PC of only a few years ago, are performing critical functions in banking and finance, healthcare, government service delivery and so on. In the banking sector, for example, they have been pressed into usage as transaction/data/user authentication mechanisms, e.g. issuing of a mobile phone SMS[2] message to a user performing a banking transaction on a desktop/laptop PC containing an authentication code, etc. A successful DoS/DDoS attack on such services is rapidly becoming a matter of concern as users depend on such systems while using vulnerable, commodity product level hand-held

[2]SMS stands for 'Short Message Service'. 'Wikipedia' explains SMS as "the text communication service component of phone, web or mobile communication systems, using standardized communications protocols that allow the exchange of short text messages between fixed line or mobile phone devices." It forms part of the internationally accepted 'Global System for Mobile Communications (GSM)' standards set developed and agreed in the mid-1980s. An example is the 'NetCode' system of Australia's Commonwealth Bank which is described by the bank as follows:

> NetCode SMS is a free service available to all NetBank customers. It replaces Personalised Identification Questions to provide added security when completing certain NetBank transactions...Every time you perform one of these transactions, a new NetCode will be sent to your mobile phone via SMS. To complete your NetBank transaction, simply key in the 6-digit NetCode in the 'Enter your NetCode' field. Each NetCode number securely matches the Commonwealth Bank's NetBank system. [6]

devices designed with little to no inherent computer security fundamentals and where functionality and ease of use take priority.

Indeed, the very nomination of the mobile phone and other devices as an 'out-of-band' security enhancement facility makes that very channel a target for DoS or allied attack. At the same time, concern has also been expressed about the vulnerability to DoS of the very 3G/GSM base service structure itself through such mechanisms as privately developed and operated base-stations use open source base station software systems. DoS/DDoS could also arise through misuse of genuine commercial femtocell systems offered by telecommunications vendors to enhance mobile services at low infrastructure cost. The Japanese company, NEC, for example stated in late 2010, in relation to India, as follows [58]:

> NEC Corporation today announced the *launch in India* of its ground-breaking 3G Femtocell solution enabling operators to offer high-quality, high-speed 3G mobile services to its enterprise and residential consumers within a few months. A trial of the NEC Femtocell solution is in progress with a mobile operator in India. NEC has 8 contracts worldwide with mobile operators for femtocell system deployments which make up nearly half a million femtocell devices.

Interestingly, the company NetGear in Australia advertises a femtocell product but states, [53]:

> **3G with Integrated ADSL2 + Modem: DVG834NH**
> WCDMA radios operate in the 1.9GHz Rx/2.1 GHz Tx bands
> Extends mobile device coverage into RF dead zones and improves reception in poor coverage areas, connecting mobile devices to the wireless network via ADSL
> Assures highest possible broadband data rates to mobile phones for downloads
> **SERVICE PROVIDER PRODUCT ONLY** – Not for general sale or ordering

However, the age of *the 'do it yourself', and even portable, femtocell* using open source software and commercial, commodity-level hardware could easily be here providing a low-cost threat to the basic channel of communication via the digital mobile/cell phone network.

Direct attack on the communicating channel is well known in the area of cryptology. In this area, a DoS attack consists of disruption of an encrypted message transmission causing failure of an encrypted message to be correctly deciphered or an attack on the actual cipher key management system in use. This type of DoS attack vector aims at making the communicating entities revert to plaintext operations which may be more easily intercepted. In these circumstances, a DoS/DDoS threat takes on new significance with a need for appropriate management policies to be in place to handle any such eventuality.

In relation to another relevant factor, the United Nations, as well as other commentators, has indicated that by 2008 50% of the world's population lived in cities. In the USA, that figure has been reported as being 80% by 2010 [75]. These cities have themselves become totally dependent upon their own information infrastructure and its connection to the national, regional and global networks for all aspects of economic, cultural, social and critical infrastructure management needs, e.g. power distribution, and mass transportation.

3.1.2.5 Overall Research Activity – 'Evidence Based Policy'

The research activities reported in this chapter have been guided by the principles clearly outlined by the former prime minister of Australia, the Hon Kevin Rudd, in a presentation given in April 2008 to leading members of Australia's Federal public service on the topic of policy development and implementation. He indicated as follows:

> Evidence-based policy.
> A third element of the Government's agenda for the public service is to ensure a robust, evidence-based policy making process. Policy design and policy evaluation should be driven by analysis of all the available options, and not by ideology. When preparing policy advice for the Government, I expect departments to review relevant developments among State and Territory governments and comparable nations overseas. The Government will not adopt overseas models uncritically. [124]

He emphasised the need to clearly identify current and future challenges and opportunities. He also stated that policy development should then be created against the perspectives of not just the government itself but also of 'all parts of the community'. That community now encompasses the 'Internet nation'.

This chapter discusses the policy and law challenges of DoS/DDoS on information infrastructures at three levels, i.e. the 'Global Information Infrastructure (GII)', the 'National Information Infrastructure (NII)' and the 'Defence Information Infrastructure (DII)'. However, as there has been much prior research into and discussion of the DII, this research has concentrated on the GII/NII aspects of the problem largely from the perspective of the private sector and the 'commercial' or non-military/intelligence government arena (Fig. 3.2).

However, there may be overlap in the assigned responsibilities for protection against DoS/DDoS between the defence/military forces, intelligence agencies, government in general at all levels and the private sector. This is exemplified by Australia's CSOC (Computer Security Operations Centre) activity within its DSD (Defence Signals Directorate), an entity within its defence department, but now charged with assistance to both 'commercial' government and owners/operators of national critical infrastructure through CERT-Australia. Another example of such overlap can be seen in the formation of 'CYBERCOM' within the USA's military establishment. However, in the US case, the boundaries of operation for CYBERCOM are yet to be defined as outlined as follows by its first commander, Army Gen. Keith B. Alexander. In a news article in September 2010 from the USA's DoD he is reported as stating as follows:

> There is confusion over who does what, the general acknowledged, so White House officials are leading an effort to sort through the needs of cybersecurity and update the policies and issues. "They are looking at the policies and authorities that need [re-]doing, and what's the right way to approach it," he said.

He went on to also state that:

> *...technology has outpaced policy and law.* The government, he added, still is dealing with laws that came out when the nation relied on rotary phones. "The laws we did 35, 40 years ago are what we have to update." [92]

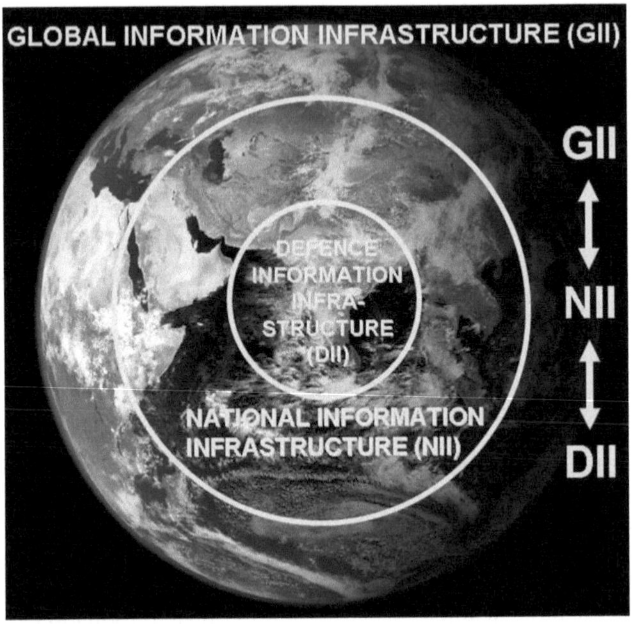

Fig. 3.2 The GII-NII-DII structure

This clearly supports the theme of this section in that it is obvious that there has been over 30 years of indecision by governments worldwide as ICT has gained a *dominant* place in ALL aspects of national and international activity, both public and private. Moreover, in many ways an internationally accepted 'monoculture' has developed in some arenas, e.g. in data networks the open TCP/IP Internet protocol set is now universal with little to no competition, at the desktop and laptop computer level, Microsoft Inc.'s 'Windows' operating system also dominates, etc. While providing benefits of ease of use with such uniformity, this monoculture factor also makes attack on those infrastructures easier with the same benefit of allowing the creation of widely used malware. At the same time massive privatisation of NCI has meant that governments themselves have to look for different processes to consider protection and administrative regulation of NCI enterprises. Normal commercial legal structures and imperatives do not indicate that corporate directors have any responsibility for national security or defence in general but this is exactly what the DoS/DDoS threat against such NCI is concerned about.

And so, given current and rapidly developing dependence on national and international information infrastructures, denial of service (DoS) to that infrastructure takes on new menace personally, locally, nationally, regionally and globally. Aside from accidental DoS, the assertion of DoS by deliberate attack on information resources exemplifies the factor of *assertion of power* over the victim and that assertion can be performed by any person, group or nation. However, the world is faced with what may be seen as three decades of *indecision by governments* worldwide in relation to any regulatory or related administrative environment, as mentioned above by General Alexander of the USA, to enable efficient and effective governance of this new environment, at least at the non-military level. Traditional government appears to have withdrawn from this sphere of activity and delegated responsibility to the private sector. However, such government-based regulatory structures did, and do, exist for the basic telecommunications industry and operators of the related infrastructure but, interestingly, outside some minor regulations related to electrical safety, regulation of the 'end-of-line' equipment (as explained below and nowadays a computer system in some form), have been non-existent. In contradiction to this, however, is the fact that governments have taken major governance roles in some industries, particularly those where 'life and limb' are seen to be highly relevant, e.g. the pharmaceutical industry, air transport, motor vehicles, drug and food sectors and swimming pool fencing. In many ways it can be clearly seen that, in general, government response waits till a sector becomes a major factor in society and attracts public and, thus, political attention. It can be argued that it has taken the last 30 years for this to occur the ICT industry and its associated service provision.

In considering the whole aspect of national information infrastructure (NII), and its protection and resilience, it is vital to understand the structure of that base, particularly when it forms the vital and underlying control and management element in nearly all national critical infrastructure (NCI). Essentially there are two major structures involved which, in many ways, have been considered for over 100 years. This information infrastructure consists essentially of:

- Interconnecting telecommunications infrastructure, now moving to a digital base and, in some countries, being reconstructed, e.g. the national broadband network (NBN) in Australia
- Connected ICT[3] products and systems, often referred to as 'end-of-line' equipment by the telecommunications industry

Both need to be considered in terms of the policy and law parameters needed for their governance. This research activity shows that almost universally only the former structure has been subject to any regulatory involvement, i.e. the telecommunications infrastructure, while the latter, the computer systems connected to that telecommunications infrastructure, have been largely ignored from a regulatory viewpoint, e.g. software product quality. Once again, the last 30 years have been vital in producing a total change in the nature of both of these elements.

In the telecommunications area the major factor that influences overall national and international information infrastructure is the growth of the 'Internet' as a general means of telecommunications in all forms, from voice to image to video

[3]The term ICT stands for 'Information and Communications Technology'. As such, it expresses the concepts of knowledge and process rather than the results of that knowledge, i.e. the artefacts or the products and systems that are created from ICT. As such, it is an expression of the knowledge and experience of humans.

to data and so on, and one that is replacing a century of base telecommunications infrastructure, commonly known as the 'circuit switched network'. At present, the Internet essentially operates on top of that prior telecommunications infrastructure, i.e. the 'circuit switched network', while in actual fact employing a different technology of 'packet switching'.

This change was heralded in 1988 by the move to interconnection of a number of data networks operational at the time. Kahn and Cerf state that, in their opinion, the Internet as we know it now really was not operational till around 1983, as they acknowledged as follows:

> Our work on the Internet started in 1973 and was based on even earlier work that took place in the mid-late 1960s. But the Internet, as we know it today, was not deployed until 1983. [105]

This interaction is summarised as follows:

> The first TCP/IP-based wide-area network was operational by 1 January 1983 when all hosts on the ARPANET were switched over from the older NCP protocols. In 1985, the United States' National Science Foundation (NSF) commissioned the construction of the NSFNET, a university 56 kilobit/second network backbone using computers called "fuzzballs" by their inventor, David L. Mills. The following year, NSF sponsored the conversion to a higher-speed 1.5 megabit/second network. A key decision to use the DARPA TCP/IP protocols was made by Dennis Jennings, then in charge of the Supercomputer program at NSF. The opening of the NSFNET to other networks began in 1988. [23]

However, it must be noted that data networks existed and operated commercially in many countries in the 1970s and 1980s, including in Australia as the 'AustPAC' service from 1982 and in the UK from 1980 as the Packet Switch Stream (PSS) service, and in some extent to the current day, based around the ITU/CCITT X.25 standard. This standard was incorporated into the lower four layers of the OSI (Open Systems Interconnection) model and should be seen as an alternative to the Internet's TCP/IP protocol structure. However, the base concepts of X.25 and the Internet's TCP/IP structures do differ in some ways, e.g. X.25's 'virtual circuit' concept.

Indeed, under the 'Open Systems Interconnection (OSI)' scheme governments worldwide adopted 'profiles', each called a 'Government OSI Profile (GOSIP)' or equivalent, that incorporated the stated OSI standard for security of systems and their interconnection, standard IS 7498-2, 'Information processing systems – Open Systems Interconnection – Basic Reference Model – Part 2: Security Architecture' [44] Policy, in some government cases, dictated that any computer and data network systems procured for public usage by that government must confirm with that GOSIP. However, even given such a policy directive, the OSI structure totally ignored along with its security architecture in favour of the overall Internet design, perceived to be a 'cheaper' model. The movement to a monoculture in network protocols and desktop/laptop operating systems, as mentioned above, presents a major security challenge. However, as mobile, wireless ICT products gain acceptance as the base interaction unit to information systems by everyone, the possibility of base system diversity arises given the limited functionality employed.

It could be noted that under the seven-layer model that formed the base for the OSi structure, the lower four layers:

- Physical
- Link
- Network
- Transport

were usually seen as being provided by the telecommunications infrastructure operators. The upper three layers:

- Session
- Presentation
- Application

were then seen as the responsibility of the end-of-line equipment supplier and service provider, i.e. the computer industry.

The marketplace, even in government procurement, essentially took over and there appears to have been no real consideration of the differences in security and resilience between the basic schemes. It should, however, be noted that computer/data communications in the 1970s was subject to quite a number of differing and often proprietary systems [37], e.g. IBM's 'System Network Architecture (SNA)' and Digital Equipment Corporation's DECNet. It could be that over 20 different schemes existed at the time. In summary, it has been pointed out that there could be at least three major periods in data network development as follows:

Period 1 – individual company systems and experiments.
Period 2 – telecommunications providers (PTT) begin to offer services based largely around the ITU's X.25 standard, while the ARPANet and like systems remain specialised and under the control of specific entities, e.g. the UK's 'JANET' system for interconnection of computer systems in the UK's universities, etc.
Period 3 – the Internet period, largely after 1988, as it became openly available.

However, equally important to the development of the national and international information infrastructure was the creation and marketing of the IBM PC, Model 5150, by that company in 1981. Essentially, since IBM made it an 'open' system, it enabled the creation of the 'PC Clone' industry and the rapid acceptance of the IBM PC as a business machine. From a DoS perspective, it is essential to examine not only the Internet protocol-based data communications infrastructure but also the hardware and software systems that are connected to it, which are still the main and easiest points of attack, outside physical or electronic attack on the telecommunications network itself. This IBM PC was based upon an operating system, DOS (Disk Operating System) supplied by Microsoft Inc. at the time, which had no security functionality or evaluation at all.

The system was meant to be, just as it is labelled, 'personal' and under an individual's control. Moreover, the original PC was not connected to any real data network at the time although very soon after its introduction add-in circuit boards were created to enable the PC to be connected to IBM SNA and other networks largely through emulation of a remote terminal, such as a member of the IBM 3270 series or simulation of a simple teletype unit or asynchronous visual display terminal such as the DEC VT100, etc.

The creation of the 'world-wide-web (WWW)' and 'MOSAIC' [67] browser program were also critical in making the combination of the Internet and the PC the force that it now is. However, like Microsoft's DOS, MOSAIC itself contained no security functionality. 'MOSIAC' is described in Wikipedia [32] as follows:

> Mosaic was developed at the National Center for Supercomputing Applications (NCSA) at the University of Illinois Urbana-Champaign beginning in late 1992. NCSA released the browser in 1993, and officially discontinued development and support on January 7, 1997.

3.2 Concepts in Policy and Legal Aspects of the DoS/DDoS Threat

Firstly, DoS has always been a major aim of any warfare activity. Indeed, even at the criminal level, threats of DoS are enacted through extortion and like activities. Thus, from warfare to criminal extortion, denial of service (DoS) has been a major activity for millennia. During World War II, for example, the French Resistance '...planned, coordinated, and executed acts of sabotage on the electrical power grid, transportation facilities, and *telecommunications networks*' [18]; in other words, the major critical infrastructures of the nation at the time living under conquest by Germany.

DoS through 'radio jamming' was, and is, an accepted military practice as demonstrated by Australia's RAAF 462 Squadron towards the end of World War II as follows:

> secret wireless jamming equipment codenamed "Airborne Cigar" (ABC) would locate and jam German flight control frequencies [35]

Essentially, radio jamming and allied tactics as practiced by 462 Squadron were all part of various operations against enemy information systems, namely radio and radar at the time, forming part of what is known collectively as 'electronic warfare'. [116] Essentially, as mentioned above, such activities demonstrate *exertion of power* over the target or victim.

This use of these 'electronic warfare' techniques to achieve DoS is also referred to by Clarke and Knake in relation to a report that '...Israel had bombed a complex in Eastern Syria, a facility being built by North Koreans.' [83] However, the important DoS note here by Clarke and Knake is that the attacking aircraft was not 'seen' by Syria's air defence radar and allied monitoring systems. They continue by stating as follows:

> 'What appeared on the radar screens was what the Israeli Air Force had put there, an image of nothing.'

Clarke and Knake clearly put the topic of cyber warfare directly into the DoS/DDoS arena when they continue to refer to the term as 'actions by a nation-state

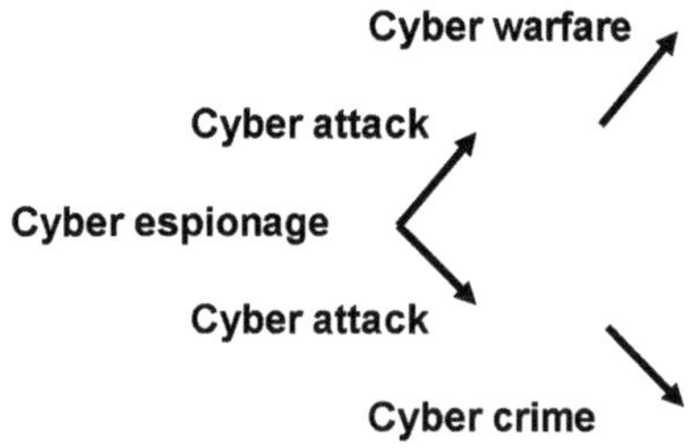

Fig. 3.3 Dimensions in cyber threats

to penetrate another nation's computers or networks for the purpose of causing damage or *disruption.*' Interestingly, the authors also clearly separate the network infrastructure from the computer systems connected to it either as managers and controllers of that network or as 'end-user' server systems, client systems or the like. This need to categorise the nature of DoS/DDoS attacks is discussed later.

3.2.1 Cyber Warfare and 'Weapons of Mass Disruption'

The terms 'cyber espionage', 'information warfare', 'cyber warfare' and 'cyber attack' have become commonly used since the mid-1990s. A number of books and monographs expound on the topic at depth [77, 108, 110, 123, 127]. Schwartau [127] defined three distinct forms of what he labelled as 'information warfare' as follows:

- Personal IW
- Corporate IW
- Nation State IW

However, it must be noted that Clarke takes a different view on the concept of IW and sees it as just an adjunct to any known and conventional form of warfare, once declared.

From a DoS/DDoS perspective, cyber warfare represents, as shown in Fig. 3.3, one extreme in the discussion in that, in a fundamental sense, acts of war aim at exactly that, 'denial of service'. At the other 'end of the spectrum', 'cyber crime' is the term used for those attacks on information systems that may be regarded as basically criminal activity, whether at a national or international level. The broad classification of such concepts is shown in the diagram.

3.2.2 Entities in Cyber/Information Warfare

Thus, in summary, this term is usually taken to refer to attacks that are carried out for political reasons, including terrorism and online protests (or 'hactivism'), which are typically directed at government and other critical infrastructure organisations.

3.2.3 Ultimate DoS/EMP Weapons

At the munitions level, the development and use of so-called electromagnetic pulse (EMP) weapons to disable all forms of electronic devices presents the ultimate in DoS activity. In this regard, a 2004 report on the topic stated as follows:

> EMP is one of a small number of threats that can hold our society at risk of catastrophic consequences. EMP will cover the wide geographic region within line of sight to the nuclear weapon. It has the capability to produce significant damage to critical infrastructures and thus to the very fabric of US society, as well as to the ability of the United States and Western nations to project influence and military power. The common element that can produce such an impact from EMP is primarily electronics, so pervasive in all aspects of our society and military, coupled through critical infrastructures. Our vulnerability is increasing daily as our use of and dependence on electronics continues to grow. The impact of EMP is asymmetric in relation to potential protagonists who are not as dependent on modern electronics. The current vulnerability of our critical infrastructures can both invite and reward attack if not corrected. [94]

However, from a DoS/DDoS perspective, it has to be noted that new vulnerabilities emerge as the density of computer chips and magnetic/electronic memory units becomes greater. This concern has long been considered in the aerospace arena as even cosmic rays may cause dense circuit failure and thus DoS. It has been reported that Intel Corp. has 'plans to tackle cosmic ray threat... because the operation of computers is through charged particles, the unpredictable hits from the rays are problematic, potentially causing the system to crash'. [49]

At the same time interference to wireless services, now taking on a critical national information infrastructure dimension, from sun radio bursts and related solar events must also be considered along with network interference as discussed below. This solar concern is summarised in the following statement from the Australian IPS (Ionospheric Prediction Service), now known as the Australian Space Weather Agency:

> Ejections of material from the sun may follow solar flares or other solar phenomena. If the material reaches the earth, it can produce magnetic and ionospheric disturbances with consequences for *HF communications,* geomagnetic surveys for minerals, the operation of long pipelines and *power lines,* and a host of other effects. [130]

This concern must be considered, to some extent at least, by those countries now seeing mobile/cell wireless communications as a lower-cost alternative for the provision of nationwide telecommunications services rather than the establishment of a fibre/wired environment. This movement is illustrated in India by the potential for over one billion cell/mobile phones coming into usage. A recent report, from a resilience viewpoint, pointed out that the associated transmission towers themselves, some 250,000 at present, must be continuously powered, often by diesel-powered generators.

> Considering the estimates predicting that India will have a billion cell phone subscribers by 2015, it is "extremely essential" that the policy mooted by the Central ministry of New and Renewable energy to energise cell phone towers using solar energy is implemented "as soon as possible" to reduce strain on the already scarce resources,... [63]

At this level of intensity, DoS from whatever source, deliberate attack, economic factors or natural phenomena, must become a major area for analysis and assessment.

Another engineering concern related to DoS, however, with the exploding deployment and usage of wireless-based products and services, is cross-network interference. This growth of wireless-based services has been recognised by APEC in 2010 as follows [60]:

> While we continue to encourage the use of all forms of information and communications technologies in a technology-neutral manner for infrastructure development, we recognize the widespread use of wireless communications as an important means to promote more efficient deployment of ICT infrastructure in the APEC region.

However, this concern has been stated by one supplier of these systems, Huawei, as follows:

> Radio interference is one of the key factors affecting the quality of wireless networks. As wireless communications technologies have developed, networks of different frequency bands and standards operating in the same region have become commonplace. Interference between wireless networks has grown more troublesome. When operators start to deploy LTE networks, how can operators address the more complex type of interference between wireless networks? [104]

Similarly, global movement to optical fibre-based data communications services means that the power service provided by the older switched service telephone exchange will disappear. This has important consequences for DoS in an emergency situation and currently the general stance is that telecommunications vendors moving to this structure need to provide customers with a battery backup unit capable of at least operating for a few hours. In Australia, the National Broadband Network Company (NBNCo) has announced that it will provide such battery-based DoS backup to premises for voice services, as indicated below [54]. However, there has been discussion as to whether or not this is a sufficient response to the resilience question. The possibility that the current situation, whereby the main telephone exchange provides power backup to at least voice communications, should continue has been canvassed. With generator capability at an exchange level, resilience could be maintained at a limited service level almost indefinitely, provided fuel is available for those generators [99].

This chapter continues by considering the more specific cases of DoS/DDoS by cyber attack and warfare rather than the wider case of electromagnetic interference concerns or electronic warfare, as mentioned above. While not specifically addressing those DoS possibilities, overall cyber threat analysis has reportedly led the international ICT consultancy and 'futures' firm, Gartner, to state:

> In the year 2015 a solar eclipse will cloak the north in darkness and a G20 nation will be rocked by a *cyber attack* as damaging as September 11. The attack may topple its stock market, or behead its government. The strike will *bombard the nation's critical infrastructure and communications networks.* [118]

Much earlier, but in a similar fashion, the bombing of Darwin, Australia by Japan in 1942 aimed, among other actions, at the destruction of telecommunications services, as follows:

> During the Second World War, the Japanese flew 64 raids on Darwin and 33 raids on other targets in Northern Australia. On 19 February 1942, 188 Japanese planes were launched against Darwin, whose harbour was full of Allied ships.... The first attack lasted approximately forty minutes. The land targets included the *Post Office, Telegraph Office, Cable Office* and the Postmaster's Residence, where postal workers were killed. (Emphasis by the author.) [29]

A bomb which hit the main Darwin post office killed the post-master and his family and six young *women telegraphists* sheltering in a slit-trench outside [129].

In simple terms, a DoS problem can arise from many sources and under many conditions.

The basic concepts in considering policy and legal aspects of the threats relate to the:

- Status of the attacking entity
- Nature of the attack itself
- Status of the entity under attack.

3.2.4 Status of the Attacking Entity

Distinction has to be made between the various forms of entity that may be the cause of DoS/DDoS. These entities may include:

- 'Acts of God', such as flood, earthquake, cyclone/hurricane, or other natural phenomena including 'sun spots', etc. nation state, as in a declaration of war that involves 'cyber warfare' activity as part of other instruments of war
- Formal military entities, as in state authorised or unauthorised attack
- State agencies, such as intelligence agencies
- Political, religious or other groups
- Organised criminal entity
- Specific criminal group or individual/individuals
- Disgruntled individuals or group of individuals, including ad hoc organisation of such individuals, competing private enterprises, and other entities

An 'act of god' is clearly represented by DoS caused by flooding in the city of Brisbane, Australia in January 2011. A major credit union, Credit Union Australia, apologised to its customers for loss of service nationwide caused by its systems being flooded in the following terms:

> Flooding and power disruptions in the Brisbane CBD are continuing to impact CUA's Brisbane Headquarters and this is currently *affecting our service availability* for the CUA internet site and online banking. CUA is working on restoring access for customers to these services as a critical priority, however we expect online banking to be unavailable for the

> remainder of today. We will continue to provide updates on our progress and if this estimate changes CUA will immediately inform customers. CUA apologises for this disruption to services and the inconvenience this is causing. [21]

However, the main policy and legal question here relates to the responsibility of corporate entities and their directors to provide what may be seen by some as 'extreme' backup and recovery (fail-soft, fail-safe) structures for critical information systems. As outlined below, there appear to be no policy or legal parameters available to really guide risk assessment and management in this regard for large-scale and critical information systems.

3.2.5 *Nature of the Attack Itself*

An actual deliberate attack may be considered as one of the following three, or possibly more, actions of relevance: Disaster caused by natural phenomena or other accident, namely 'Act of God',

- An 'Act of War',
- An 'Act of Terrorism'
- A 'Criminal Act'.

These actions need to be combined with the status of the attacking entity and the victim in order to determine appropriate policy and legal responses, if they exist.

3.2.6 *Status of the Entity under Attack*

The status of the 'victim' is also crucial in determining what policy and/or legal response applies. In this sense, the 'victim' may be any of the following or a combination of them:

- Military establishment, infrastructure or allied entity
- Government entity of national security significance and interest, e.g. office of the prime minister (India and Australia), etc.
- Individual State or Province administration, e.g. the State of Queensland, Australia, the State of Gujarat, India, etc.
- Owners/operators of systems of national security interest and/or of national critical infrastructure (NCI)
- Publicly and privately, or jointly, owned and/or operated
- Major commercial enterprises
- Small to medium businesses
- Religious/political entities
- Individuals/families, and others

Each of the above, as well as their combination, dictates the entity which can be held responsible for maintaining resilience against attack, i.e. maintaining sufficient 'cyber-defence' mechanisms and processes, and/or responding to an actual attack, i.e. endeavouring to determine the origin and nature of the attack as well as threatening or activating a counter-attack or allied 'information operation' against the attacker once determined.

3.2.7 *Status of the Entities of Concern*

Policy, law and regulation are three terms that are often used to define those processes that are relevant to efforts needed to develop, propagate and enforce procedures that contribute to order in society. Moreover, they can operate at a number of levels, individually or together, in nation states, regional associations of nations, international organisations, national and trans-national corporations, and so on.

Policy may be further separated into two different aspects, i.e. public policy and private policy. In the public arena, policy may lead to enactment of legislation that enhances law and then creates associated regulatory instruments. Alternatively, it may be used in general as indications of 'best practice', or the like, in legal cases where specific legislation does not exist. At the private sector level, enterprises are normally expected to have created and to administer security policy against clear understandings of the risks faced by the enterprise and the formulation of an associated risk assessment document or its equivalent. Moreover, the so-called industry self-regulation may consist of a set of standardised security policy statements for a particular industry or other enterprise coupled with a set of specific processes and procedures to be followed, i.e. a set of regulations. In turn, this set of regulations may then become the subject of public policy and then legislation through appropriate governmental processes. At a minimum, such industry developed and propagated 'regulations', usually without any legally or really enforceable penalty, can become vital evidence for submission in any legal proceedings.

Law may separately cover both 'actors' in any security situation, i.e. the perpetrator and the victim. It involves the understanding of the legal environment in which a nation 'lives'. Both India and Australia share a common inheritance of British Common Law which enables analysis of the approaches to the emerging concerns of threats to the national critical infrastructure (NCI) and allied national information infrastructure (NII) which has become a base for the NCI itself. Moreover, both nations also share a similarity in the structure of their military and intelligence enterprises, again inherited from a common English background. Legal concerns also cover the identification of those enterprises charged with the protection of national interests and the identification and 'bringing to justice' of

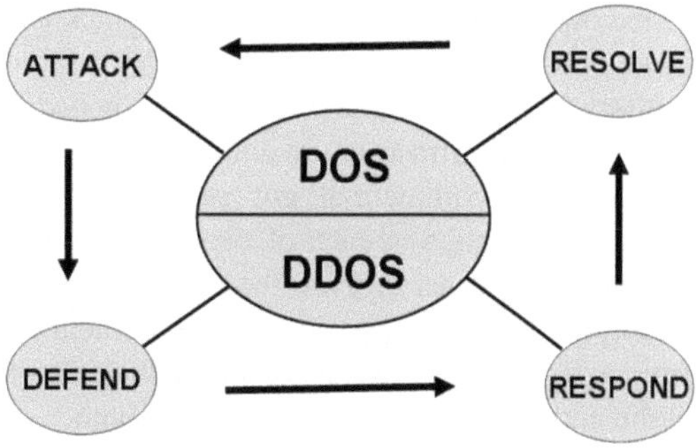

Fig. 3.4 Dimension in DoS/DDoS threat

any alleged perpetrator. In this sense, the range of relevant organisations can be large, from a small contracted security service to safeguard a business's information infrastructure to a nation's military forces.

3.2.8 Dimension in DoS/DDoS Threat

From both policy and legal perspectives, four distinct actions may be readily determined in relation to any DoS/DDoS matter. These four concerns are illustrated in Fig. 3.4 and include:

- Attack
- Defence
- Response
- Resolution

An *attack* may be mounted from any source, internally, locally, nationally or internationally, and normally refers to the organisational and technical aspects of the activity. *Defence* may include all the necessary technical and management processes needed to detect, protect and analyse the attack. Moreover, such systems as 'intrusion detection systems (IDS)' and 'intrusion prevention systems (IPS)' may form a vital part of the defence activity aimed at preventing the attack from being successful in penetrating the target system or from successfully operating. These defence systems need to be flexible and be able to react to changing attack technologies and circumstances as well as catering for vulnerabilities in the target system itself which may be known or which gradually emerge over time. *Response* refers to the ability, or inability, of the attacked system to provide any form of active countermeasure, including any form of retaliation, against the attacker or even to attempt to use such counter measures, including 'probes' of the attached network or apparent attacking computer system, to more reliably determine the source of the attack. It also includes the capacity to collect and store any necessary forensic data that may be needed for any later legal or related activity. In this regard, the status of the ownership of the system under attack is of critical importance. For example, if we consider the scenarios of a military information system within a designated military installation, a control system owned and operated by a state or privately owned critical infrastructure entity, such as those relevant to the national electrical power grid, a 'commercial' government information system, a private company, and an individual, then in each of these situations there may be marked differences in the authority to act or even the technical competence to act in a reliable manner. Resolution refers to the actions of settlement of the matter in hand which may vary from a declaration of a state of war in the case of a nation to attempts at legal redress at the national or international level or both.

3.2.9 Dimensions in the Attack Scenario

From policy and legal perspectives, the target, mode and perpetrator of an attack are all important aspects of any reaction that is to occur at a private sector, national government or international level. In this regard, the following terms need to be clearly separated and understood as being relevant to the overall policy making concern:

- Vulnerability
- Threat
- Accident
- Observation
- Attack
- Penetration
- Compromise
- Damage

3.2.9.1 Vulnerability

A vulnerability in any ICT product or system first of all has to exist in either a recognised or unrecognised form. Moreover, this dual possibility exists in relationship to both the vendor and user of a system. From a vulnerability viewpoint, responsibility usually lies with the system vendor to make good any necessary 'repairs'. The same can also apply to the owner/user of any ICT system where 'applications' have been purchased or developed and those application systems are in turn offered for sale either directly or in the form of a service. Protective measures would normally be in place to protect a system from exploitation of any such vulnerability, and such deployment and effective management of any such countermeasures would normally be considered to be part of the obligations of those responsible for the operation of the company or public enterprise. At the product level, vulnerabilities fit into the class of concerns related to product quality and 'fitness for purpose'.

Consumer laws may also apply. Simply put, if the ICT product or system were fully protected then denial of service via malicious attack would be highly improbable.

Legal questions may arise if a product is sold by a manufacturer/vendor with full knowledge of any such vulnerabilities without any such vulnerabilities being recognised by the customer. The question, from a policy viewpoint, then arises as to exactly who is responsible for ensuring that ICT products and systems contain required security functionality, particularly resilience against denial of service attacks, and that such functionality has been demonstrated as being robust and sufficient for the protection purpose needed, e.g. protection against DoS attacks. The international standard IS15408, 'Information technology – Security techniques – Evaluation criteria for IT security' [103], sets out such parameters under what is known as the 'Common Criteria' for ICT products and systems [79] but its existence and usage, outside the military and intelligence community environment, is minimal at best.

Similar assurance standards do exist at the national level, e.g. with NIST in the USA, and these often apply to specific industries such as the electrical power industry as discussed later.

3.2.9.2 Threat

A threat exists if a vulnerability can be exploited maliciously or can result in DoS or loss by some accidental or other means. Natural disasters can easily be recognised as a threat to any ICT system along with unpredicted failure of any critical software or hardware component. An example of this occurred in Australia in October 2010 in the case of the Virgin Blue airline, reported as follows:

> The failure of a hardware component in the outsourced Navitaire New Skies system disrupted the travel plans of 25,000 passengers and cost the airline millions of dollars as staff were forced to revert to manual processing.
>
> Reservation systems are the heart and lifeblood of airlines and even a partial loss can pose a serious problem.
>
> But in this case the primary check-in system stopped working and it took Navitaire 21 hours to switch to a backup system instead of the few hours it should have taken, forcing the airline on to a last-resort failsafe using laptops. It would take almost two weeks before Virgin Blue's reservations system was back up and fully operational. [85]

The threat here was the possibility of failure of a key component of the system which, in turn, presented a possible vulnerability. Large-scale disruption with associated human misery can be the result of enterprises losing the operation of ICT systems on which they now have become fully dependent. Their ICT systems are now critical infrastructure. Indeed, in the case of one of the earliest airline reservation systems, American Airlines 'SABRE' system developed by IBM, it has been stated as follows:

> In 1959, when American created Sabre, it could not have predicted that the information it gathered about flights and customers – and the way it gathered that information – would be more valuable than the flights themselves. [107]

This realisation is also referred to by Carr in the following way:

> Building and running computer systems had become as important to the success of American Airlines as flying planes and pampering passengers. [78]

Some 45 years after SABRE, the interesting phenomenon is one of 'outsourcing' whereby such mission and corporate critical systems as the enterprise information system is 'outsourced' to a third party. The information system and its associated computers are viewed as a 'cost centre' to be minimised where possible and no longer seen as the 'core' of the business. This has even led to 'offshoring' whereby that critical information system may even be subcontracted to an enterprise in another nation. These phenomena need to be seen as a progression with major implication for the resilience and overall security of an enterprise's most valuable asset, its information.

The threat, in this case, is one of the apparent lack of provision of what could be seen as necessary backup and recovery equipment capable of taking over all information activities rapidly, commonly known as a '*fail-soft* scheme'.

In many critical systems where DoS cannot be tolerated such fail-soft systems became common in the 1970s and 1980s in the era of the mainframe computer system, e.g. in on-line gaming and wagering systems. Such 'cut-over' to backup fail-soft systems was usually estimated to occur in seconds or minutes, in the worst case. Tandem Computers Inc., founded in 1974, was an example of a supplier of such systems that exemplified this approach to resilience against DoS. Its entry in Wikipedia states as follows:

> Tandem Computers, Inc. was an early manufacturer of fault-tolerant computer systems, marketed to the growing number of transaction processing customers who used them for ATMs, banks, stock exchanges and other similar needs. These systems use a number of independent processors and redundant storage devices to provide high-speed 'failover' in the case of a hardware failure, an architecture that is named NonStop. [137]

The point of note here is that recognition of a threat, namely accidental or deliberate loss of service, could be ameliorated by choice of an appropriate ICT product, system or service with proven lower levels of vulnerability or even a product or system that has clearly stated security functionality that has been independently evaluated. Essentially, this demonstrates that not all ICT products have the same level of quality assurance or set of vulnerabilities, as is the case in most other industries, e.g. car and truck industry, etc. The question as to whether or not current client-server systems built on commodity hardware and software components present a real threat to national critical infrastructure, including military requirements, has to be considered. Incorporation of these commodity products into vital systems with high resilience to DoS must be considered against what may seem to be reasonable 'commercial' risk assessment, which is then used in turn to determine investment levels in system resilience and security – a problem referenced by Clarke at the start of this paper.

However, commercial risk assessment does not adequately consider a ‘binary’ situation, such as total DoS, particularly where the end result may be very large scale or even catastrophic. Risk functions in the commercial arena have been built around more complex, but essentially continuous functions in loss assessment whereas DoS is completely different and involves large-scale anomalies which business, unlike the military, is unused to handling.

This ‘virgin’ case presents two interesting policy and legal aspects in DoS studies as follows. Firstly, the relevant information system was ‘outsourced’ to another party even though it forms the very ‘heart’ of the business, cf. the American Airlines SABRE system. Secondly, the resilience and robustness of the outsourced system is unknown to that airline’s customers, unlike the situation with the actual aircraft where specific and legally binding safety and security standards apply and are available for perusal. No security and resilience parameters for the ICT system, normally specified by the enterprise wishing to outsource critical information infrastructure, are known by an airline’s customers who depend upon it totally and who may suffer from DoS in a most direct fashion, e.g. stranded at an airport.

The threat from human attackers has to be considered in a different light. It has to be evaluated against a number of parameters which include the following attributes of the attacker or attackers:

- Technical knowledge at an advanced level
- Access to education and training, or access to those with that knowledge
- Ability to create appropriate ‘cyber-weapons’ to enable a DoS/DDoS attack, e.g. virus/Trojan or allied software products and systems
- Access to ‘cyber-munitions’ to actuate an attack, i.e. the actual data sequences and message content in the required format that need to be ‘fired’ from the ‘cyber weapon’
- Financial resources to purchase each of the above or to ‘sub-contract’ or outsource the DoS/DDoS attack itself
- Motivation to perform the attack

Now that technical knowledge may become even easier if all connected devices become members of one ICT ‘monoculture ecosystem’, i.e. one using Microsoft Corp software and firmware products as a base. This was alluded to by Mr S. Ballmer of Microsoft in his inaugural speech at the 2011 CES conference and exhibition in January 2011 as follows:

> Support for system-on-a-chip means that Windows will be *everywhere on every kind of device* without compromise; all the performance and flexibility of Windows on low-power, long-lasting devices. [71]

Essentially, what is indicated is support for a situation where a commodity-level software structure is incorporated into all devices linked into the international Internet via mobile, wireless services as well as traditional fixed line structures. The attack vector via such embedded systems now becomes an even bigger opportunity.

3.2.9.3 Accident

A DoS situation can occur through either accidental/unintentional or deliberate causes. A DDoS activity must almost always be considered to be deliberate, even if its actual occurrence is triggered from unsuspecting systems themselves. These systems, usually 'Trojan'-infected systems forming a 'botnet', have themselves been subject to deliberate attack in order for the attacking 'Trojan' program to be inserted. As mentioned above, the policy and legal parameters in the accidental situation would appear to fit into the arena of individual, enterprise or government responsibility. In the case of a public company, e.g. such responsibility is set out by law in relation to the responsibilities of directors. Such responsibility applies whether or not natural phenomena are involved, (flood, fire, earthquake, etc.), or if system reliability is the problem. A recent example of this latter problem is an outage of Microsoft's 'Hotmail' email services reported as follows:

> A service bug that left a group of Windows Live Hotmail users without access to new messages and entire folders for days has been explained and remedied against future instances. Writing on the Windows Team Blog, Mike Schackwitz of the Hotmail team says the problem stemmed from an error with an automated script that Microsoft uses to test the service for errors in every day usage. [112]

With massively increased system complexity, DoS caused by unknown interactions in software systems can be expected to be a major DoS problem in the near future and a problem that needs to be addressed in policy formation. However, at the simple physical level, the dependence of the global Internet on undersea cables[4] is almost total and is a major concern should such infrastructure be accidentally or deliberately disrupted.

3.2.9.4 Observation

What Clarke has called 'cyber-espionage' [95], the detailed observation of a potential DoS attack victim, forms the base for any attack itself. In turn, such cyber-espionage can in principle become a topic of interest in relation to both 'cyber warfare' and 'cyber crime', activities which it would normally precede.

---

[4]The 'tech-exclusive' website, on 27 April 2010, reported as follows under the banner headline 'India faces Disrupted Internet Service Due to Undersea Cable Issue'

> Getting slow download speed? Well things will continue like this for the next 3 to 4 days. The slow internet speed is to be blamed on an undersea cable repair going on. The SEA-ME-WE4 undersea cable got cut a few weeks ago and is now being repaired. This cable links Europe, the Middle East and South Asia region. The worst hit country is India. This single cable carries around 89% of the traffic load of the above mentioned region. The major internet players – Bharti Airtel and Tata Communications – get their bandwidth from the same undersea cable. This is not the first time that an undersea cable cut has disrupted the internet speeds in India. In 2008, an undersea cable cut led to India losing around 50% to 60% of its internet bandwidth. [122]

3.2.9.5 Attack

An attack on a system may be considered to be an observed or unobserved attempt to compromise that system in some way. It may or may not be successful. In policy and legal terms, there are many parameters that must be considered surrounding an attack in order to determine what, if any, policies or laws apply, e.g. the origin of the attack whether internal, local, national or international, the level of the attack, the ability of a system to withstand an attack and prevent penetration and location of any countermeasure system from within the target system itself to its network connections.

The nature of the attack, in the case of ICT systems and services, can be quite varied but can achieve the same result. For example, the accepted attack 'vector' of 'flooding' the target system with unwanted messages or data packets so as to prohibit it from performing its normal tasks is well known. However, equally successful may be attack via the deprivation of the Internet's 'domain name system (DNS)' to the attacked system in such a way that the attacked system can no longer address any entities on the Internet. Alternatively, the vital 'routing tables' that permit traffic on the Internet to be correctly received from and sent to necessary computer systems may be disrupted. The attack, therefore, may vary in its level of sophistication. Essentially, there are two major types of deliberate DoS/DDoS attacks possible on an ICT system, namely:

- 'Flooding' attacks
- 'Vulnerability exploitation' attacks

This separation of types is discussed by De Villiers from the legal perspective with the conclusion that differing legal parameters may apply [86].

3.2.9.6 Penetration

A successful attack culminates in the penetration of the target system. The nature of that penetration can take various forms from gaining access to confidential data, either owned by the enterprise itself or held in trust for other parties, to physical damage caused to systems that are controlled by that target computer system. Supervisory control and data acquisition (SCADA) systems exemplify the last case and present a real concern in relation to the safety and security of national critical infrastructure (NCI) such as the electrical power grid, water and sewage reticulation, etc.

3.2.9.7 Compromise

Following successful penetration of the system, an attacker may wish to violate programs and/or data that form the base of the targeted system. This could include placing 'back-door' code into computer programs in such a way that critical data are sent to the attacker for use in planning larger-scale attacks aimed at DoS, for example. It may be the more simple act of defacing pages in a website so as to attempt to destroy confidence in that website by its users. This form of 'psychological operation' achieves its DoS aim by simply causing users to distrust the system and thus to avoid using it.

3.2.9.8 Damage

The attack may succeed in its intent by finally destroying or modifying data and programs in the target system so as to render the system inoperable or for some other purpose. It may be sufficient for the attack to be successful if it just succeeds in slowing down system response so as to force users to another enterprise, etc. In the case of 'cyberwar', Clarke summarised the situation in the ABC interview referenced above in the following way:

> Cyber war can do many of the same things that regular war can do. It can blow up electric power grids and cause blackouts, it can cause trains to derail, it can cause pipelines to explode, refineries to explode. In other words, this is what people don't normally get, cyber war isn't about ones and zeros killing each other in some theoretical fourth dimension, it's about making things in the real world blow up. [83]

Damage can thus extend way beyond the actual computer and data network that is attacked. DoS extends to the major, critical services of a nation. However, all this was reported in 1997 in the USA in the 'Report of the President's Commission on Critical Infrastructure Protection, October 1997' in the following terms:

> In the cyber dimension there are no boundaries. Our infrastructures are exposed to new vulnerabilities – cyber vulnerabilities – and new threats – cyber threats. And perhaps most difficult of all, the defenses that served us so well in the past offer little protection from the cyber threat. Our infrastructures can now be struck directly by a variety of malicious tools. [115]

This was again emphasised in the January 2011 interview with an IBM Vice-President for strategy in security solutions as follows:

> The electric grid, oil and gas refineries and water facilities could top the list of targets by cyberterrorists hell bent on causing destruction or chaos in 2011, according to IBM, which issued a list of cybersecurity predictions for 2011. Cybercriminals can target embedded devices found in a number of critical infrastructure systems to cause harm or societal conflict, said Kristin Lovejoy, vice president of strategy for IBM Security Solutions. IBM is also predicting an increase in mobile threats posed by smartphones and tablet devices, an increase in global compliance mandates making business across borders more complex and a renewed emphasis on security being built into software and systems, rather than being bolted on. In this interview, Lovejoy addresses mobile and virtualization security, and the government's role in securing critical infrastructure... What is the government's responsibility or capacity to mandate security in and around those systems?" The pattern I'm seeing arising in the marketplace is very close to our concept of 'secure by design.' Security must be an intrinsic element of the service that is being delivered and to that extent these critical infrastructure industries need to think about it. They are going to be mandating that this happens less through direct prescriptive legislation and more through

> the procurement process. If I'm a utilities provider and I want a contract with a local government to provide utility services, in order for me to do so, I need to adhere to a number of security principles before I can respond. [135]

This statement from IBM offers support for any concern for the threat imposed by an ICT system 'monoculture' already mentioned above. A suggestion from India for the creation of its own national base software system product set can be seen as a response to this threat situation as discussed below.

3.2.10 *Defence and Response Status*

Defence and response reflect two differing philosophies in relation to DoS/DDoS attack. Any entity normally aims at providing sufficient *defence* mechanisms and services to enable attacks to be repelled or their effects minimised. Such provision of defence services may be considered a normal part of any management obligations on an information system. This involves such concepts as 'secure by design' as well as secure through management and operational processes, etc.

The term response is used often but there is no clear definition of what a response to DoS/DDoS attacks constitutes. The term 'response' might include both passive and active resistance to a perceived threat or an actual attack. In all probability, active resistance or response, such as a 'counter-attack', may not be legally possible although being technically feasible. This would certainly be the case if the source of the attack were not clearly and reliably identified. There are a very limited number of alternatives open to a victim.

Firstly, the system being attacked may be shut down but this alone may cause severe damage to users of a critical infrastructure, e.g. the electrical power grid. In other instances, the traffic may be diverted away from the network or systems under attack, if that is possible, or the load switched to another 'surrogate' system. However, any such action may cause damage or harm to another network. Apart from the technical difficulties, there are the uncharted legal waters if a retaliatory strike is mounted on the perceived attacker who may or may not be the real attack source. In the latter case, as for the use of so-called botnets, the direct attack systems may belong to and be operated by unknowing users themselves who have not been able to detect the 'capture' of their computer or its ongoing attack actions.

There is much discussion in regard to appropriate laws and policies that can help in countering DoS attacks. These discussions have included the following topics.

3.2.11 *Due-Care by ISPs and Websites*

The Wordnet Dictionary defines 'due-care' as:

> The care that a reasonable man would exercise under the circumstances; the standard for determining legal duty.

Due-care is a powerful legal theory of liability because all members of society are liable if they negligently harm another. This includes ISPs and websites related to critical information infrastructure. ISPs and websites could be held liable when they transmit malicious code that infects users' computers. Reasonable care might include taking reasonable security measures to detect and stop a DoS attack. ISPs in particular are well placed to detect attacks; in a sense they are the portals of such attacks. Once an ISP detects an attack it can terminate service from its customers sending out the malicious code. This policy statement is discussed in the Australian situation later in this chapter.

3.2.11.1 Standardisation of Security and Risk Assessment

The International Standards Organization (ISO) in the year 2000 issued ISO 17799, its 'Code of Practice for Information Security Management'. ISO 17799 addresses the topic in terms of policies and general good practices. If ISO 17799 is to become a useful tool in adapting due-care, it must first be incorporated with the ISP policy. These minimum security standards could establish a legal standard of care owed by ISPs and others.

3.2.11.2 Policies

The following topics, e.g., may be included in any policy statement by a relevant entity:

- There should be a policy for ISPs and large networks to employ antivirus protection that is frequently updated.
- It should be assured that all software used on the Internet pass a test of minimum security standards. Computers sold in the country should come pre-loaded with firewall protection.
- ISPs should keep records for a minimum amount of time so that attacks can be back-traced.
- ISPs and large networks should promptly terminate service to users who send malicious code, and giving firms legal immunity from liability for such action.
- ISPs should verify email senders' 'sent from' addresses.
- Parents of 'script kiddies' should be financially responsible for their children's DoS damage.

Thus a set of 'best practice actions' may be the method used to resolve many concerns in relation to DoS/DDoS threats, as follows:

Firewall Usage

Firewalls work by blocking attempts by individual hackers to intrude into your computer. They can therefore be effective against hacked denial of service attacks but they do not help against a distributed denial of service attack. A hacker can enter your computer by exploiting vulnerabilities in the computer's operating system or its applications.

Patching

A computer virus requires action by the user in order to spread. This action typically involves opening an email or an email attachment. Worms, on the other hand, can enter with emails but then spread automatically by themselves, without any action by the user. Worms are best deterred by diligently using software patches, as described below. Viruses can best be deterred by using up-to-date anti-virus software. As soon as a software firm discovers a new vulnerability, it develops a corrective program for it, called a patch. Prompt application of the patch will protect against that worm.

Shutdown

Whenever the user's computer is not being used for an extensive period it should be shutdown as opposed to letting it run in power-saver mode. Although the computer appears to be shut down in power-saver mode, a hacker could gain access to it.

User Actions and Controls

As noted above, most viruses are spread by email, and require the user to open the message or its attachment. Email with unfamiliar sender addresses or names should be deleted. The most vulnerable part of the email is the attachment. The user should never open an attachment to an unrecognised email address. Most Internet browsers, firewalls, and intrusion detection systems have security settings. If the user finds that his or her computer is under frequent attack, a higher security setting may provide better protection. The downside of increasing security settings is that they may keep out wanted communications.

If prevention efforts have failed and an attack has been made on your computer or system, there are still best practices that can be undertaken to limit the resulting damage.

3.2.11.3 Data Back-Up

Important data should be regularly backed up. Optical discs or an external hard drive are convenient mechanisms for this. Once the data is backed up, thought must be given to the physical location where the back up data will be stored. It is desirable to store the back up data a reasonable distance from the original data. Several firms located in New York's Twin Towers had their back up data stored in the other tower. After the 9/11 attack, that back up data was lost. Most firms, organisations, and even some large families use a network of computers. One computer may become infected before the others. A prompt removal of that computer from the network may save the others from infection. It is useful to have an extra computer available and ready to take the place of an infected machine.

Contacting Law Enforcement Agencies

Promptly contacting law enforcement may be required for insurance purposes. It is also important so that law enforcement can secure evidence that may be used in a criminal prosecution of the hacker who perpetrated the attack.

So, response to a DoS/DDoS situation may involve a number of actions, including:

- Migration of all information systems activity to a 'backup' facility, e.g. in the case of natural disaster and malicious attacks
- Active attempts to determine the origin of the attack
- Counter-attacks on the attacker, once reasonably determined, in order to stop or hinder the attack
- Counter penetration of the attacking systems in order to hinder further attacks or to deflect current attacks
- Passive 'attack' or 'probing' so as to observe the attacking system in order to be able to predict future attack situations, etc.

The nature of the entity performing the response must also be considered. These entities may belong to any of the following groups:

- The attacked entity itself
- A contracted company charged with providing protection services, e.g. 'outsourced' information security systems and services
- A computer emergency response team (CERT) within the attacked entity itself or acting on behalf of the entity through a form of contract or subscription service or government provided service
- The appropriate Internet service provider (ISP) responsible for the connections used by the attacked entity, which may even be the ISP of the attacked system itself, or an appropriate telecommunications service interface supplier
- Police and law enforcement acting on behalf of the attacked entity, with associated international affiliations, e.g. Interpol
- A government body, including any national intelligence or defence-related organisation
- A military or designated other government establishment

However, the responsibility of the ICT industry itself, through the products, hardware and software, and services it produces and sells must also be held responsible for the DoS/DDoS protection of its customers. IBM researchers clearly stated the matter in the following way in 2002 [88]:

> ...hardware on which applications run must be secure, as must the operating system and run time environment in between, while offering a reasonable API for application developers...
>
> ...applications cannot be more secure than the kernel functions they call, and the operating system cannot be more secure than the hardware that executes its commands...

3.3 International Perspectives

The following is a list of some of the international groups who have various roles to play, some in the policy area, some in the law area and some encompassing both. The list is not exhaustive and is presented as an indication of the overall global significance of activities in the cyber security arena.

3.3.1 ITU/UN: The International Telecommunications Union and the United Nations

In 1865 the 'International Telegraph Union' was formed in Paris, France. The name was changed to the 'International Telecommunications Union' in 1932. The organisation exists to 'coordinate the establishment and operation of global telecommunication networks and services. It consists of Member States and sector members representing public and private companies and organisations with an interest in telecommunications.' [52]

The ITU claims a major role in all aspects of security in telecommunications arenas and in cybersecurity, in particular, with the following statement:

> A fundamental role of ITU, following the World Summit on the Information Society (WSIS) and the 2006 ITU Plenipotentiary Conference, is to build confidence and security in the use of Information and Communication Technologies (ICTs). At WSIS, Heads of States and world leaders entrusted ITU to take the lead in coordinating international efforts in the field of cybersecurity, as the sole Facilitator of Action Line C5, 'Building confidence and security in the use of ICTs'. In response, ITU Secretary-General, Dr. Hamadoun I. Touré launched the Global Cybersecurity Agenda(GCA), which is a framework for international cooperation aimed at enhancing confidence and security in the information society. [28]

For instance, in 2008, the ITU drafted in Geneva recommendation X.tb-ucr, *Trace back use case and requirements,* looking at ways to identify the source address of packets sent via the Internet Protocol (IP). This was aimed at preventing flood-type DDoS attacks in which the origin of the attack packets had been deliberately falsified or 'spoofed'.

The ITU is also involved in developing a 'Botnet Mitigation Toolkit', which attempts to increase awareness of botnets and their associated criminal activities. Based on the Australian Internet Security Initiative (AISI) the toolkit draws on existing resources and takes into account the specific needs of international member states with a view to both raising awareness of the threats posed by botnets and helping to develop effective policy to counteract them.

3.3.2 *ICANN/IANA*

The 'Internet Corporation for Assigned Names and Numbers (ICANN)' [25] and its associated enterprise the 'Internet Assigned Numbers Authority (IANA)' [24] have roles to play in the overall security and resilience of the Internet. In particular ICANN maintains a 'Plan for Enhancing Internet Security, Stability, and Resiliency'. As national and international telecommunications move to packet switched networks and the Internet protocol set in particular, the Domain Name System (DNS) plays a critical role in the operation of the network. ICANN, through IANA, manages the content of the root domain, the pool of unallocated IP addresses (IPv4 and IPv6) and the Internet protocol's numbering system. The World Wide Web (WWW) and associated 'web applications' thus depend upon it for their very operation. [24]

This dependency was revealed by the distributed denial of service (DDoS) attacks against the 13 root servers that formed the backbone of the Internet's Domain Name System (DNS) system in 2002, and again in 2007 [47, 48, 56, 90, 133]. Although they had only a brief and limited effect, these attacks represented a serious threat to the viability of the Internet. Accordingly, the ICANN Security and Stability Advisory Committee (SSAC) regularly issues advisories on DNS Distributed Denial of Service (DDoS) attacks[5]. ICANN is also encouraging the deployment of Domain Name System – Security Extensions (DNSSEC), which is designed to ensure that DNS data has not been modified in transit through the use of digital signatures, establishing a chain of trust originating from the root servers [100].

3.3.3 *Organisation for Economic Cooperation and Development (OECD)*

The (OECD) was formed in 1961 out of the Organisation for European Economic Cooperation and Development (OEEC), which had been established in 1947 to administer the 'Marshall Plan' for the reconstruction of Europe after World War II. Its mandate is to help its 34 member countries achieve sustainable economic growth and improved standards of living, and so contribute to the development of the world economy. It assists governments by monitoring statistical data, trends, analyses and forecasts, and by identifying 'best practice' in developing their own domestic and international policies [36].

---

[5] A contributing author, Caelli, is the Chair of the SSAC for auDa, the Australian Domain Name Authority.

In 2008 [119], the OECD recognised that 'the Internet has, in a short space of time, become fundamental to the global economy.' As such, the OECD has taken a major interest in all aspects of Denial of Service (DoS) related to the Internet and has been organising international workshops on all matters related to denial-of-service threats and their mitigation.

Both Australia and India are members of the OECD.

3.3.4 APEC

'Asia-Pacific Economic Cooperation (APEC)' consists of some 21 member 'economies' having been formed after a suggestion by former Prime Minister of Australia, Mr Bob Hawke in 1989. Its main goal is to create greater prosperity for the people of the Asia-Pacific region [68].

Of interest is its stated aim to enhance 'human security'. A key part of its 'Strategic Action Plan for 2010–2015' is to 'Promote a Safe and Trusted ICT Environment' to protect consumers and their personal data, and to ensure a safe and secure online environment by collaborating with all relevant stakeholders in undertaking cyber security initiatives against perceived threats.

After the 9/11 attacks on the USA, the APEC Leaders issued a Statement on Counter-Terrorism, urging the reinforcement of APEC activities to protect critical infrastructure. In 2002, Telecommunications and Information Ministers of the APEC economies set out a Program of Action, designed to secure information and communication infrastructure. They made a commitment to:

> Endeavor to enact a comprehensive set of laws relating to cybersecurity and cybercrime that are consistent with the provisions of international legal instruments, including United Nations General Assembly Resolution 55/63 (2000) and Convention on Cybercrime (2001) by October 2003.

By 2003 the e-Security Task Group had created a 'Cybercrime Legislation & Enforcement Capacity Building Project'. And in 2010 APEC produced a major statement in its 'Key APEC Documents 2010' under the broad heading of an ambition to 'Promote a Safe and Trusted ICT Environment' as follows:

> We recognize that our society is becoming increasingly dependent on ICT, which means that online threats can have a major social and economic impact. The safety, security, trustworthiness, and reliability of ICT in the APEC region are crucial to ensuring ICT's important contribution to sustainable growth.
>
> We emphasize the need for enhanced measures to address malicious online activities. We also recognize that a safe and trusted ICT environment is facilitated, in part, through *strengthened consumer protection measures*, which include effective policies that protect personal information protection practices and promote the security of networked systems.
>
> We therefore encourage each economy to enhance mutual cooperation on countering malicious online activities, to engage in efforts to increase cybersecurity awareness and to share information on protecting ICT. These efforts need to align with efforts by and in collaboration with industry partners, the Internet technical community and all other relevant stakeholders including Internet Service Providers (ISPs), telecom operators as well

> as regional and other international organizations. Such efforts will foster a more secure online environment that protects ICT networks and users and secures access to information in an appropriate manner.

Of particular note here is the emphasis on 'consumer protection'. This acknowledges that ICT, and particularly the computer hardware and software industries, operates in an environment of little to no governmental regulation, except for aspects of the telecommunications services industry, unlike most other industries. Moreover, the consumer or 'end user' of ICT systems is the least knowledgeable in relation to security services and mechanisms on commodity-level computing systems. Overall APEC has designated six major areas for cooperative development of cyber security activities:

- Development of legal structures
- Information sharing and cooperation
- Guidelines
- Public awareness
- Education and training
- Wireless security

3.3.5 Others

3.3.5.1 FIRST – Forum for Incident Response and Security Teams

While not directly involved in policy formation, it could be argued that FIRST has a role to play in the legal domain. It describes its mission as follows:

> Forum for Incident Response and Security Teams (International) (FIRST) brings together a variety of computer security incident response teams from government, commercial, and educational organizations. FIRST aims to foster cooperation and coordination in incident prevention, to stimulate rapid reaction to incidents, and to promote information sharing among members and the community at large [17].

From a legal viewpoint, FIRST creates and issues so-called best practice documents that may be citable as evidence in any proceedings. India's CERT-In and Australia's CERT-Australia and AusCERT are members of FIRST.

3.3.5.2 Interpol

Interpol is the world's largest international organisation of police forces with 188 members. It maintains a working group related to all forms of Information and Communication Technology (ICT)-related criminal activity with members from its representative organisations worldwide [20].

3.3.5.3 IETF

The term 'IETF' stands for the 'Internet Engineering Task Force'. Its stated ambition is summarised as follows:

> The goal of the IETF is to make the Internet work better. The mission of the IETF is to make the Internet work better by producing high quality, relevant technical documents that influence the way people design, use, and manage the Internet [64].

From a policy and legal perspective, the construction, role and function of the IETF can be seen as being outside the considerations of international regulatory or related bodies, governments or states. Moreover, its activities are stated as being governed by 'cardinal principles' which may be summarised as follows:

- *Open process* – any interested person can participate in the work, know what is being decided, and make his or her voice heard on the issue. ...
- *Technical competence* – ... issues where the IETF has the competence needed to speak ... and ... is willing to listen to technically competent input from any source.
- *Volunteer Core* – our participants and our leadership are people who come to the IETF because they want to do work that furthers the IETF's mission of making the Internet work better'.
- *Rough consensus and running code* – We make standards based on the combined engineering judgement of our participants and our real-world experience in implementing and deploying our specifications.
- *Protocol ownership* – when the IETF takes ownership of a protocol or function, it accepts the responsibility for all aspects of the protocol, even though some aspects may rarely or never be seen on the Internet [66].

An Internet RFC explains the term DoS in a more specific sense, for that network, as follows:

> Denial-of-Service (DoS) attack is an attack in which one or more machines target a victim and attempt to prevent the victim from doing useful work. The victim can be a network server, client or router, a network link or an entire network, an individual Internet user or a company doing business using the Internet, an Internet Service Provider (ISP), country, or any combination of or variant on these [96].

It should be noted that this RFC does not consider the case of the actual user computer systems that are attached to the Internet and thus only examines those sub-systems that make up the Internet itself. It should also be noted that, in relation to the processes and procedures used in creating Internet standards, 'it is difficult to linearise a complicated and interlocked process [76].'

As of January 2011, the latest RFC document is numbered RFC 6093, indicating the huge set of standards and related documents that comprise the underlying protocols and formats that are required or used on the Internet. The 'standards' set alone has, as at 8 January 2011, some 70 official entries [19]. This sheer size and complexity contributes markedly to concern in relation to DoS/DDoS matters from an Internet perspective as it is highly unlikely that any information systems manager or non-expert user can appreciate the structures and vulnerabilities inherent in the global network.

3.3.5.4 IFIP

The 'International Federation for Information Processing' (IFIP), established in 1960 under the then auspices of United Nations Educational, Scientific and Cultural Organization (UNESCO), is an international organisation of ICT professionals. It represents Information Technology (IT) profession societies from some 56 countries [22]. It has two international technical committees (TC) interested in policy and law for information security and assurance. These are:

- TC-9 – ICT and Society
- TC-11 – Security and Protection in Information Processing Systems[6]

3.4 Regional Perspectives

The following outlines two regional entities involved in all aspects of cyber security.

3.4.1 EC/EU

In 2005, the Council of the European Union (EU) adopted Framework Decision 2005/222/JHA on attacks against information systems ('the Framework Decision'). This decision was a step in the EU's fight against cyber crime. Framework decisions were introduced by the 1997 Amsterdam Treaty. Under Article 34 of the amended Treaty on European Union, the Council may, by unanimous decision, adopt a framework decision for the purpose of the approximation of the laws of Member States.

Network and information security belongs to the core of the European Commission's policy regarding the information society. Tackling cyber crime is also an issue under the EU's 'third pillar', i.e. co-operation between Member States in the field of justice and home affairs. The Framework Decision had two main objectives:

1. Creating a common set of legal definitions and criminal offences across the EU and
2. Improving the effective prosecution of offenders by setting out minimum rules with regards to penalties, as well as rules with regards to the judicial co-operation between Member States.

The 2005 framework decision has been regularly updated and replaced with new directives. The Framework Decision also aimed at combating 'denial of service' (DoS) attacks. The information systems security interests of the European Community and European Union (EC/EU) are largely entrusted to the European

[6] A contributing author, Caelli, has been a member of TC-11 since 1984.

Network and Information Security Agency (ENISA). It has the following stated aim [1]:

> …the Agency's Mission is essential to achieve a high and effective level of Network and Information Security within the European Union. Together with the EU-institutions and the Member States, ENISA seeks to develop a culture of Network and Information Security for the benefit of citizens, consumers, business and public sector organisations in the European Union. ENISA is helping the European Commission, the Member States and the business community to address, respond and especially to prevent Network and Information Security problems. ENISA is as a body of expertise …

3.4.2 NATO

The 'North Atlantic Treaty Organization (NATO)' is described in Wikipedia [34] as follows:

> The North Atlantic Treaty Organization or NATO …; French: Organisation du traité de l'Atlantique Nord (OTAN), also called the (North) Atlantic Alliance, is an intergovernmental military alliance based on the North Atlantic Treaty which was signed on 4 April 1949.

In 2010, cyber-security was nominated as a key priority for the organisation, as follows:

> NATO mobilises for cyber warfare [114]. NATO leaders meeting in Lisbon are set to enshrine cyber security as one of the 28-nation alliance's priorities. There are as many as 100 attempted cyber attacks on the military force every day, IT experts have said. In 1989, before the Internet revolution, Suleyman Anil was the lone man in charge of the security of NATO's IT system, armed with a single computer. Two decades later, with the threat of cyber attacks on the rise, Anil oversees two teams tasked with protecting the networks of the alliance's political headquarters in Brussels and operations command in Mons, Belgium.

This statement largely applies to the information systems and networks that belong to the military members of NATO but these may impact upon the general stance of the NII in the member countries.

3.5 National Perspectives

3.5.1 Australia

3.5.1.1 Policy

Historical Background – Australia Linked to England via India

Australia's history in relation to both postal and telecommunications systems has been summarised as follows [45]:

> From the earliest days of European settlement in Australia, communication services have been seen as the almost exclusive responsibility of government. For the first decades of British governance no reliable mail service existed on the new continent. Messages, both between the colonies and internationally, were sent primarily by an ad hoc system of favours and paid messengers. It was not until 1821 that the first regular postal service began, initially operating only within New South Wales. Even then there were no truly reliable postal services within the new colonies until 1832, when Tasmania established the first post office as a Government Department. The other colonies quickly followed and soon Australia's fledgling mail network was entirely run and regulated by government. This began a pattern that was to dominate the communication industry for the next 150 years – almost complete government control of Australia's communications services ...
>
> However, in a country as large and isolated as Australia a communication system that relied on horseback could only go so far. When the telegraph first appeared in Europe in 1844, the young country was quick to adopt the new technology. Morse code was brought to Australia in 1853 by Samuel McGowan, and by 1859 telegraph cables linked Melbourne, Adelaide, Sydney and even Tasmania. By the mid-1860s all regional centres in the south east of the country were part of a virtually instantaneous communications network owned, maintained and managed by government. The final and most significant breakthrough was made in 1872, when Sturt's crossing of the Northern Territory enabled the establishment of Australia's first international telecommunications system, a telegraph link to Asia. This in turn linked Australia to the European and American lines, and the great southern land finally ended its isolation from the rest of the world.
>
> Over the next few years *Australia's dependence on the new telecommunications industry rapidly grew*. The population quickly embraced all new technological developments in what would become the historical norm. In the final years of the nineteenth century Australia sent more telegraphs per capita than any other nation. Telephones quickly followed the telegraph, and in 1882 the first public telephone exchange, based in Sydney, made personal communication available to the average Australian, just six years after Alexander Graham Bell took out his patent.

International telecommunications has been a major concern in Australia since before federation of the colonies. That ideal has been described as follows [3]:

> From 1858 the prospect of a cable being laid from Britain to Australia began to generate competition within the Australian colonies for landing rights. In 1869 the Queensland colonial government established a cable across the southern intersect of Cape York peninsula in preparation for a connection with a then envisaged cable, linking Australia to England via India, Singapore and Java. In the event the South Australian colonial government successfully out manoeuvred Queensland in a bid to secure landing rights with an offer to finance the cost of a telegraph from Port Augusta to Port Darwin (at that time South Australia administered what was to become the Northern Territory). A contract was let in 1870 and, following immense physical hardship and considerable logistical challenge, the Overland Telegraph (the Port Augusta to Port Darwin line) was completed on 22 August 1872. (*with the cable to England being restored to operation in October of that year* following a four month breakage in it. The cable had been initially brought ashore at Darwin in November 1871, with Australia's first international telecommunications message being received on 19 November.

The significance of DoS was a concern even in the nineteenth century as expressed above. The main cable path actually went through Mumbai (Bombay) India as follows, showing an early connection between what were to become Australia and India [5]:

> The All-Sea Australia to England Telegraph, supplied by Telcon, was opened in 1872. It was operated in two sections, *Bombay to Singapore* by the British India Extension Telegraph

> Company and *Singapore to Adelaide* by the British Australian Telegraph Company, both under Pender's control.

Moreover, the problem of DoS through loss of international cable connection, now by fibre-optic cables, is a major concern as both Australia and India develop their 'digital economies'. This takes on specific significance in the area of the offering of information and computer-based services on an international basis. Even by 1901 a similar concern for DoS threats was obvious [15].

Defence

The topic of policy related to Australia and the response to 'cyber-war' have been examined by Waters et al. in their monograph 'Australia and Cyber-Warfare' [134]. In a foreword to this volume a former Minister for Defence in Australia, the Hon. Kim Beazley, AC, Australia's ambassador to the USA as of January 2011, commented on the need to rethink national defence and security/military requirements in the following way:

> Politics has distorted what is really important in the Australian debate on our future defence needs. We are obsessed with platforms and personnel numbers. Over the last decade, the Australian Government has burnished its popular security credentials by junking any serious study of platform needs and acquiring capabilities based on immensity with big dollar signs attached, thereby seeking to impress public opinion with size and cost whilst saying little of relevance about modern and future warfighting

In expanding on the overall theme of cyber actions he went on, in the same foreword, to issue a clear warning as follows:

> George Orwell said during the Second World War that we sleep safe in our beds because rough men do violence in the night to those who would wish us harm. The rough men are now joined by the "geeks" of both genders.

He was sceptical about Australia's ability to conduct, or even maintain interest in the whole area of cyber warfare, or whatever other term it goes by. He specifically referenced 'network-enabled operations' in a military sense and the problem of any intelligence agency or service moving to 'warfighting operations'.

Dudgeon [87], in the above monograph, quotes Australian 'defence doctrine' in relation to a definition of the national information infrastructure. He quotes that definition from an *Australian Defence Doctrine Publication (ADDP) 3-13 – Information Operations (2006)* as follows:

> ...compris[ing] the nation wide telecommunications networks, computers, databases and electronic systems; it includes the Internet, the public switched networks, public and private networks, cable and wireless, and satellite telecommunications. The NII includes the information resident in networks and systems, the applications and software that allows users to manipulate, organise and digest the information; the value added services; network standards and protocols; encryption processes; and importantly the people who create information, develop applications and services, conduct facilities and train others to utilise its potential.

The incorporation of those obviously involved in training, education and research in this definition is interesting and in line with some similar definitions related to the nature of national security significant matters in the USA, e.g. the encryption export regime.

Ball [70] continues in the same monograph by proposing a 'cyber-warfare centre'. He sees it as essential component for Australia to be able to create an '...architecture for achieving and exploiting Information Superiority and Support (IS&S) beyond around 2020.' Elsewhere this concept, or ideas and structures similar to it, has been referred to as the development and deployment of a 'cyber range' [117]. He finds that there are many organisations, both inside and outside Australia's defence structures, involved in some aspects of the cyber warfare topic but that they are "poorly coordinated" and 'not committed to the full exploitation of cyber-space for either military operations or IW more generally'. A similar concern may be expressed about the need for a so-called 'cyber range', a facility for both military entities as well as owner/operators of NCI, where they may test and validate the resilience, and overall security, of ICT products, systems and services that are to be incorporated into mission-critical information networks and systems, e.g. SCADA systems.

Government-Federal (Non-military)

Development and implementation of CIP/NIIP policy is largely concentrated in the:

- Department of the Prime Minister and Cabinet (Australia) (PMC)
- Australian Government Attorney-General's Department (AGD)
- Department of Broadband, Communications and the Digital Economy (Australia) (DBCDE)

However, there are numerous crossover points in the area which have been seen as being a weakness in the whole structure.

AGD

The *Critical Infrastructure Protection Branch* is charged with the development and administration of policy in the CIP arena. On 30 June 2010 the Australian Government released its *Critical Infrastructure Resilience Strategy*. An important note is the emphasis placed on 'resilience' which the ministerial introduction to the strategy summarises as follows:

> The responsibility for the continuity of critical infrastructure is shared by all governments and by owners and operators ...It has a strong focus on business-government partnerships [10].

The strategy also notes the importance of coordination with appropriate Australian State and Territory entities. This is particularly appropriate where critical infrastructure may be owned, or jointly owned, by that State or Territory. To this end,

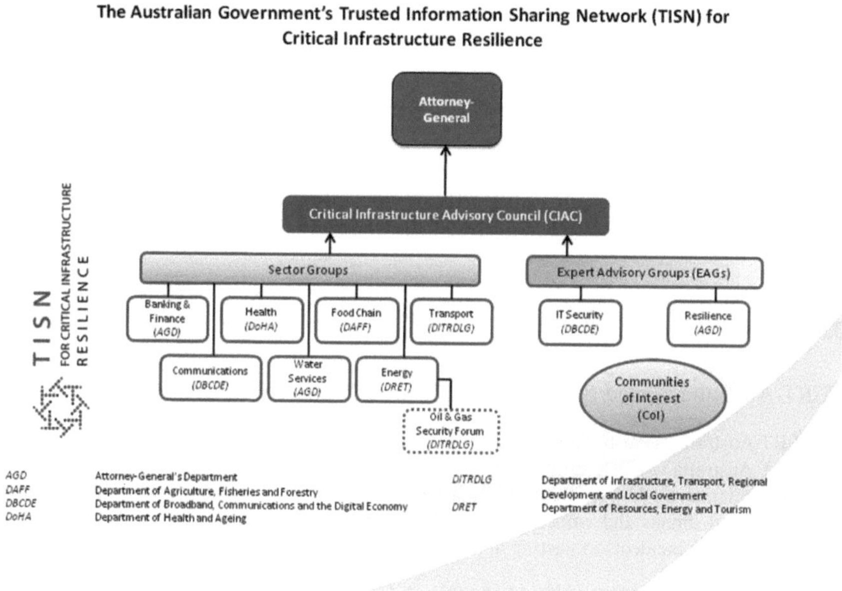

Fig. 3.5 Australia's TISN for critical infrastructure resilience

> the Australian Government has established the Trusted Information Sharing Network (TISN) for Critical Infrastructure Resilience (CIR) as its primary mechanism to build a partnership approach between business and government for CIR [40].

The TISN consists of seven industry groups and two technical advisory groups as shown in Fig. 3.5. These are as follows (Jan 2011):

Sector groups:

- Banking and finance
- Communications
- Energy
- Food chain
- Health
- Transport
- Water services

Expert advisory groups:

- Resilience
- IT security (ITSEAG)[7]

---

[7]The author, Caelli, was a foundation member of this group and served on it until Jan 2011.

Communities of Interest (COI)

Communities of interest are formed from time to time as deemed fit by the sector groups with interest from those sector groups, the expert groups and other parties. An example is the supervisory control and data acquisition (SCADA) Security COI, of major significance to many NCI owners and operators particularly in the electrical power, water and sewerage, transportation and allied areas. The TISN and its role are discussed by Christensen et al. in a 2010 paper [81].

3.5.1.2 CERT-Australia

CERT-Australia started operations on 28 Jan 2010. It describes itself as follows:

> CERT Australia is Australia's official national computer emergency response team (CERT). CERT Australia works to ensure that all Australians and Australian businesses have access to information on how to better protect their information technology environment from cyber based threats and vulnerabilities. CERT Australia is the initial point of contact for cyber security incidents impacting upon Australian networks [4].

CSOC – Computer Security Operations Centre

CERT Australia describes the CSOC as follows:

> Cyber Security Operations Centre (CSOC) provides the Australian Government with all-source cyber situational awareness and the ability to facilitate operational responses to cyber security events of national importance.

In this regard, CSOC, within the Australian Department of Defence, provides assistance to CERT-Australia. CSOC itself was established in response to the 2009 Australian Government's Defence White Paper [91].

CSOC describes itself as follows:

> The Cyber Security Operations Centre (CSOC) is a Defence Signals Directorate capability that serves all government agencies. The CSOC has two main roles:
>
> - It provides government with a comprehensive understanding of cyber threats against Australian interests; and
> - It coordinates operational responses to cyber events of national importance across government and critical infrastructure [11].

Essentially, CSOC responds to governmental requirements while CERT-Australia is the contact point for the private sector involved in ownership or operation of NCI.

DBCDE

This department also plays a role relation to DoS/DDoS and threats to the security and resilience of the NII. The following activities, among others, are of note:

ITU – International Telecommunication Union

> The Department coordinates Australia's participation in the ITU. Australian organisations – both Government and private – participate in many of the specialist ITU meetings.

TISN/ITSEAG

> The Department participates in the Australian Government's Critical Infrastructure Resilience (CIR) activities through the Trusted Information Sharing Network for Critical Infrastructure Resilience (TISN)…. Under the TISN, the Department provides secretariat support to the Communications Sector Group (CSG) and the IT Security Expert Advisory Group (ITSEAG) and its working group, the Supervisory Control and Data Acquisition systems Community of Interest (SCADA COI) [9].

PMC

The Department of the Prime Minister and Cabinet has a number of groups and coordinating functions relevant to DoS/DDoS threat to the NCI and to Australia in general.

National Security Science and Technology (Australia) (NSST)

> The NSST Branch within the National Security and International Policy Group provides a national focus for science and innovation aimed at enhancing Australia's national security [33].

National Security and International Policy Group

> The National Security and International Policy Group provides advice on Australia's foreign, trade and treaty matters, defence, intelligence, non-proliferation, counter-terrorism, law enforcement, border security and emergency management matters; coordinates security-related science and technology research matters; and plays a coordinating leadership role in the development of integrated, whole-of-government national security policy.

Government: State Level

Similarly appropriate policy and coordination functions at the State level usually fit within the responsibilities of the office of each State Premier, e.g. in the State of Queensland this is the *Department of the Premier and Cabinet*. It describes its function as follows [42]:

> The Department of the Premier and Cabinet is the agency that deals with the 'bigger picture' aspects of government. As our name suggests, it's our job to support and advise the Premier and Cabinet so that the government's priorities can be met.

Private / 'Not-for-Profit' Sector

AusCERT:

AusCERT was one of the first '*Computer Emergency Response Teams*' to be created worldwide, being formed in 1993. It operated as the 'national CERT' till 2010 at which time CERT-Australia, incorporating the previous 'Gov-CERT', took over that role. It describes itself as follows:

> AusCERT is the premier Computer Emergency Response Team (CERT) in Australia and a leading CERT in the Asia/Pacific region. AusCERT operates within a worldwide network

> of information security experts to provide computer incident prevention, response and mitigation strategies for members and assistance to affected parties in Australia. As a not-for-profit, self-funded organisation based at The University of Queensland, AusCERT relies on member subscriptions to cover its operating costs [2].

AusCERT is a member of both the *Forum for Incident Response and Security Teams (FIRST)* and the *Asia Pacific Computer Emergency Response Team (APCERT)*.

The Internet Industry Association:

In a November 2010 press report, Mr P Coroneos, Chief Executive of Australia's Information Industry Association (IIA)[8], was quoted as stating that essentially the problems facing the security and, in particular, the resilience of the global information infrastructure '... are not necessarily within the control of government. The infrastructure is now privately held, Government's role must move more to a coordinating role [98].'

Now, interestingly, this statement of public and potentially regulatory policy is at odds with the role and intervention of government in numerous other industries, such as the pharmaceutical manufacture and distribution sector, air transport, motor vehicle manufacture and distribution, trucking and transport industry, healthcare, and so on, where those industries are also owned and controlled within the private sector.

Thus, the now accepted policy of 'industry self-regulation' or private/public 'joint coordination' of the regulatory regime in relation to the ICT product and service sector is strangely at odds with those other sectors in numerous national economies. The reasons for this are interesting and unusual, including successful lobbying of governments by the ICT products and systems industry against regulation over many years. However, that approach is by no means universal, as the case of India demonstrates, as detailed below.

IIA's 'iCODE':

In June 2010, the IIA announced its new voluntary code of practice for Internet Service Providers (ISPs) in Australia with the overall aim of trying to improve the overall security of internet usage for all users. It has become known as the 'icode'. In relation to the DoS threat, this code includes the provision for an ISP to monitor the activity of a connected system that comes to its attention as having been possibly comprised and which is now behaving as a 'zombie' system in a larger 'botnet' used for DDoS attacks. It permits the ISP, after a process of due notification to the end-user and associated remedial actions, to stop the connection of the 'zombied'

---

[8]The Internet Industry Association (IIA) is an industry organisation governed by its own constitution and by a board of directors. It maintains an operational executive group including a Chief Executive Officer based in Australia's Federal Capital, Canberra. Its home page on the world wide web is at URL http://www.iia.net.au/.

system to the Internet. The remedial actions are not detailed, e.g. if a system is only connected to the Internet via a low-speed 'dial-up' connection then download of such large data collections as operating systems 'patches', virus signature files, etc. is impossible and the end user has no real ability to perform any remedial action based on those requirements. The problem will be one of determining what reasonable actions can be taken prior to any threat of disconnection.

Interestingly, this icode '…recognises both (ISPs) and consumers can and must share responsibility for minimising the risks inherent in using the internet. [57]' However, and unlike the situation that applies in almost all other industry policy statements, it says nothing about any responsibilities of the manufacturers, importers, suppliers or vendors of the actual hardware or software systems involved, e.g. the PC, internet connectable mobile phone, 'personal assistant', computer operating system and Internet 'browser', such as the Apple iPad, iPod and iPhone. Microsoft's Windows operating system, 'Internet Explorer' or Office application software products, and so on.

Indeed, unless such products and systems are 'secure by design' then it may be that actions taken by a services vendor, such as an ISP, or even the inexpert end user may be of minimal value. In particular, as end users move to specific 'appliance level' ICT products for their normal information requirements, and these products also find their way into corporate and governmental systems, e.g. the mobile/cell phone, the portable tablet computer, such as Apple Inc.'s *iPad* product, etc., the ability of the end user or even the ISP to effect remedial action may be extremely limited.

3.5.1.3 Law

Response to DoS/DDoS

There appears to be no *specific* legislation in Australia that governs the protection of or response to DoS/DDoS on Australia's NCI or NII. Georgiades et al [93] point out that:

> There is no formal definition of the terms 'cyber-attack' nor of 'cyberwar/ cyberwarfare'. At the same time, neither is there any definition of what comprises a 'cyber weapon' or 'cyber munitions'. This makes distinguishing between attacks that constitute a crime, an act of terrorism or of warfare extremely difficult.

Moreover, the terms used are changing in meaning and emphasis at a rapid rate along with the status of the underlying technology and products used. The terms also do not readily relate to formal and accepted legal terms and their meanings.

The well-documented and discussed *Maroochy Case* [46] in Queensland, Australia, highlights the distinctions needed in any legal discussion. The case involved essentially the illicit interference with the monitoring and control systems of a sewerage system in the Maroochydore area of the State which affected the associated pumping stations. The case involved the release of sewerage into the local waterways with the resultant risks to the health of the community as well as financial loss to the local government entity responsible for the provision of such sewerage services. The attack on the system involved the misuse of the wireless control and monitoring systems of the pumping stations by the accused. At that time, and even today, 'cyber-crime' suffers from no clear and consistent statutory declaration. At the time the accused attacker was charged and convicted prior to the amendments to the Criminal Code Act of 1995, as discussed below. In the case of R vs Boden, the State of Queensland's Criminal Code was used. Today, if a similar criminal act were to be performed then prosecution under the terrorism provisions of the Criminal Code could be expected to be used. Specifically, sub-section 2 of section 100.1 of the ACT would be the most probable citation. The Act defines a 'terrorism' act as follows:

> . . . an action or threat of action where:
>
> 1. the action falls within sub-section (2) and does not fall within sub-section (3); and
> 2. the action is done or the threat is made with the intention of advancing a political, religious or ideological cause; and
> 3. the act is done or the threat is made with the intention of:
>
>   a. coercing, or influencing by intimidation, the government of the Commonwealth or a State, Territory or foreign country, or a part of a State, Territory or foreign country; or
>   b. intimidating the public or a section of the public.

Now sub-section (2) continues by defining those actions that are covered under section 100.1. These actions include those that may:

- Cause serious physical harm to a person
- Cause serious damage
- Cause a person's death
- Endangers a person's life, except for the person actually taking that action
- Creates a serious risk to the health or safety of the public or a section of it
- Seriously interferes with, disrupts or destroys an electronic system

The section then goes on to nominate examples of such 'electronic systems' that include such infrastructures as those used for and vital to telecommunications, financial systems, government service delivery, essential utilities, transportation or any 'information system' in general.

As mentioned above, a form of collaboration exists between government and the private sector which now accounts for the majority of ownership and operation of NCI. The theme of 'best practice' is the current philosophy although, at law, that theme may or may not assist in any ensuing legal actions. Essentially, in Australia, the legal approach to the threat of DoS/DDoS may be summarised as a combination of:

- Common law doctrine, covering liability and redress
- Specific Federal and State legislative instruments

This applies in both the cases of deemed responsibility of the NCI owner/operators to safeguard their enterprises against such attacks as well as any recourse that may be available should such activity occur and/or be successful. Australian Federal legislation, as pointed out by Christensen et al., *governs national security, terrorism and espionage and includes:*

- *Criminal Code 1995 (Cth) s. 100, Schedule 10.7 Computer Offences;60*
- *Telecommunications Act 1997 (Cth) (for enforcing and surveillance) and obligations of ISP providers and carriers;*
- *Cybercrime Act 2001 (Cth), amending Criminal Code 1995 (Cth), Crimes Act 1914 (Cth) and Customs Act 1901 (Cth);*
- *Spam Act 2003 (Cth);*
- *National Security Information (Criminal and Civil Proceedings) Act 2004 (Cth);*
- *Anti Terrorism Act 2004 (Cth);*
- *Corporations Act 2001 (Cth); and*
- *Privacy Act 1988 (Cth).*

It should also be noted that the *Telecommunications (Interception and Access) Act 1979* may also be relevant. At the individual Australian State level, e.g. Queensland, the appropriate criminal laws may also apply, e.g. Criminal Code (Qld) 1899 and Summary Offences Act 1953 (SA).

The current legal framework for *responding* to DoS attacks is provided under the Australian Criminal Code Act 1995; which has provisions related to computer offences and acts of terrorism. This arises from the nature of such attacks and more specifically, the target of those attacks.

If a DoS attack occurs on a 'critical information infrastructure', the attack may invoke the terrorism provisions in accordance with national security interests. This means that depending on the nature of the attack and the consequence or the impact of the attack, the terrorism provisions potentially apply. This is problematic, given that such attacks may originate from anywhere in the world, because of the fact that international legal response is not explicit and is obscured under international laws and conventions. Australia's Criminal Code Act 1995, provides that any activities that disrupt or interfere or destroy an electronic system may be included under the terrorism provisions. Specifically terrorism is defined under section 100.1 of the Criminal Code Act 1995, to mean 'an action or threat of action where the action is done or the threat is made with the intention of advancing a political, religious or ideological cause or it is done with the intention of coercing or influencing by intimidation the government of the Commonwealth or State or Territory or intimidating the public.'

Further actions under this section also include:

- Serious harm or serious damage, causing death or endangering a person's life,
- Creating a serious risk to the health or safety of the public or a section of the public
- More interestingly, seriously interfering with or disrupting or destroying an electronic system which includes an information system, telecommunications

system, financial system or a system used for the delivery of essential government services or an essential public utility or a transport system

Consequently, a DoS attack that may be categorised under any of these instances potentially captures the terrorism provisions regardless of whether it is a private corporation or not. If the target is deemed to be a critical information infrastructure then, plausibly, the terrorism provisions will be invoked.

However, any legitimate and legally available response alternative to private corporations which operate and own critical infrastructures is faced with uncertainty because the attacked entity could itself be potentially liable under the computer offences provisions of the Criminal Code Act 1995! In this respect, the potential overlap between the various provisions do not assist in clarifying the legal responses to DoS attacks. Given the above concerns, the overall problems of policy and law must be considered. Policy parameters may exist at the enterprise, regional or industry sector or the national and general public level. However, legal structures may limit the defensive posture that may be taken under any protection policy.

The associated problem of responsibility for protection against DoS/DDoS by NCI owner/operators is also a difficult concern. The *Corporations Act (2001)* may apply for cases where it may be claimed by affected parties that 'due diligence' has not been taken in the protection of the corporate infrastructure. The concept of agreed and acknowledged 'best practice' in protection may be important in any ensuing legal actions. Moreover, corporations may also be liable for employees' actions or negligence where those best or acknowledged and agreed practices have not been followed. Section 180 of the Corporations Act (2001) requires that company officers keep themselves well informed and make decisions in the best interests of the company. It should be noted here that the basic dichotomy faced by such corporate officers is a fundamental one, i.e. performing their functions to the best interest of the company and its shareholders does not imply that those actions are in the best interest of overall national security!

Those suffering from a DoS/DDoS attack on NCI may have civil recourse under associated contracts of service or tort. In turn, the affected NCI operator may be able to claim for damages from the attacker, if he or she can be reasonably, or even definitely, identified. Contracts, such the End-User Licence Agreements (EULA), may, however, contain exclusion clauses that limit any liability. The recent adoption of national consumer protection legislation may, however, place some constraints on those clauses.

In summary, failures in the legal framework situation may be summarised as follows:

- Lack of defined responsibility on behalf of NCI owner/operators to incorporate appropriate levels of resilience into their systems capable of preventing, or at least lessening the effects of, a DoS/DDoS attack, e.g. use of the international standard IS-15408 'Common Criteria' for the evaluation of the security stance of products is not required
- No requirement to report such attacks to those affected or to any other authority, e.g. police

- Lack of certainty as to what response actions or steps the owner/operator may take to mitigate against or even attempt to stop any attack, particularly where the attack may originate from another jurisdiction or country.

In summary, evidence points to the fact that market failure has occurred in the cybersecurity area in the computer products and systems industry, hardware and software, and government intervention is necessary to increase the resilience of basic computer components in national information infrastructures as being discussed by India, as below. However, where almost all ICT products and systems are fully imported the question as to how such an approach may be taken arises and clearly points to the need for *an international effort* in the area. The *Common Criteria Recognition Arrangement (CCRA)* may be an appropriate vehicle. The CCRA, with 26 nations participating, including *both Australia and India*, has the following aims [38] that appear fully aligned with the needs for DoS/DDoS protection for the NII:

> The Participants in this Arrangement share the following objectives:
>
> 1. To ensure that evaluations of Information Technology (IT) products and protection profiles are performed to high and consistent standards and are seen to contribute significantly to confidence in the security of those products and profiles
> 2. To improve the availability of evaluated, security-enhanced IT products and protection profiles
> 3. To eliminate the burden of duplicating evaluations of IT products and protection profiles
> 4. To continuously improve the efficiency and cost-effectiveness of the evaluation and certification/validation* process for IT products and protection profiles

3.5.2 *India*

3.5.2.1 Background to the ICT Sector in India

A key characteristic of the development and deployment of information and communication technologies is its elimination of time and space as barriers to service delivery. From the introduction of the of the earliest telegraph to the roll-out of modern mobile networks, the history of the ICT sector in India not only mirrors this inexorable trend but also captures the momentum of the transition of India to an emerging technological super-power. However, concomitant with this growth in the use of ICT is an increased dependence on its continuous availability of what is now critical infrastructure. Hence the harmonisation of policy, legal and regulatory framework is a must for CIIP (Critical Information Infrastructure Protection).

From one perspective, and as discussed in the 'History of Indian Telecommunications', one of the key drivers for the introduction of telecommunications services in India was the need to satisfy the commercial imperatives associated with trade [102]. For example, in 1851 the first telegraph service was established. It was a 21-mile link connecting Kolkata, then the seat of the British colonial government, with

Diamond Harbor, a trading post of the British East India Company. However the telegraph, and later the telephone which was introduced in India in 1882 had another key function. These technologies, particularly when they were integrated with the existing postal service in 1883, were key tools of command and control that were essential to enabling the British to maintain law and order in the country [102].

The history of the development of the telephone sector in India reflects the tension that played out elsewhere in deciding whether the emerging service was best delivered via a public or private enterprise. As is described in another historical perspective on Indian telecommunications, the 'History and growth of Calcutta telephone' [74], in 1880 two telephone companies, namely The Oriental Telephone Company Ltd. and The Anglo-Indian Telephone Company Ltd., approached the Government of India to establish telephone exchanges in India. Citing the view that telephones were a Government monopoly and that the Government itself would commence the work, permission was refused. This decision was reversed in 1881 and the Government eventually granted a license to the Oriental Telephone Company Limited of England to build and operate telephone exchanges at Kolkata, Mumbai, Chennai (Madras) and Ahmedabad. With the exception of Ahmedabad, these were opened in 1882.

Continuing the story from the 'History and growth of Calcutta telephone', the telephone system remained under private ownership until 1941 when all the shares of the private companies were purchased for Rs 117 lakhs. In April 1943, the Indian Posts and Telegraphs Department assumed direct control of the telephone system in Calcutta, Madras and Bombay. Hence, at the time of independence from Britain in 1947, post, telegraph and telephone services were exclusively in the domain of the state. This state of affairs meshed with Jawaharlal Nehru's post-independence socialist policies, the cornerstone of which was state-run, state-owned monopolies in various sectors, including Indian telecommunication. However, while all the major cities and towns in the country were linked with telephones during the British period, the total number of telephones in 1948 was only around 80,000 [102].

From the time of independence until the mid-1980s, development and growth in the telecommunications sector was heavily impeded by the grossly over-bureaucratised and highly insular Ministry of Posts and Telegraphs. In particular there was an ingrained perception that telephones were a 'luxury' rather than a 'necessity', i.e a 'status symbol' as opposed to a vital 'instrument of utility', for a modern economy. Consequently it has been estimated that by the mid-1980s India had only 2.5 million telephones and 12,000 public phones for a population of 700 million. Only 3 percent of India's 600,000 villages enjoyed telephone services [102].

However, as in a number of other countries, a view emerged in India that provision of world-class telecommunications infrastructure and information technologies was a key element in rapid economic and social development of the country. The winds of change were blowing. The first step in the revolution occurred in 1985 when control of telephony was transferred to a new entity called the Department of Telecom and thereby clearly delineating the difference between telephony and the older postal and telegraphy services. This presaged a series of reforms that opened the sector to competition from private providers. For example, an important first step in this direction was announcement of the National Telecom Policy in 1994 (NTP 94) which set ambitious targets such as telephone-on-demand and the provision of universal service (i.e. connecting all villages). This was followed by a second major policy initiative the 'New Telecom Policy 1999' [132]. A key component of this policy was the idea of the Universal Service Obligation (USO) in order to widen the reach of telephony services in rural India, thereby bridging the gap between urban and rural services. Such changes contributed to a dramatic growth in investment, and particularly foreign direct investment, in telecommunications infrastructure. The result was a complete transformation of telecommunications in India such that by the end of the millennium, India could boast that its telecommunications network was the ninth largest in the world [128].

Figure 3.6 shows the history of Indian Telecommunication.

3.5.2.2 Policy, Legal and Regulatory Framework

Currently, the telecommunications services sector in India can be broadly divided into three categories: basic (fixed wired line and wireless in local loop), mobile (GSM/CDMA) and internet services. As with other countries, India has sought to provide a policy and legal framework that sustains vibrance and innovation in a sector that is vital to the development of the economy. At the same time, the Government of India (GoI) has endeavoured to create a framework that also protects the interests of the key stakeholders and in particular those of the nation as a whole as well as those of the individual consumer [128]. These objectives are captured in the mission statement for the Telecom Regulatory Authority of India (TRAI) which is the body responsible for the oversight of telecommunications regulations in India. In essence their mission is:

> to ensure that the interests of consumers are protected and at the same time to nurture conditions for growth of telecommunications, broadcasting and cable services in a manner and at a pace which will enable India to play a leading role in the emerging global information society. [131].

Managing the inherent tensions in attempting to balance all of these competing objectives is part of the fabric of the Indian telecommunications sector.

Rohan Raichaudhuri [121] provides a more detailed insight into these tensions and specifically the legal and regulatory framework of the telecommunication market in India.

As noted by Raichaudhuri, the Indian telecom industry is governed by the DoT (under the Ministry of Communications and Information Technology). The DoT along with the Telecom Commission are responsible for all matters relating to, amongst others, policy formulation, licensing, wireless spectrum management, administrative monitoring, research and development, standardisation and validation of equipment as well as private investment in the sector.

As has been indicated previously, the policy and legal frameworks must also be adaptive to the changing telecommunications environment. Raichaudhuri also

Fig. 3.6 India's telecommunications development

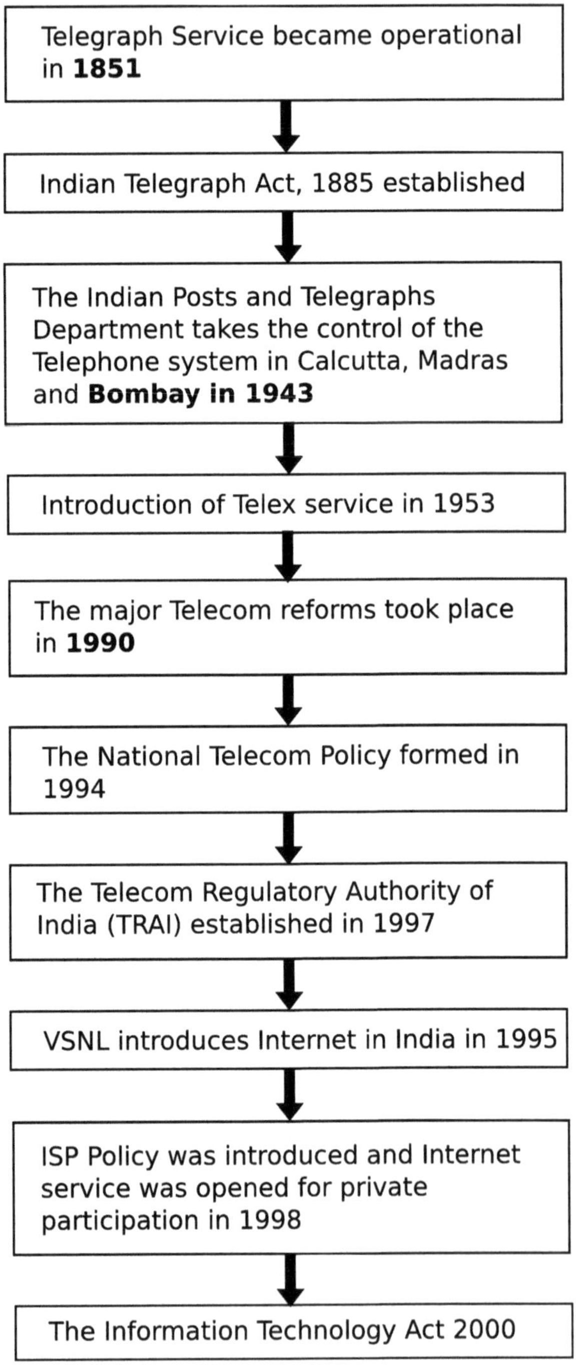

identifies the primary role of the TRAI to look into matters relating to, amongst others, the need and timing for introduction of new service providers and the terms and conditions of license to them, ensure technical compatibility and effective interconnection between different service providers, regulate arrangements of revenue sharing amongst the service providers, ensure compliance of terms and conditions of license as well as recommend revocation of license for non-compliance, protect the interest of the consumers of telecommunication service, conduct periodical survey to monitor the quality of services provided by the service providers.

Raichaudhuri further notes that the TRAI Act, 1997, as amended in 2000, set up the Telecom Disputes Settlement and Appellate Tribunal (TDSAT), a dispute settlement body, in order to protect the interests of the various service providers as well as consumers of telecom services. The TDSAT is empowered to adjudicate on any dispute between a licensor and a licensee, two or more service providers, a service provider and a group of consumers and to hear and dispose of appeals against any direction and even a decision or order of the TRAI.

As in other countries, the diversity of the telecommunications sector has spawned the need for specialist regulatory bodies to cover specific aspects of the industry. One in particular is that of wireless communication and the management of the wireless spectrum. In India this is the responsibility of the Wireless Planning and Coordination (WPC) Wing of the Department of Telecom [8].

Because of the lack of much-needed quality infrastructure in India, the GoI has also formulated several guidelines to encourage the private sector to participate in central sector public-private partnership projects, under the Public Private Partnership Appraisal Committee, with the Ministry of Finance, GoI as the nodal ministry.

Rohan Raichaudhuri also identifies other allied legislation that affects the delivery of ICT-based services. These include the Information Technology Act, 2000, Communication Convergence Bill, 2001, Indian Telegraph Rules, 2008, TRAI Act, 1997, and several other subordinate rules and regulations.

3.5.2.3 Cyber Laws in India [13, 27, 30, 31, 39, 41, 50]

Inventions, discoveries and the development of new technologies with their associated artifacts, systems and services widen scientific, technological and engineering horizons but pose new challenges for the law and governance requirements. Information technology, generally consisting of computer and data networks, with the associated recognition of so-called cyberspace, has opened new dimensions in human interconnectivity but has also created governance and regulatory problems. It is now recognised that there is an urgent need to solve many of these problems even though the legislative processes take a long time and are totally at odds with the fast pace of change in the IT sector as well as in the applications of such IT and its products. Solutions to the problems of adequate governance, statutory or otherwise, can provide answers to some problems. Such legislative solutions are broadly known as 'Computer Laws', 'Information Technology Laws' or simply 'Cyber Laws'.

The Indian government has enacted its 'Information Technology Act 2000' which is itself based on the UNCITRAL Model Law (United Nations Commission on Trade Related Laws). Beyond this, India has not enacted many other relevant legal provisions. For example, problems in the area of Intellectual Property Rights (IPRs), have been covered by amendments to the relevant 'Copyright and Patent Act'.

However, the single most important legislative measure taken may be considered to be the 'Information Technology Act 2000' and later amendments to it. This Act provides for a number of conditions that are particular to the use of IT products and systems as follows:

- Documents and contracts in electronic form are now legally recognised as such, in a similar fashion to similar legislation in Australia.
- Documents, data records and the like in electronic form are admissible as evidence in any court of law.
- The concept of a 'digital signature' is now accepted as such for signing purposes; appropriate penalties are set out for such actions as propagation of obscene materials and for forms of cyber-related crime.
- A 'cyber regulation advisory committee' and a 'cyber regulation appellate tribunal' are both established.
- Conditions and terms for electronic filing and maintenance of 'e-records' are both enhanced.

In a similar way, India has amended four pertinent and related Acts, as follows:

- The Indian Penal Code, 1860
- The India Evidence Act, 1872
- The Bankers' Book Evidence Act 1891
- The Reserve Bank of India Act, 1934

These four Acts were amended by Section 91 to 94 of the IT Act. In summary, as in Australia, India's IT Act has positioned an electronic record at par with any paper record. One main reason for the enactment of the IT Act itself was the enormous growth of India's IT software and services sector, as explained elsewhere in this chapter. Indian companies and individuals were now competing with other global entities in this sector. Thus, in order to attract such IT business to India, there was an urgent need for an appropriate level of legislative provisions to address the concerns of both European and American companies and their governments in respect of provisions related to data privacy and security as well as other assurances related to such transnational contracts.

In summary, a need for such a binding Act was perceived against a number of factors including:

- Global penetration and use of the Internet, causing a major change in social, political and economic factors worldwide

- Complexity of various legal issues that arose from major differences across nations relevant to the control and regulation of the telecommunications industry and other e-commerce activities
- Massive growth in the IT services sector in India

At the same time as the IT Act was being decided, another act, the so-called Communication Convergence Bill 2001, was under consideration in India. The aim here was to fully harness the benefits of converged and newly converging technologies, products, systems and services, particularly the areas of telecommunications, information technology/computers and broadcasting. In 2000, a committee was set up to consider such a 'Communications Convergence Bill' but sharply divided opinion existed among experts about the desirability of having such an act [101]. Essentially, such disagreement has put this bill into 'cold storage', meaning that enabling legislation is unlikely in the near future. However, some of its proposed provisions have been incorporated into the 'IT Act', by amendment. (The 'Communication Convergence Bill', with complete text of the relevant report of the Parliament of India has been published [7].)

An expert committee was also set up to consider amendments for the IT Act. It made its recommendations, with proposed amendments, in 2005. These were introduced, in modified form, as the 'Information Technology (Amendment) Bill 2006'. This bill was further modified and passed by the parliament on 23 December 2008. After the assent of the president, it was notified on 5 February 2009 as the Information Technology (Amendment) Act 2008 (Central Act no. 10 of 2009). The new Act has incorporated some important provisions of the original Communications Convergence Bill. Sections 91 to 94 have been omitted in the amending Act, but in view of Section 6A of the General Clause Act, these amendments in the respective Act will continue. The first two Acts have been further amended by the amending Act.

In summary, the following policy areas are addressed in India's relevant 'cyber' laws:

- Violation of Intellectual Property Rights (IPR) and associated remedies;
- Violation of information systems other than IPR related crimes and associated remedies
- Interception of telecommunications and data networks, including the banning of such activities coupled with associated monitoring, related to the freedom of the Indian citizen
- Liabilities relevant to intermediary entities, such as telecommunications/internet service providers
- Computer forensics
- Collection and admissibility of computer evidence
- Awareness, training and enforcement activities
- International cooperation

3.5.2.4 Social/Economic Benefits/Impacts of the Act

The social and economic benefits of the IT Act may be summarised as follows:

- Email will now be a valid and legal form of communication in India, where such evidence can be duly produced and approved in a court of law.
- Electronic commerce may now be carried out with confidence under the legal infrastructure provided by the Act.
- 'Digital signatures' may now be used to carry out online/e-commerce transactions and such digital signatures have legal validity and sanction in law.
- Enabling of the entry of companies into the business of being 'Certifying Authorities' for the issuance of 'Digital Signature Certificates'.
- It enables a file to be created and managed in any form, an application to be developed or any other document maintained by any office, authority, body or agency owned or controlled by the appropriate Government to exist in electronic form as may be prescribed by the appropriate Government.
- The Act now allows Government to issue notifications via the 'world wide web', heralding a move towards e-governance.
- The IT Act also addresses important issues of information security which are critical to the success of electronic transactions. Under the IT Act, 2000, it is now possible for corporate entities to have a statutory remedy in case anyone breaks into their computer systems or networks and causes damage or copies data. (The remedy provided by the Act is in the form of access to monetary damages, not exceeding Rs 1 crore.)
- Some recent commentary on the amendments to the Information Technology Act has claimed that the amendments are little different from already existent powers via the 'Indian Telegraph Act' of 1885, and thus civil-liberties concerns may be mistaken. This is not correct. Section 69 of the Information Technology (Amendment) Act, 2008, passed by the Parliament on India on 23 December 2008, is far more intrusive than the Indian Telegraph Act of 1885, which was drafted to protect the interests of the British Raj. Under the new IT Act, any Government official or policeman will be able to listen in to all phone calls, read SMS messages and emails, and monitor the websites visited. A warrant from a magistrate to do so will not be required.
- Until the passage of the amended IT Act, 'phone tapping' was governed by Clause 5(2) of the Indian Telegraph Act of 1885, which said that 'On the occurrence of any public emergency, or in the interest of the public safety, the Government may, if satisfied that it is necessary or expedient so to do in the interests of the sovereignty and integrity of India, the security of the State, friendly relations with foreign States or public order or for preventing incitement to the commission of an offence, for reasons to be recorded in writing, by order, direct that any message or class of messages to or from any person or class of persons, or relating to any particular subject, brought for transmission by or transmitted or received by any telegraph, shall not be transmitted, or shall be intercepted or detained, or shall be disclosed to the Government making the

order or an officer thereof mentioned in the order'. Other sections of the act mention that the government should formulate 'precautions to be taken to prevent improper interception or disclosure of messages'. Many calls have been made, both inside and outside Parliament, to formulate rules to govern the operation of Clause 5(2) but from 1885, no government has formulated any such precautions, since all governments have wanted to retain the right to spy on their opponents unfettered.

- Section 69 of the amended IT Act drops all references to public emergency or public safety, meaning that government powers have been vastly extended.
- The earlier IT Act of 2000 mentioned only decryption of messages; interception and monitoring were not mentioned. Section 69 of the new IT Act enhances scope from the 2000 version to include interception and monitoring. It also broadens the scope of surveillance to include the investigation of any offence, whether cognizable or not.
- In view of the many reported incidents of the tapping of the phones of politicians, the Supreme Court, in the PUCL case, laid out procedures and guidelines to protect citizens against the arbitrary exercise of power by the government. However, this judgement, as well as the relevant sections of the Telegraph Act, have become infructuous or not fruitful with the passage of the amended IT Act, since the latter has overriding effect.

However, what of the safeguards in the act?

The Indian Government has not formulated any safeguards in relation to Section 5 of the Telegraph Act since independence. Moreover, it is unrealistic to expect it to formulate any safeguards under Section 69 (2) of the amended IT Act – especially in view of the prevailing terrorism situation. Until suitable safeguards are in place, Section 69 of the Information Technology (Amendment) Act of 2008 appears to be in violation of Article 21 of the Constitution which states 'no person shall be deprived of his life or personal liberty except according to procedure established by law.' [113]

3.5.2.5 Indian Telegraph Act

The Indian Telegraph Act, 1885, is a law in India that governs the use of telegraphy, telephones, communications-related products and services, radio, telex and fax in India. It gives the Government of India exclusive privileges of establishing, maintaining and working telegraphs. It also authorises the government to tap phone lines under appropriate conditions. The latest amendment is dated 2004.

3.5.2.6 Telecom Reforms in India [14, 26, 62]

There is an acknowledged need for this Act and related policies. These requirements may be summarised as follows:

- Recognition by the Government of India that provision of a world-class telecommunications and national information infrastructure is the key to rapid economic and social development of the country.
- Development and provision of a comprehensive and forward-looking telecommunications policy which creates a suitable framework for further development of the IT services industry.
- Governance over the use of telegraphy, telephones, communications systems, radio, telex and facsimile systems in India.
- Availability of infrastructure for electronically transferring and accessing information perceived as critical for hastening the realisation of economic, social and cultural benefits, as well as for conferring a competitive advantage for all entities in the nation.

The first step in this direction in India was the announcement of the National Telecommunications Policy in 1994 (NTP 94). This provided for the opening up of the telecom sector to competition in basic telephony services as well as value-added services such as cellular mobile services, radio paging, VSAT services, etc. It also defined certain important objectives, including the availability of telephone on demand, the opening up of long-distance telephony, provision of world-class services at reasonable prices, improving India's competitiveness in global markets with the promotion of exports, attractive FDI and stimulation of domestic investment, ensuring India's emergence as a major manufacturing and export base for telecommunications equipment and the universal availability of basic telecommunications services to all villages. It was agreed that the quality of telecommunications services should be of world standard. Resolution of consumer complaints and disputes, and the development of a public interface will receive special attention. The objective will also be to provide the widest permissible range of services to meet customer demands at reasonable prices.

3.5.2.7 Telecom Regulatory Authority of India (TRAI) Act 1997

The Telecom Regulatory Authority of India or TRAI (established in 1997) is the independent regulator established by the Government of India to regulate telecommunications business in India. [131]. Its aims may be summarised as follows:

- To regulate telecom services, including fixation/revision of tariffs for telecom services which were earlier vested in the Central Government.
- Facilitate competition and promote efficiency in the operation of telecommunication services so as to facilitate growth in such services.
- Protect the interest of the consumers of telecommunication service.
- Monitor the quality of service and conduct a periodical survey of any such service provided by the service providers.
- Inspect the equipment used in the network and recommend the type of equipment to be used by the service providers.

- The TRAI Act was amended by an ordinance, effective from 24 January 2000, establishing a Telecommunications Dispute Settlement and Appellate Tribunal (TDSAT) to take over the adjudicatory and disputes functions from TRAI. TDSAT was set up to adjudicate any dispute between a licensor and a licensee, between two or more service providers, between a service provider and a group in TRAI.

3.5.2.8 New Telecom Policy 1999

The most important milestone and instrument for telecommunications reform in India is the 'New Telecom Policy 1999' (NTP 99) [132]. It laid down a clear roadmap for future reforms, contemplating the opening up of all the segments of the telecom sector for private sector participation. It recognised the need for strengthening the regulatory regime as well as restructuring the departmental telecom services to that of a public sector corporation so as to separate the licensing and policy functions of the Government from that of being an operator. It also recognised the need for resolving the prevailing problems faced by the operators so as to restore their confidence and improve the investment climate. Key features of the NTP 99 include [26]:

- Strengthening of the Regulator.
- National long-distance services opened to private operators.
- International long-distance services opened to the private sector.
- Private telecom operators licensed on a revenue sharing basis, plus a one-time entry fee; resolution of problems of existing operators envisaged.
- Direct interconnectivity and sharing of network with other telecom operators within the service area was permitted.
- Department of Telecommunication Services (DTS) was corporatised in 2000.
- Spectrum Management made transparent and more efficient.
- Provision of Internet access to all district headquarters by the year 2000.

3.5.2.9 Internet Service Providers (ISPs)

Internet service was opened for private participation in 1998 with a view to encourage growth of the Internet and to increase its penetration. The sector has seen tremendous technological advancement over a short period of time and has necessitated taking major steps to facilitate technological ingenuity and the provision of various services. The Government, in the public interest in general, consumer interest in particular and the proper conduct of telegraph and telecom services, has decided to issue new guidelines for the granting of a license for the provision of Internet services on a non-exclusive basis. Any Indian company with a maximum foreign equity of 74% is eligible for such a grant of license.

3.5.3 Some Conclusions

Policy, as well as the legal and regulatory framework in India for telecommunications and the telecom sector, were initially monopolistic and based on 'command and control' policy and legal philosophies during the 1880s. After India's Independence it continued to be colonial in nature and run and managed by the Government until the 1980s. During this period these facilities were considered as a luxury and penetration was limited to only high income groups and senior government officials. The main reason for this poor performance until the 1980s was, among other factors, an official view that a telephone was a 'luxury' rather than a 'necessity'. This resulted in a slowdown of India's telecommunications industry until the 1980s. While India was focusing on improving delivery of telegrams, other countries were adopting and implementing digital telephony and facsimile transmission technologies. Statistics in relation to the history of Indian telecommunications, as indicated in the this chapter, show that there has been a mammoth revolution in this sector during the 1980s. During this period, a dramatic policy shift from consideration of telephony as a luxury item to a necessity occurred. Basic telephony services penetrated into villages across India, and PCOs (Public Call Offices) had mushroomed all across India including remote, rural, hilly, and tribal areas as well [136]. The emergence of PCOs satisfied the strong Indian socio-cultural need of keeping in touch with family members. Therefore, telephone expansion in India serves a strong socio-cultural function for its users, in addition to any commercial use.

Revenues of the Department of Telecommunications (DoT), the state-run telecommunications operator, increased fivefold during the 10 years period from 1988 to 1998. Over the next several years, India planned to add four to five million digital telephone lines to increase its tele-density from 2.5 per 100 people in 1999 to 7 per 100 by 2005 and to 15 by 2010. This has resulted in a lowering of costs, improvement in service and facilities and deep penetration of telephony and Internet services throughout India including rural and urban sectors covering the entire strata of the Indian economy. At present, even a rickshaw puller, auto wala, maid servant, service sector person, etc., all have access to and are using telephony services. This is indeed a revolution and the future is very bright with the development of 4G services and the convergence of related technologies. However, while these technologies, products, systems and services provide convenience, they also pose problems for security and privacy, including the resilience of that infrastructure once everyone becomes dependent upon it. In order to protect any national and international CIIP (namely Critical Information Infrastructure Protection), these problems have to be tackled by harmonisation of policy, legal and regulatory frameworks across the globe.

From another Indian perspective, the basic parameter related to DoS for NCI/NII policy development may be stated as follows:

> The IT infrastructure has become an integral part of the critical infrastructures of the country. The IT infrastructures interconnected computers, servers, storage devices, routers,

> switches, and wire line, wireless, and hybrid links increasingly support the functioning of such critical national capabilities as power grids, emergency communications systems, financial systems, and air traffic-control networks. *The operational stability and security of critical information infrastructure is vital for economic security of the country.* [12]

There has also been recognition that DoS/DDoS threats must be answered with a next generation of secure basic software and firmware structures as reported recently and as follows:

> The Defence Research and Development Organisation (DRDO) is working on creating a futuristic computing system, including India's own operating system, said V.K. Saraswat, Scientific Adviser to the Defence Minister and DRDO Director-General. Talking to journalists after inaugurating the DRDO Transit Facility here on Saturday, Dr. Saraswat said: 'We do not have our own operating system. Today, various bodies, including banks and defence establishments, need security. Having our own operating system will help us prevent hacking of our systems.' [59]

The important point in policy that appears to be understood here is that a computer operating system and its associated hardware structures are the base for any security scheme that is required, including any defence against DoS/DDoS and penetration of a computer. Simply put, security must be a design parameter and be built into any product or system. Adding it in later may not result in any real improvement and may actually result in a false sense of security, e.g. a computer application cannot be more secure than the underlying computer hardware, operating system, 'middleware' and software libraries that are used. Moreover, it is implied in India's mission statement that such structures are not commercially available in the usual way through normal ICT product and system supplier channels. This attitude is supported by a 2001 paper by P. Loscocco, USA's National Security Agency (NSA) and S. Smalley, NAI Labs, in the following way:

> End systems must be able to enforce the separation of information based on confidentiality and integrity requirements to provide system security. Operating system security mechanisms are the foundation for ensuring such separation. Unfortunately, existing mainstream operating systems lack the critical security feature required for enforcing separation: mandatory access control ... Instead, they rely on discretionary access control (DAC) mechanisms. As a consequence, application security mechanisms are vulnerable to tampering and bypass, and malicious or flawed applications can easily cause failures in system security. *DAC mechanisms are fundamentally inadequate for strong system security.* [111]

Moreover, it is worth noting that the original intention of the USA's Trusted Computer System Evaluation Criteria (TCSEC), known as the 'Orange Book' and finally fully released in 1985, clearly aimed at trying to get the computer industry to enhance the security of its products and systems, a matter of vital significance in relation to DoS/DDoS, but expressed in terms of confidentiality at the time, in the following way:

> ... the criteria have been developed to serve a number of intended purposes:
> * To provide a *standard to manufacturers* as to what security features to build into their new and planned, commercial products in order to provide widely available systems that satisfy trust requirements (with particular emphasis on preventing the disclosure of data) for sensitive applications. [43]

The evidence is simply that the 'Orange Book', and other similar publications in what has become known as the USA's 'Rainbow Series' did not achieve that aim at all. The current 'Common Criteria', as international standard IS15408, now covers this area and both India and Australia are signatories to the 'Common Criteria Recognition Agreement (CCRA)'.

3.5.4 International and Other Nations

3.5.4.1 International

The DoS/DDoS threat on information systems is an international threat and as such an international legal response is required. However, the base for any such arrangement is only being discussed now. The following instruments have been referred to as possible bases for such legal arrangements:

- Mutual Legal Assistance Agreements (MALT)
- United Nations Convention on the Law of the Sea (UNCLOS)
- Antarctic Treaty System
- Nuclear Non-proliferation Treaties
- Law of Armed Conflict

At the cyber conflict level, Wingfield [138] has acknowledged that there are numerous problems in considering the international legal situation in relation to DoS/DDoS under conditions of 'information operations' or 'cyber attack', depending upon the nature of the attack, the victim and the perpetrator. He proposes that three distinct legal problems emerge, as follows:

1. Reliable attribution to a person, organisation/enterprise or nation
2. Determination of the loss suffered by the victim, including first- and second-order effects
3. Lack of understanding of any applicable international laws

It has been pointed out by both Wingfield and Clarke that there is a marked difference between cyber attacks under differing conditions. For example, Clarke sees such attacks as a normal adjunct to any condition of war, along with any other means of conducting any war. The situation is a state of *jus in bello* and use of cyber attack fits under normal rules of war. However, cyber attack during times of peace, *jus ad bellum*, affords a totally different perspective and set of legal standards. The situation described on 15 Jan 2011 [73] by the *New York Times* in relation to likely sources of and responsibility for the 'STUXNET' computer 'Trojan' allegedly aimed at the disruption, i.e. DoS, of nuclear fuel centrifuges in Iran would fit into the latter situation.

3.5.4.2 Other Nations

Both the USA and the UK are examples of nations which have set out major discussion documents and programs in all aspects of DoS/DDoS against their NCI and NII, in particular. An extensive literature exists based up on the relevant policy documents and legal analysis. Detailed examination of the approaches being adopted by these two countries is a major project and is outside the sphere of this general discussion paper. However, the entry of Australia into bilateral arrangements with both the USA and UK in all matters of cybersecurity is of note. The January 2011 communiqué [126] issued following the third 'AUKMIN' ministerial conference in Australia between Australian and UK ministers exemplifies emerging arrangements between nations. The communiqué stated as follows [126]:

> While globalisation and interconnectedness bring new opportunities for growth and development, both countries recognised that new threats to the security of their citizens and their interests are also emerging. National interests are now affected, more than ever before, by events which take place beyond the borders of our own nations. Australia and the UK acknowledged the value of cooperating with third countries in assessing and addressing strategic and security challenges. Australia and the UK are committed to working together in concrete and practical ways to shape a more secure environment and advance common interests with respect to outer space and cyber security.

This supports the statements by the UK's Minister for the Armed Forces at Chatham House, London, on Tuesday 9 November 2010 [97].

He clearly stated as follows:

> As Minister for the Armed Forces, I am concerned with how we should defend ourselves against those who would use cyber space to do harm, and how we best use the new technologies to further our national security...... we must recognise that the threats in cyber space do not just come from malicious mischief makers or organised criminality...The consequences of a well planned, well executed attack against our critical networks could be catastrophic. The fact that cyber security has been identified as one of the top national security threats for the UK over the next five years indicates both the likelihood of such an attack and the level of impact. Without doubt, man has brought the capacity for war to cyber space too.

This speech clearly identified the dual nature of DoS/DDoS threats against the NII in that it saw the dual nature of criminal activity, on one side, and determined, state-sponsored attack on the other. He emphasised the inherent asymmetry in the cyber challenge in that the low cost of attack is unequally matched by the high cost of defence. He stated this fact as follows: 'cyber is a powerful asymmetric tool for warfare'.

3.6 Summary and Conclusions

The emergence of total dependence by nation states, encompassing all private and public sector enterprises, on a national information infrastructure (NII) has been clearly identified in the first decade of the twenty-first century. Moreover, the dependence of other national critical infrastructure (NCI) on that same NII has also been seen as a major concern internationally. The need for bilateral and multilateral work in the area of protection of that NII has been recognised globally and the project reported in this chapter demonstrates a bilateral approach to the problem on behalf of Australia and India. The problem of the role and function of nation states in the regulation of the international digital network infrastructure, now almost totally based around the Internet engineering structures, has been identified as an area of research needing much further attention. At the same time, movement of end-user equipment to a mobile, wireless environment, while maintaining connection to the global Internet, has also been seen as an area of major policy concern, particularly related to the security and resiliency of the overall network infrastructure. The three concepts of global, nation and defence information infrastructure have been identified and the problem of assignment of responsibility for the protection of those systems has been discussed. Policy and law parameters need to be carefully identified and examined in order for appropriate levels of response to be carried out. In this regard, policy and law concerns have been identified in relation to so-called 'end-of-line' equipment, namely end-user systems, and the actual connection networks themselves. In this regard, it has been important to consider the background and history to the development of national and international telecommunications infrastructure as well as the development of any associated information and communications technology industry. The topics of both 'cyber' and 'electronic' warfare have both been canvassed in this chapter but problems of definition and relevance have been identified. In this regard, the problem of identification of cyber threats has been covered as well as the responsibility for the installation and management of appropriate countermeasures in connected computer systems, including those computers that form the base for the actual networks themselves. Indeed, from determination of attack parameters, response requirements will need to be determined against bodies of law that exist and against perceived needs for the development of necessary policy, law and regulatory instruments. In particular, the problem of adequate levels of response to attack has been seen as a major concern requiring much further research work. The chapter has reviewed a number of international and national organisations working in the area of policy research in relation to DoS/DDoS and related cyber security concerns. It should be noted here that in relation to the actual international Internet itself, response capacity has seen as being concentrated in a number of private sector and voluntary professional organisations as well as in governments. This change is reflected in the global movement to privatisation of national critical infrastructure enterprises such as electrical power generation and distribution, telecommunications services and the like. The history of this movement has been seen as playing a vital role in the analysis of policy, law and regulations relevant to the sector. In India, that history has been examined in detail as it plays a major part in the development and enactment of associated instruments in India, e.g. the 'IT 2000' Act and its later amendments.

The chapter concludes that much further research is needed into overall social, political, legal and regulatory aspects of the move by nations and private sector enterprises alike to total dependence on the national and global information infrastructure. It perceives that the age of the 'Digital Economy' has indeed arrived, as predicted in the mid-1990s. This fact requires that appropriate responses are needed to the requirements of governance of that reality.

3.7 Summary of Research Questions in Policy and Law: DoS/DDoS in NII Systems

Research questions:

1. Self-regulation policy
   1.1 Will ICT product, systems and services industry self regulatory policy work to mitigate against DoS/DDoS threats against the national information infrastructure and critical national infrastructure in general?
   1.2 Is there evidence for or against this in other industries of similar criticality to the nation?
2. Regulation of the ICT and ICT Services Industry
   2.1 Is there any evidence for or against the introduction or strengthening of a regulatory or legal environment for the ICT products and systems as well as the ICT services industry?
   2.2 Could any regulatory environment be effectively and efficiently managed by government?
3. Directors' responsibilities
   3.1 Should the law impose an obligation on enterprise managers and directors to implement DoS prevention and mitigation measures in information systems? If so, how should such laws and regulations be formulated?
   3.2 Should directors of private sector, public sector or joint public/private entities owning, controlling or managing national critical infrastructure have any national security obligations above the normal commercial obligation imposed by current law?
4. Military role
   4.1 As in the protection of the 'physical space' of a nation, should military enterprises be charged with and funded to perform the same protection functions for 'cyberspace'?
   4.2 If so, how should this scheme be arranged and the military-private sector interface be managed, e.g. is there a need to reconsider the concept of Government and/or military security clearances and classification of relevant documents and data?

4.3 Is a 'cyber' force an appropriate scheme? If so, how should it be created, e.g. as a separate military entity, as an adjunct to a traditional force such as an army, navy or air force, etc.?

5. Government role

5.1 Is it necessary for the government to take an intervention position in relation to DoS/DDoS threats relevant to CIP and NIIP?

6. Education and training

6.1 Should research, education and training in information systems security, with an emphasis on resilience to DoS/DDoS, be given emphasis in government policy and be appropriately funded?
6.2 How should any such education program be arranged, e.g. publicly funded tertiary education and/or private sector enterprises, specialised institutions or division of current institutions? Should such education and training be regarded as being of national security interest with limitations on enrolment in designated courses, government input into curricula taught and assessment of teachers?
6.3 Is a 'cyber range' a useful concept and worthwhile investment for training, education and research purposes as well as for ICT security product and system testing and evaluation?

References

1. About ENISA. 2011. http://www.enisa.europa.eu. Accessed 17 Jan 2011.
2. Australian Computer Emergency Response Team. 2011. http://www.auscert.org.au/. Accessed 12 Jan 2011.
3. Australian Telegraphic History. 2011. http://en.wikipedia.org/wiki/Australian_telegraphic_history. Accessed 24 Jan 2011.
4. Australia's National Computer Emergency Response Team. 2011. http://www.cert.gov.au/. Accessed 21 Feb 2011.
5. Cable & Wireless Worldwide. 2011. http://www.answers.com/topic/cable-wireless-jamaica-limited. Accessed 24 Jan 2011.
6. Commonwealth Bank – Security. 2011. http://www.commbank.com.au/help/faq/netbank/security.aspx#whatisnetcodesms. Accessed 17 Jan 2011.
7. Communication Convergence Bill 2001. 2011. http://164.100.24.208/ls/committeeR/communication/39.pdf. Accessed 25 Feb 2011.
8. Communication Division. 2011. http://planningcommission.nic.in/sectors/commun.html. Accessed 25 Feb 2011.
9. Communications Critical Infrastructure Resilience. 2011. http://www.dbcde.gov.au/online_safety_and_security/Communications_critical_infrastructure_resilience. Accessed 10 Jan 2011.
10. Critical Infrastructure Resilience. 2011. http://www.ag.gov.au/www/agd/agd.nsf/Page/Nationalsecurity_CriticalInfrastructureProtection. Accessed 10 Jan 2011.
11. Defence Signals Directorate. 2011. http://www.dsd.gov.au/infosec/csoc.html. Accessed 10 Jan 2011.

12. Department of Information Technology. 2011. http://www.mit.gov.in/content/overview. Accessed 25 Feb 2011.
13. Department of Information Technology, Ministry of Communications and Information Technology, Government of India. 2011. http://www.mit.gov.in/content/overview. Accessed 23 Dec 2010.
14. Department of Telecommunications. 2011 http://www.dot.gov.in/. Accessed 25 Feb 2011.
15. Eastern Telegraph Co's System and Its General Connections. 2011. http://upload.wikimedia.org/wikipedia/commons/a/a5/1901_Eastern_Telegraph_cables.png. Accessed 25 Jan 2011.
16. Facebook. 2011. http://www.facebook.com/press/info.php?statistics. Accessed 1 Jan 2011.
17. Forum of Incident Response and Security Teams. 2011. http://www.first.org. Accessed 11 Jan 2011.
18. French Resistance. 2010. http://en.wikipedia.org/wiki/French_Resistance#cite_ref-source1_3-0. Accessed 31 Jan 2010.
19. IETF STD Index. 2011. http://www.rfc-editor.org. Accessed 8 Jan 2011.
20. Information Technology Crime. 2011. http://www.interpol.int/public/TechnologyCrime/default.asp. Accessed 11 Jan 2011.
21. Information Update for CUA Customers. 2011. http://www.cua.com.au/. Accessed 13 Jan 2011.
22. International Federation for Information Processing. 2011. http://www.ifip.org. Accessed 11 Jan 2011.
23. Internet. 2011. http://en.wikipedia.org/wiki/Internet. Accessed 5 Jan 2011.
24. Internet Assigned Numbers Authority. 2011. http://www.iana.org/about/. Accessed 11 Jan 2011.
25. Internet Corporation for Assigned Names and Numbers. 2011. http://www.icann.org/. Accessed 11 Jan 2011.
26. Invest India Telecom. 2011. http://www.dot.gov.in/osp/Brochure/Brochure.htm. Accessed 25 Feb 2011.
27. IT Act of India 2000. 2011. http://www.cyberlawsindia.net/Information-technology-act-of-inida.html. Accessed 4 Feb 2011.
28. ITU Activities Related to Cybersecurity. 2011. http://www.itu.int/cybersecurity/. Accessed 10 Jan 2011.
29. The Japanese Bombing of Darwin and Northern Australia. 2010. http://www.cultureandrecreation.gov.au/articles/darwinbombing/. Accessed 30 Dec 2010.
30. Kerala Police, India. 2011. http://www.keralapolice.org/newsite/hitech_need_cyberlaw.html. Accessed 4 Feb 2011.
31. Legal Services India. http://www.legalserviceindia.com/cyber/cyber.htm and http://www.legalserviceindia.com/article/l323-Cyber-Crimes-&-Cyber-Law.html. Accessed 18 Jan 2011.
32. MOSAIC (Web Browser). 2011. http://en.wikipedia.org/wiki/Mosaic_(web_browser). Accessed 18 Jan 2011.
33. National Security Science and Technology Branch. 2011. http://www.dpmc.gov.au/nsst/index.cfm. Accessed 10 Jan 2011.
34. NATO, Wikipedia. 2011. http://en.wikipedia.org/wiki/NATO/. Accessed 17 Jan 2011.
35. No. 462 Squadron RAAF. 2011. http://en.wikipedia.org/wiki/No._462_Squadron_RAAF. Accessed 5 Jan 2011.
36. Organisation for Economic Cooperation and Development. 2011. http://www.oecd.org. Accessed 11 Jan 2011.
37. Packet Switched Network. 2011. http://en.wikipedia.org/wiki/Packetswitched_network. Accessed 14 Jan 2011.
38. The common criteria recognition arrangement: Purpose of the arrangement. http://www.commoncriteriaportal.org/ccra/. Accessed 11 Jan 2011.
39. The Information Technology (Amendment) Bill 2006. 2011. https://www.prsindia.org/uploads/media/Information_Technology/1168510210_The_Information_Technology__Amendment__Bill__2006.pdf. Accessed 4 Feb 2011.

40. TISN for critical infrastructure resilience. http://www.tisn.gov.au/. Accessed 6 Jan 2011.
41. Uncitral model law. http://www.uncitral.org/. Accessed 4 Feb 2011.
42. Who we are, Department of the Premier and Cabinet, Queensland. http://www.premiers.qld.gov.au/about-us/who-we-are.aspx. Accessed 17 Jan 2011.
43. Trusted computer system evaluation criteria. Technical report DoD 5200.28-STD, USA Department of Defense, 26 Dec 1985. (Supersedes CSC-STD-001-83, dated 15 Aug 83).
44. ISO 7498-2: 1989 Information processing systems – Open Systems Interconnection – Basic Reference Model – Part 2 Security Architecture. Technical report, International Organization for Standardization, 1989.
45. History of communications in Australia. 2001. http://www.abs.gov.au/ausstats/abs@.nsf/Previousproducts/1301.0Feature%20Article432001?opendocument&tabname=Summary&prodno=1301.0&issue=2001&num=&view=. Accessed 25 Jan 2011.
46. (2002). R vs Boden. *QCA 164*.
47. SSAC advisory SAC008 DNS distributed denial of service (DDoS) attacks. Technical report, ICANN Security and Stability Advisory Committee, Mar 2006. http://www.icann.org/en/committees/security/dns-ddos-advisory-31mar06.pdf. Accessed 4 Feb 2011.
48. DNSSEC@ICANN 2008. Signing the root zone: A way forward toward operational readiness. http://icann.org/en/announcements/dnssec-paper-15jul08-en.pdf. Accessed 4 Feb 2011.
49. Intel plans to tackle cosmic ray threat. 2008. http://news.bbc.co.uk/2/hi/technology/7335322.stm. Accessed 9 Jan 2011.
50. The Information Technology (Amendment) Act, 2008. 2008. http://www.mit.gov.in/sites/upload_files/dit/files/downloads/itact2000/it_amendment_act2008.pdf. Accessed 4 Feb 2011.
51. Changing landscape of DDos threats and protection. 2009. http://verisigninc.com/assets/whitepaper-ddos-threats-protection-forrester.pdf. Accessed 23 Feb 2011, March 2009.
52. *United Nations Handbook: 2009–2010*: An Annual Guide for those Working with and within the United Nations. United Nations Publications, 308–314.
53. 3G with Integrated ADSL2+ Modem DVG834NH. 2010. http://www.netgear.com.au/au/product/routers-and-gateways/3gfemtocells/dvg834nh?gclid=CKuP9OzpzKYCFQHVbgod8VV-Gw. Accessed 18 Jan 2011.
54. Corporate plan, 2011–2013. 2010. http://www.nbnco.com.au/. Accessed 9 Jan 2011.
55. (2010, Dec). Critical ICT Infrastructure Protection in India. 2010. cyberlawsindia.blogspot.com/2010/12/critical-ict-infrastructure-protection.html. Accessed 2 Feb 2011. http://cyberlawsinindia.blogspot.com/2011/01/critical-infrastructureprotection-in.html. Cited at 11 Jan 2011.
56. ICANN's First DNSSEC Key Ceremony for the Root Zone. 2010. http://www.icann.org/en/announcements/announcement-2-07jun10-en.htm. Accessed 4 Feb 2011.
57. icode to commence 1 December 2010. http://www.iia.net.au/. Accessed 22 Dec 2010, Oct. 27 10:14 2010.
58. NEC rewrites the rules for 3G rollout in India. 2010. http://www.nec.com.au/News-Media/Media-Centre/Media-Releases/NEC-rewrites-the-rules-for-3G-rollout-in-India.html. Accessed 4 Jan 2011.
59. Saraswat: "DRDO working on India's own computer operating system". *The Hindu*, 9 Oct 2010. http://www.thehindu.com/sci-tech/science/article821933.ece. Accessed 25 Feb 2011.
60. The eighth apec ministerial meeting on the telecommunications and information industry (telmin 8). 2010. http://www.soumu.go.jp/main_content/000087132.pdf. Accessed 23 Feb 2010.
61. A New Year wish: more digital visionaries. *The Australian*, 01 Jan 2011. http://www.theaustralian.com.au/news/a-new-year-wish-more-digital-visionaries/story-e6frg6n6-1225979630797. Accessed 1 Jan 2011.
62. Mondaq IT & Telecoms. 2011. http://www.mondaq.in. Accessed 25 Feb 2011.
63. Solar energy use in cell towers can reduce co2 emission: Study. 2011. http://www.businessstandard.com/india/news/solar-energy-use-in-cell-towers-can-reduce-co2-emissionstudy/122349/on. Accessed 16 Jan 2011.
64. The Internet Engineering Task Force. 2011. http://www.ietf.org. Accessed 11 Jan. 2011.

65. Ahmed, M. 2011. Driving the engine of success. *The Australian*, 26, 08 Jan 2011.
66. Alvestrand, H. 2004. A mission statement for the IETF: RFC 3935. Technical report, IETF Network Working Group.
67. Andreessen, M. 2011. Mosaic – The first global web browser. http://www.livinginternet.com/w/wi_mosaic.htm. Accessed 9 Jan 2011.
68. APEC Secretariat. 2010. APEC at a glance. http://publications.apec.org/file-download.php?filename=211_sec_APECatAGlance_F.pdf&id=1077. Accessed 23 Feb 2011.
69. Australian Government. 2006. Denial of service/distributed denial of service MANAGING DoS ATTACKS. http://www.dbcde.gov.au/__data/assets/pdf_file/0011/41312/DoS_Report.pdf. Accessed 23 Feb 2011.
70. Ball, D. 2008. *Australia and cyber-warfare*, Chap. 6: An Australian Cyber-warfare Centre, 119–148. ANU, Canberra.
71. Ballmer, S. 2011. Remarks by Microsoft CEO Steve Ballmer at 2011 International CES Las Vegas, 5 Jan 2011. http://www.microsoft.com/presspass/exec/steve/2011/01-05CES.mspx. Accessed 7 Jan 2011.
72. Beckstrom, R. 2010. The new internet nation state. Internet Corporation for Assigned Names and Numbers (ICANN) As Prepared for delivery to the United Nations, Department of Economic and Social Affairs, Multi-stakeholder Consultation on Enhanced Cooperation New York. http://www.icann.org/en/presentations/beckstrom-speech-united-nations-14dec10-en.pdf. Accessed 9 Jan 2011.
73. Broad, W., J. Markoff, and D. Sanger. 2011. Israeli test on worm called crucial in Iran nuclear delay. *New York Times*. http://www.nytimes.com/2011/01/16/world/middleeast/16stuxnet.html. Accessed 18 Jan. 2011.
74. BSNL. 2011. Calcutta telephones. http://www.calcutta.bsnl.co.in/history_ctd.shtml. Accessed 25 Feb 2011.
75. Caelli, W. 2010. Towards the digital city: Information assurance and resilience policy. In *Proceedings of the Hong Kong Institute of Engineers (HKIE) Conference "The Way to the Digital City – Sustainable Public Services and Solutions"*, Hong Kong – SAR, PRC, 1–2 Nov 2010.
76. Carpenter, B. 2009. The IETF process: An informal guide. http://www.ietf.org/about/process-docs.html. Accessed 9 Jan 2011, Oct 29 2009.
77. Carr, J. 2010. *Inside cyber warfare: Mapping the cyber underworld*. Sebastopol: O'Reilly.
78. Carr, N. 2009. *The big switch: Rewiring the world, from Edison to Google*. New York: W.W. Norton & Co.
79. CCRA. 2011. Common criteria. http://www.commoncriteriaportal.org/. Accessed 23 Feb 2011.
80. Chabrow, E. 2011. 5 obstacles to infosec reform in 2011 : Some players have changed, but the issues remain the same. *GovInfo Security*, 7 Jan 2011. http://blogs.govinfosecurity.com/posts.php?postID=844&rf=2011-01-10-eg. Accessed 9 Jan 2011.
81. Christensen, S., W. Caelli, W. Duncan, and E. Georgiades. 2010. An Achilles heel: Denial of service on Australian critical information infrastructure. *Information and Communications Technology Law* 19(1): 61–85. http://dx.doi.org/10.1080/13600831003708059
82. CISCO. 2004. Distributed denial of service threats: Risks, mitigation, and best practices. Cisco Systems white paper. https://www.info-point-security.com/open_downloads/alt/cisco_wp_ddos.pdf. Accessed 23 Feb 2011.
83. Clarke, R. and R.K. Knake. 2010. *Cyber war: The next threat to national security and what to do about it*. New York: HarperCollins.
84. CNN. 1998. Heading off an electronic Pearl Harbor. *CNN Tech*. http://articles.cnn.com/1998-04-06/tech/9804_06_computer.security_1_hackers-national-security-cybercrime?_s=PM:TECH. Accessed 2 Jan 2010.
85. Creedy, S. and F. Foo. 2010. Virgin brand feeling a bit blue after computer glitch. *The Australian*, 25, 9 OCtober 2010.
86. de Villiers, M. 2007. *Distributed denial of service: Law technology and policy*. Sydney: University of New South Wales, School of Law.

87. Dudgeon, I. 2008. *Australia and cyber-warfare*, chapter 4: Targeting Information Infrastructures, 59–83. Canberra: ANU.
88. Dyer, J.G., M. Lindemann, R. Perez, L. van Doorn, S.W. Smith, and S. Weingart. 2001. Building the IBM 4758 secure coprocessor. *IEEE Computer*, 57–66, Oct 2001.
89. Dyson, E. 1997. *Release 2.0: A design for living in the digital age*. Viking/Penguin.
90. Evers, J. 2007. New shield foiled Internet backbone attack. http://news.cnet.com/2100-7349_3-6166107.html. Accessed 4 Feb 2011.
91. Fitzgibbon, J. 2009. Defending Australia in the Asia-Pacific century: Force 2030. Defence white paper. http://www.defence.gov.au/whitepaper/. Accessed 10 Jan. 2011.
92. Garamone, J. 2010. Cybercom chief details cyberspace defense. http://www.defense.gov/news/newsarticle.aspx?id=60987. Accessed 13 Jan 2011.
93. Georgiades, E. Crisis on impact: Responding to cyber attacks on critical information infrastructures. In preparation.
94. Graham, W.R., J.S. Jr. Foster, E. Gjelde, R.J. Hermann, H.M. Kluepfel, R.L. Lawson, G.K. Soper, L.L. Jr. Wood, and J.B. Woodard. 2004. Report of the commission to assess the threat to the United States from electromagnetic pulse (EMP) attack : Volume 1: Executive report. Technical report. http://empcommission.org/docs/empc_exec_rpt.pdf. Accessed 2 Feb 2011.
95. Hall, E. 2010. Former White House security advisor warns of cyber war. The World Today, ABC Radio National http://www.abc.net.au/worldtoday/content/2010/s3086792.htm. Accessed 7 Dec 2010.
96. Handley, M. and E. Rescorla. 2006. Internet denial-of-service considerations: RFC 4732. http://tools.ietf.org/html/rfc4732. Accessed 22 Nov 2010.
97. Harvey, N. 2011. 2010/11/09 – Meeting the cyber challenge. http://www.mod.uk/DefenceInternet/AboutDefence/People/Speeches/MinAF/20101109MeetingTheCyberChallenge.htm. Accessed 4 Jan 2011.
98. Hilvert, J. 2010. Cybercrime response a win for self-regulation. http://www.itnews.com.au. Accessed 20 Dec 2010.
99. Hopewell, L. 2011. NBN batteries pose disaster issue: Experts. http://www.zdnet.com.au/nbn-batteries-pose-disaster-issue-experts-339308546.htm. Accessed 13 Jan 2011.
100. ICANN. 2011. DNSSEC @ ICANN Signing the root zone: A way forward toward operational readiness. http://www.icann.org/en/announcements/dnssec-paper-15jul08-en.pdf. Accessed 23 Feb 2011.
101. ICT-India. 2011 ICT regulation toolkit. http://www.ictregulationtoolkit.org/en/PracticeNote.1222.html. Accessed 25 Feb 2011.
102. IndianetZone. 2008. History of Indian telecommunications. http://www.indianetzone.com/42/history_indian_telecommunications.htm. Accessed 25 Feb 2011.
103. ISO. 2011. Information technology – security techniques – evaluation criteria for it security – part 1: Introduction and general model. http://www.iso.org/iso/iso_catalogue/catalogue_ics/catalogue_detail_ics.htm?csnumber=50341. Cited at 5 Jan 2011.
104. Jingfei, S. 2009. Solution–Mitigating interference between LTE and 2G/3G networks. Technical report.
105. Kahn, R. and V. Cerf. 2000. Al Gore and the Internet. http://amsterdam.nettime.org/Lists-Archives/nettime-l-0009/msg00311.html. Accessed 5 Jan 2011.
106. Kitney, D. 2011. The web whirlwind. *The Weekend Australian*, 1, 8 Jan 2011.
107. Koch, C. 1995. Mutable markets. *CIO Magazine* 9(5): 68–76.
108. Kramer, F.D., S.H. Starr, and L.K. Wentz, eds. 2009. *Cyberpower and national security*. Washington, DC: Center for Technology and National Security Policy.
109. Lemos, R. 2002. Security czar: Button up or get hacked. *CNET News*. http://news.cnet.com/2100-1001-840335.html. Accessed 31 Dec 2010.
110. Libicki, M. 2007. *Conquest in cyberspace: National security and information warfare*. Cambridge: Cambridge University Press.
111. Loscocco, P. and S. Smalley. 2001. Integrating flexible support for security policies into the Linux operating system. In *2001 USENIX Annual Technical Conference Boston*, 25–30 Jun 2001. USENIX.

112. Lowensohn, J. 2011. Hotmail's recent message loss hiccup explained. *CNET News*. http://news.cnet.com/8301-17939_109-20027726-2.html?tag=nl.e776. Accessed 8 Jan 2011. http://news.cnet.com/8301-17939_109-20027726-2.html?tag=nl.e776. Cited at 8 Jan 2011.
113. Maheshwari, V. 2011. Legal service India.com. http://www.legalserviceindia.com/articles/art222.htm. Accessed 25 Feb 2011.
114. Mallet, P. (2010, Nov 18). NATO mobilises for cyber warfare. http://www.physorg.com/news/2010-11-nato-mobilises-cyber-warfare.html: Accessed 10 Jan 2011.
115. Marsh, R.T. 1997. Critical foundations: Protecting America's infrastructures – The report of the President's Commission on Critical Infrastructure Protection. Technical report, The White House, Washington, DC. http://www.fas.org/sgp/library/pccip.pdf. Accessed 4 Jan 2011.
116. Australian War Memorial. 2011. 462 squadron raaf. *Australian War Memorial*. http://www.awm.gov.au/units/unit_11167.asp. Cited at 5 Jan 2011.
117. Ohlden, A. 2011. Northrop Grumman opens UK cyber security test range For evaluation of threats on large scale networks. http://www.science20.com/newswire/northrop_grumman_opens_uk_cyber_security_test_range_evaluation_threats_large_scale_networks. Accessed 26 Jan 2011.
118. Pauli, D. 2010. 2015, the year of cyberwar: Gartner. http://www.zdnet.com.au/2015-the-year-of-cyberwar-gartner-339307693.htm. Accessed 6 Dec 2010.
119. Perset, K. and S. Paltridge. 2008. The future of the internet economy. Policy brief, OECD. http://www.oecd.org/dataoecd/20/41/40789235.pdf. Accessed 4 Feb 2011.
120. Preimesberger, C. 2010. 2010 saw the dawn of nation-state cyber wars: Citrix CTO. *eWeek*. http://www.eweek.com/c/a/Cloud-Computing/2010-Saw-the-Dawn-of-NationState-Cyberwars-Citrix-CTO-430619. Accessed 31 Dec 2010.
121. Raichaudhuri, R. 2011. "INDIA CALLING" – The rise of the Indian telecommunications industry. http://www.indialawjournal.com/volume2/issue_3/article_by_rohan.html. Accessed 25 Feb 2011.
122. Rajesh. 2010. India faces disrupted internet service due to undersea cable issue. http://www.tech-exclusive.com/india-faces-disrupted-internet-service-due-to-undersea-cable-issue/. Accessed 22 Feb 2011.
123. Rattray, G. 2001. *Strategic warfare in cyberspace*. Cambridge: The MIT Press.
124. Rudd, K. 2008. Address to Heads of Agencies and members of senior executive service. http://www.apsc.gov.au/media/rudd300408.htm. Accessed 2 Feb 2011.
125. Rudd, K. and S.M. Krishna. 2011. Australia-India foreign ministers' framework dialogue. foreignminister.gov.au/releases/2011/kr_mr_110120.html. Accessed 20 Jan 2011.
126. Rudd, K., S. Smith, W. Hague, and L. Fox. 2011 Australia-United Kingdom ministerial consultations: Joint communiqué; Sydney, 18 Jan. 2011. http://foreignminister.gov.au/releases/2011/kr_mr_110118a.html. Accessed 19 Jan 2011.
127. Schwartau, W. 1996. *Information warfare*, 2nd edn. New York: Thunder's Mouth Press.
128. Singh, H., A. Soni, and R. Kathuria. 2003. *Telecom policy reform in India*, chapter 4. World Bank and Oxford University Press.
129. Stanley, P. 2002. Remembering 1942: The bombing of Darwin, 19 February 1942. Australian War Memorial presentation. http://www.awm.gov.au/atwar/remembering1942/darwin/transcript.asp. Accessed 3 Feb 2011.
130. Thompson, R. 2010. The sun in action. http://www.ips.gov.au/Educational/2/1/1. Accessed 9 Jan 2011.
131. TRAI. 2011. Telecom Regulatory Authority of India. http://www.trai.gov.in/Default.asp. Accessed 25 Feb 2011.
132. TRAI. 1999. New telecom policy 1999. http://www.trai.gov.in/TelecomPolicy_ntp99.asp. Accessed 25 Feb 2011.
133. Vixie, P. 2002. Securing the edge. http://www.icann.org/en/committees/security/sac004.txt. Accessed 4 Feb 2011.
134. Waters, G., D. Ball, and I. Dudgeon. 2008. *Australia and cyber-warfare*. Canberra: ANU Press.

135. Westervelt, R. 2011. IBM predicts rising mobile threats, critical infrastructure attacks in 2011. http://searchsecurity.techtarget.com/news/article/0,289142,sid14_gci1525624,00.html?track=NL-102&ad=806829&asrc=EM_NLN_13112830&uid=5074973. Accessed 4 Jan 2011.
136. Wikipedia. 2011. Public call office. http://en.wikipedia.org/wiki/Public_call_office. Accessed 25 Feb 2011.
137. Wikipedia. 2011. Tandem Computers Inc. *Wikipedia*. http://en.wikipedia.org/wiki/Tandem_Computers. Cited at 5 Jan 2011.
138. Wingfield, T. 2009. *Cyberpower and national security*, Chapter 22: International law and information operations. National Defense University Press.
139. Zakaria, F. 2008. *The post American world*. New York: W.W. Norton & Company.

Chapter 4
DDoS Testbed

D. Schmidt and S.M. Shalinie

4.1 Introduction

Testing for denial of service vulnerabilities, the effects of attacks and mitigation strategies all require the construction of a dedicated testbed facility. Although the tools for launching such attacks are widely available, for example the Stacheldraht, Trinoo and Phatbot tools [15, p. 87], and although the attacks themselves are well understood, the simulation on a small scale in the laboratory of a large and complex system (the Internet) is fraught with difficulty. Recent research suggests, however, that large botnets, consisting of purportedly millions of machines, may in fact be much smaller than previously estimated, or at least at any one time use only a small fraction of their total size [20, 25]. Nevertheless, the requisitioning, installation and maintenance of thousands of machines, along with their physical routers, high-speed fibre backbones and complex network topologies, renders live experiments at the real world scale wholly impractical. Attacks on the live Internet likewise cannot usually be carried out for legal reasons. This leaves open only the possibility of carrying out experiments into those aspects of distributed denial of service that can be readily simulated under normal laboratory conditions.

A DDoS testbed needs to provide the experimenter with a number of services: it must be capable of coordinating and replaying attacks, it must measure the effects of the attacks on the target, it must provide a means to reconfigure the testbed structure, and if the facility is to be shared, provide a login and job scheduling facility.

D. Schmidt (✉)
Information Security Institute, Queensland University of Technology, Brisbane, Australia
e-mail: schmidda@qut.edu.au

S.M. Shalinie
Thiagarajar College of Engineering, Madurai, Tamil Nadu, India
e-mail: shalinie@tce.edu

S.V. Raghavan and E. Dawson (eds.), *An Investigation into the Detection and Mitigation of Denial of Service (DoS) Attacks: Critical Information Infrastructure Protection*,
DOI 10.1007/978-81-322-0277-6_4,

The rest of this chapter investigates various aspects of these needs: Sect. 4.2 surveys existing testbed designs, Sect. 4.3 examines the goals of DDoS testbed design, and Sects. 4.4 describes the hardware and software designs of the Indian and Australian testbeds. Section 4.5 describes some experiments carried out on the two testbeds and Sect. 4.6 looks forward to future work.

4.2 Existing Testbed Designs

Denial of service testbeds are constructed either deliberately to provide a stable research facility or ad hoc to carry out particular experiments. There is no consensus on how best to construct a DDoS testbed. There are basically three strategies: simulation, emulation and live experimentation. Each has advantages and disadvantages. Sometimes these approaches are combined and in one case virtual machines have been used instead of physical machines to create a large botnet [9].

4.2.1 *Simulation*

In simulation the network and its devices (target and attackers) are all modelled in a live software system, such as the ns-2 [10], OMNET++ [14] or Opnet [7] simulators. The attraction here is the freedom to specify any topology, any combination of target and attackers. One serious drawback with this approach, however, is that simulated routers and operating systems do not adequately model the bottlenecks from experimental nodes, such as CPUs, buses, devices and drivers under the stress imposed by a DDoS attack [11, 16, 29]. For that reason simulation is sometimes used in combination with emulation as in Netbed [29] and SWOON [11].

4.2.2 *Emulation*

In emulation the network topology is realised through virtual LANs and soft-routers, while the target and attack nodes are represented by physical machines [4, 5, 18, 29]. This produces a higher degree of realism than with simulation, and operates at a much higher speed. It allows researchers to recreate experimental topologies without rewiring the testbed, and with the Emulab software it allows the facility to be shared. Against these benefits, however, must be ranged the increased cost of maintaining typically 200–300 machines, and the questionable realism of soft routers versus physical routers.

Our own experiments with commonly used soft routers, when stress-tested by a Smartbits 600 traffic generator, showed that they perform poorly in comparison with their hardware counterparts, even under modest load. The reason appears to

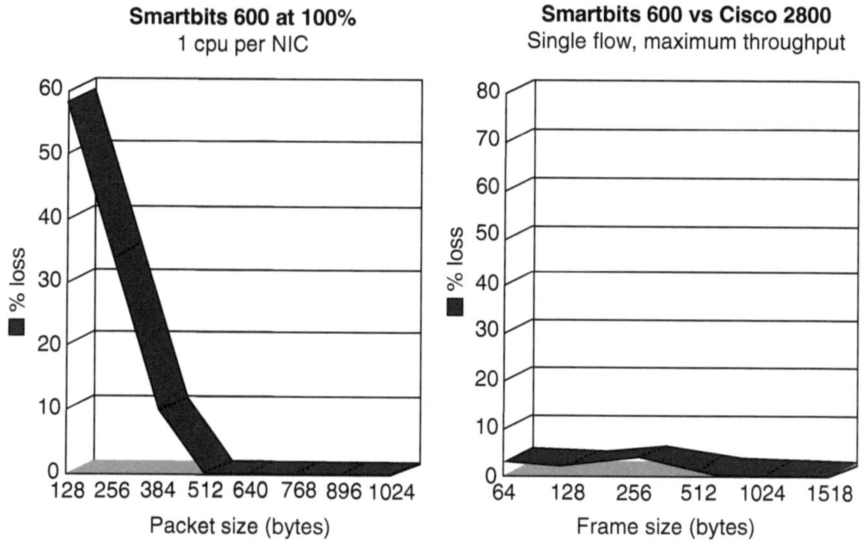

Fig. 4.1 Soft Router versus hardware router performance

be the handling of the interrupts raised by commodity network interface cards [21], with which the CPU is unable to cope, particularly with small packet sizes. Figure 4.1 (left) shows the best performance we could obtain from the Vyatta soft router after trying a variety of different interface cards and configuration parameters. By contrast, the Cisco 2800 hardware router (right) showed few dropped packets at full 1 GB load even for the smallest packet sizes, with no special configuration. Similar results were also obtained for several other Linux soft routers.

4.2.3 Live Experimentation

The favoured strategy for ad hoc testbeds is the construction of local isolated networks of real computers (or virtual machines) and real routers. The largest such DDoS testbed appears that constructed by Calvet et al. [9], which has around 2,000 virtual machines. For the most part, however, cost is not a problem; such rigs can be cobbled together from existing inventory. Like simulated and emulated test environments a single ethernet network cannot possibly mimic the performance of a section of the Internet but it can test how real networking equipment and computers perform under simulated load. Even the small scale of such testbeds can be overcome in several ways: by the use of virtual machines as attack nodes and by more intelligent software traffic generation. Probably the major drawback of this approach is that in order to reconfigure the testbed layout, rewiring is required.

4.3 Designs of the Indian and Australian Testbeds

Because of geographic separation two testbeds have been built, one in QUT, Australia, and the other in Tamil Nadu, India. The QUT testbed is an isolated facility, consisting of a small number of computers connected via a 1 GB network. The Indian testbed is an MPLS cloud distributed among eight nodes at various academic institutions in Tamil Nadu. Interconnection speeds between the nodes vary between 2 and 10 MB. Unlike the Indian testbed, which is accessible to all the universities connected to the MPLS cloud, the Australian facility is not shared. These differences between the underlying hardware and usage enable the testing of different aspects of denial of service. Much of the testbed software is, however, shared.

4.3.1 Design of the Indian Testbed

The Testbed Application (TBA) is a web-based application that accepts 'traffic source' and 'traffic sink' applications from the user, who schedules them to be run on the nodes and target-machine, respectively. Users can have individual accounts on the TBA.

A *traffic source* is a C, Python or Java program that runs on *node* machines and generates network traffic. This program also communicates with the TBA and submits information about the traffic it generates.

A *traffic sink* is a server application that runs on the target machine, and accepts data generated by the traffic sources. This application communicates with the TBA and submits information on the traffic it receives.

A *node* is a machine in the test environment that can participate in an experiment conducted by the TBA. The TBA runs applications on a node using SCP and SSH. Nodes will be registered with the TBA and the TBA's public key will be considered 'authorised' by the Node's SSH daemon.

The *test environment* is a private network of *nodes*, including the target machine and the machine serving the TBA. Users can create a test scenario that will be conducted in the test environment at a scheduled time, and specify traffic source and traffic sink applications. There are also predefined sample scenarios for testbed beginners and custom scenarios for students and researchers, which allow the source and sink programs to be modified.

Figure 4.2 shows at a high level how tests are carried out and the results monitored.

At the scheduled time, the TBA pushes the traffic sink and traffic source applications to the target machine and the nodes, respectively, and starts them remotely. The traffic source and traffic sink applications submit data to the TBA via the web service. The data submitted by the web service is a loosely coupled key-value pair. The TBA does not try to understand the meaning of this data. However, it

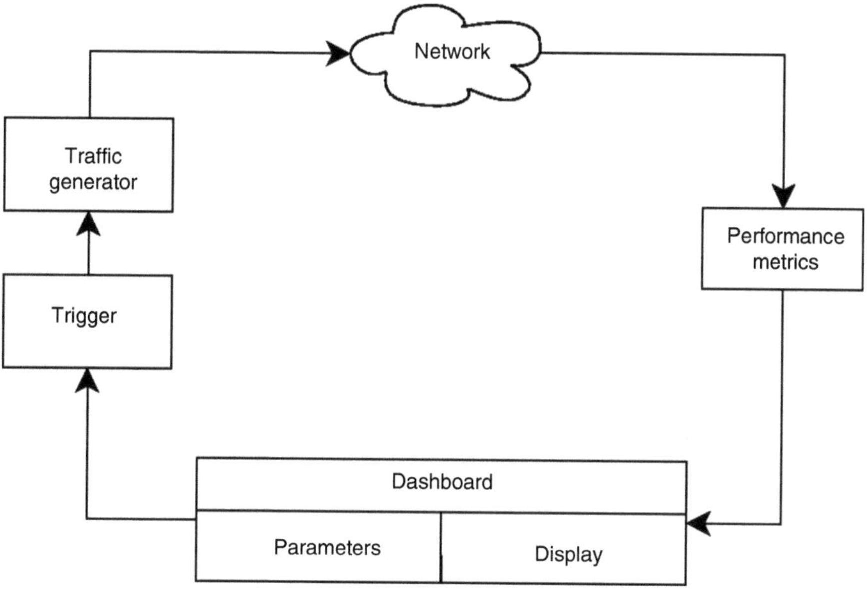

Fig. 4.2 Indian testbed flow of control

simply associates the data to the respective test sequence and files it away in the DB. The web service of the TBA also provides ways to extract the stored test sequence data so that higher-level applications can be written to analyse it.

The web interface displays the status of the source/sink nodes. This enables users to identify busy or free nodes. Running multiple programs on the same node may cause overloading which may be reflected in the results.

4.3.2 Design of the QUT Testbed

Unlike the Indian testbed, the QUT testbed does not have a fixed layout, the idea being that different experiments will require physical reconfiguration. To compensate for the inability to specify the network topology in software (an advantage of the emulation approach), dynamic discovery of suitable hosts on the network to act as attack and target machines is performed when running an experimental scenario. Figure 4.3 shows one configuration of the testbed hardware. The static features of the design are:

1. Partitioning of the network into separate attack and monitoring sections by the use of twin interface cards. The attack network carries attack and background traffic only. The monitoring network carries command and control instructions

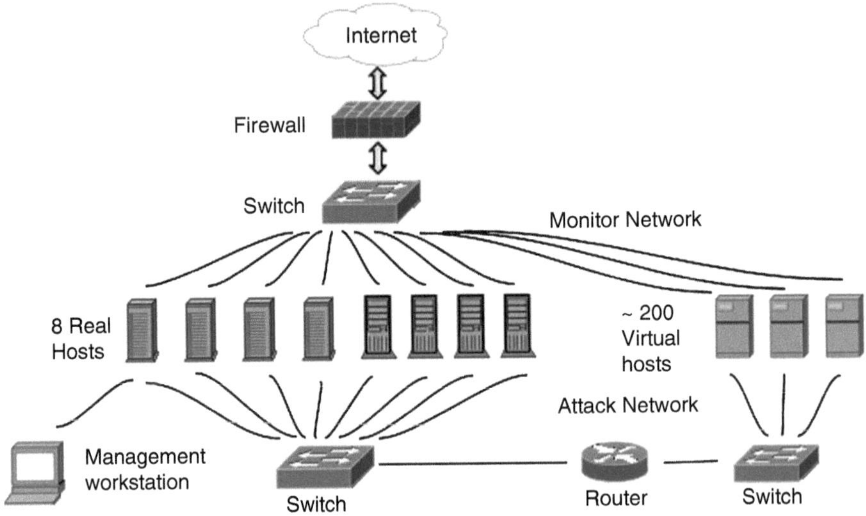

Fig. 4.3 QUT testbed physical structure

and SNMP monitoring requests. Keeping the two separate ensures that when the attack network is saturated, monitoring and control can still take place, so long as some CPU resources are still available on the target [4, 9].

2. Isolation via a firewall that lets port 80 Internet traffic only. This is to facilitate maintenance of the computers, installation of software etc.
3. A single workstation controls the launching and monitoring of the attack.
4. A virtual machine server provides numerous hosts to act as attackers or targets that supplement the limited number of physical nodes.

4.4 Elements of Testbed Software

Although the QUT testbed software is different, it shares many features with the Indian testbed, as noted below. Instead of a monolithic approach, the QUT software is designed as a set of reconfigurable tools that can be deployed equally on either testbed. There are three basic classes of tools:

1. *Measurement* of the effects of an attack on the target and/or network devices
2. *Control* of an experiment's components and their execution
3. *Generation* of background and attack traffic

4.4.1 Measurement

Mircovic et al. [17] argue that current measurements of denial of service are incomplete and imprecise, that they often rely on secondary characteristics such as statistics of dropped traffic on an interface, and rarely measure the actual effects on the service under attack. This gathering of secondary statistics is due to two problems:

1. How do you measure the responsiveness of a service that is being denied?
2. How do you measure application-specific data when most available statistics about operating system performance relate to system-wide parameters?

The metrics proposed by Mircovic et al. [17] boil down to four values: throughput, request response time, data loss and resources (CPU and memory). Data loss, throughput and response time are closely interrelated, and measurement of any one is probably sufficient indication of the others. CPU and memory usage on the other hand are different aspects of denial of service, and their relationship to throughput and data loss is more complex. However, we would agree that the key need is for direct measurement of the service under attack, not its indirect effects on the system as a whole.

Accordingly we have devised a method for directly measuring CPU and memory usage, responsiveness and throughput of an application under attack [22]. This uses SNMP (Simple Network Management Protocol) as the information-gathering protocol, but rather than requesting information about system-wide or interface statistics it uses a custom MIB (Management Information Base) to gather statistics about a specific process bound to a specific port. Because it is a custom software module, the MIB can also be extended to measure anything that is desired. Using SNMP has the advantage that any standard SNMP tools for monitoring can be used, all that is required is to know the OID (Object Identifier) of the parameter being measured.

The problem of how to measure a service under attack can be solved by assigning the highest priority to the SNMP daemon process. In this way, even when under heavy attack, queries on the monitoring interface can still be answered.

Another advantage of SNMP is that it is also supported by router manufacturers so that monitoring of network devices using the same protocol and tools becomes possible.

4.4.2 Control

A DDoS attack is usually controlled by a single computer, whose identity is protected by several layers of indirection. For example in the Waledac botnet the botnet owner hides behind a layer of 'protectors', which relay commands via a network of 'repeaters' [9], as shown in Fig. 4.4. A traditional botnet using IRC

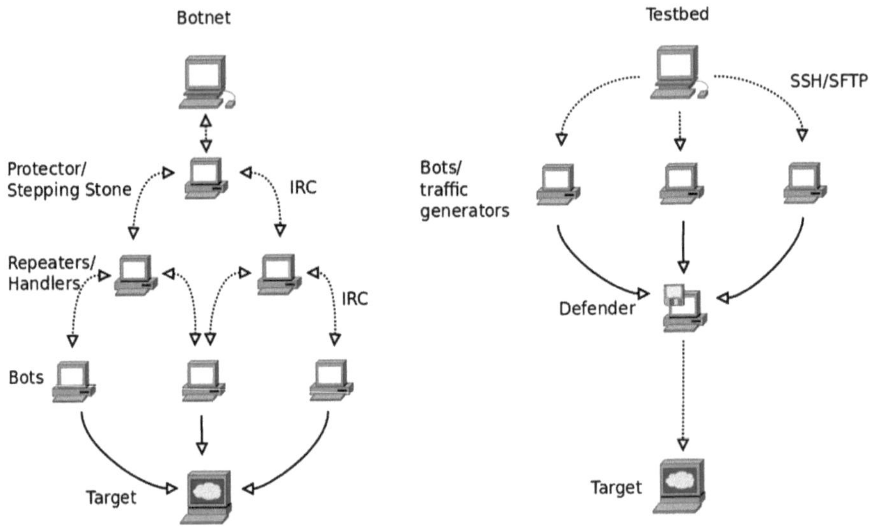

Fig. 4.4 Schematic botnet (*left*) and testbed attack structure (*right*)

(Internet Relay Chat) as its command and control protocol may similarly use some of the bots as 'stepping stones' and 'handlers' to protect the identity of the botmaster [15, p. 212f]. In a DDoS testbed, hiding the identity of the command and control server by such additional layers is not needed unless actual botnet control software is being tested.

Instead of IRC, the communication protocols for both testbeds are SSH for commands and SFTP or SCP for file transfer. As on the Indian testbed control commands are issued from the command and control workstation, which only requires the existence of suitable user accounts on the agents of the testbed. These are dynamically discovered before the script is run, then the same command via SSH can be issued simultaneously to all registered bots, which may be virtual or physical. The agents can be loaded with background traffic generators as well as attack scripts, or they may act as defence nodes. For realism, the target is usually a physical machine (Fig. 4.4). As on the Indian testbed, the traffic generators or attack programs can be anything that runs on the command line.

4.4.3 *Traffic Generation*

The generation of benign and attack traffic is a crucial part of DDoS experimentation. Simple flood-type attacks, such as SYN flood, ICMP flood and UDP floods can be pumped 'onto the wire' at high speed without caring much about how the target service will cope, because interaction with the attacker is practically zero. Attacks like http get flood, and modern application-level attacks, on the other hand, require a complete TCP connection (the exchange of SYN, SYN-ACK and ACK packets) as well as further data once the connection is established.

There are two types of traffic that need generating to provide a realistic simulation of an attack workload:

1. Background traffic, of which there are two types:
   a. Traffic present on the network but not directed to the target
   b. Benign traffic directed towards the target
2. Attack traffic

Tools such as NTGC [27], Harpoon [24] and D-ITG [8] may be used to generate realistic TCP flows between a source and sink (usually not the target) and to provide background traffic during a DDoS experiment. But in spite of the realistic range of packet types, inter-packet times, destination port number, and distribution etc., they cannot stress the server as it would in handling genuine benign traffic. On the DETER testbed the same gap can be seen: the SEER tool generates background traffic of the NTGC/Harpoon/D-ITG type [23], but there is no workflow generation facility for interacting directly with the target in a benign way.

4.4.4 IP-Aliasing

One way to generate background and attack traffic would thus appear to be to use workload generators, such as httperf [19] or Apache flood [3]. These normally only generate a load from a single IP address, whereas a realistic workload would come from many IP addresses at once. There is also a need for more flexibility in the type of traffic load generated. The usual way to simulate multiple IP addresses is to spoof the source address, but spoofed addresses cannot communicate with the target and hence cannot carry on meaningful TCP connections. One way around this problem is to use IP-aliasing, such as used by curl-loader [12]. In this technique one interface can be aliased many thousands of times, and each alias assigned a unique IP address in any desired range. Maintaining a connection for each such address is only limited by memory, and any kind of interaction – benign or attacking – can be carried on for each connection.

This technique thus has the effect of multiplying the effective size of the testbed. A mere half a dozen machines can then realistically simulate a powerful botnet. We have developed our own tool, called bot-loader, which can create 10,000 aliases in less than a second, then generate traffic for each one that can realistically load a server. Each type of interaction is provided by pluggable modules, which can be instantiated many thousands of times each, and which communicate in a random sequence with the target. In the real world, attack and benign interactions going on at the same time must be forced onto the wire in some sequence, and it is this sequence that can be directly simulated using a simple randomised sequential

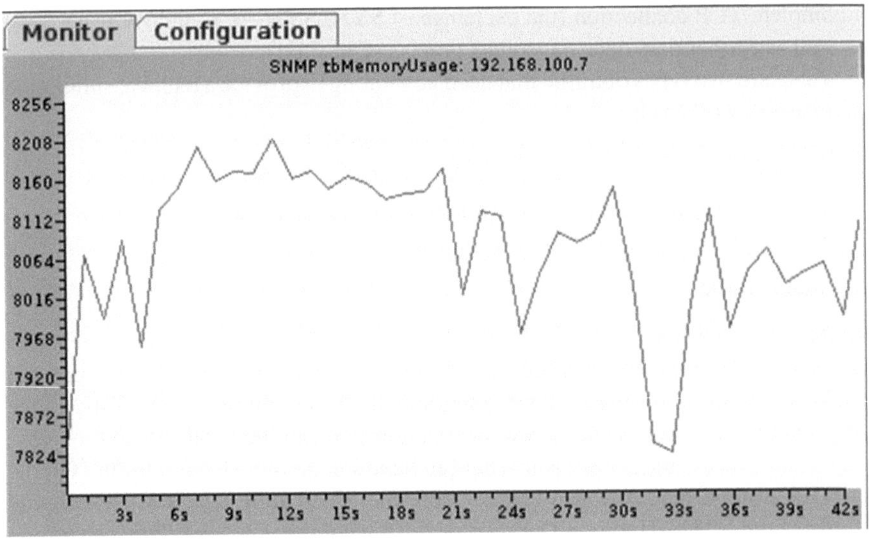

Fig. 4.5 SNMP monitoring tool

approach. Bot-loader is still under development, but this approach does indicate that a realistic DDoS testbed may be built using modest physical means, and hence the DETER type of large emulated testbed may not be necessary.

4.4.5 Monitoring Tool

A visual Java-based monitoring tool that listens issues and displays the results of any numerical SNMP get requests can be used on either testbed, as shown in Fig. 4.5. The frequency of requests is once per second, and the graph dynamically adjusts the scale so that the top and bottom values are both displayed. Because SNMP does not support floating point numbers, percentages are expressed as integers times 100. Historical values for each parameter/IP-address combination are saved in a local file for later analysis.

4.5 Experimental Results

Details of experiments already carried out on the Australian testbed include the effects of DoS attacks against web services (Chap. 7), and the detection of high rate flooding attacks using change-point analysis (Chap. 5). Here we describe an experiment in mitigation of high rate flooding attacks carried out on the Indian testbed.

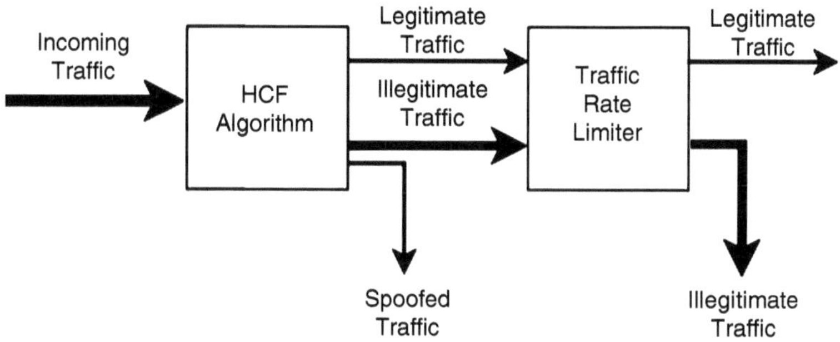

Fig. 4.6 Block diagram of the proposed work

4.5.1 A TTL-Based Hop Count Filter Scheme

Hop count filtering (HCF) is a technique that uses the Time-To-Live field of the IP header [13, 83]. This was originally designed to specify the maximum lifetime of IP packets on the Internet. During transmission, each intermediate router decrements the TTL value by one before forwarding it to the next-hop router. The final TTL value when a packet reaches its destination will thus be set to the initial TTL less the number of intermediate hops. Most modern OSs use only a few selected initial TTL values: 30, 32, 60, 64, 128 and 255 [1]. Since the maximum number of hops in practice is around 30, this allows the hop-count for a particular IP address to be reliably inferred in most cases.

The DDoS mitigation technique described here is a hybrid of this HCF approach and application layer-based bandwidth rate limiting to detect and filter high-rate traffic flows. As well as HCF at the network layer, traffic limiting is applied at the application layer, as shown in Fig. 4.6. The HCF technique involves the construction of a table correlating source IP addresses with expected hop-counts under normal conditions. This method can be deployed at the victim without the assistance of routers. During a DDoS attack, randomly spoofed attack packets can then be filtered based on their source IP addresses. To reinforce this approach, attackers using genuine IP addresses are also subject to traffic limitation at the application layer [30].

Based on the TTL (time to live) field of the IP header, hop count is computed as follows:

The hop count inspection algorithm [13, 26], as shown in Fig. 4.7, extracts the source IP address and the final TTL value from each IP packet. The algorithm infers the initial TTL value and subtracts it from the final TTL value to obtain the hop count. Then, the source IP address serves as the index into the table to retrieve the correct hop count for this IP address. If the computed hop count matches the stored hop count, the packet has been 'authenticated', otherwise it is classified as

Fig. 4.7 Hop count filtering algorithm

```
For each incoming packet
  Extract the final TTL T_f and Source IP S;
  Infer the initial TTL T_i
  Calculate the hop count H_c = T_f - T_i
  Index S to get the stored hop count H_s;
  If (H_c = H_s)
    Packet is legitimate
  Else
    Packet is spoofed
```

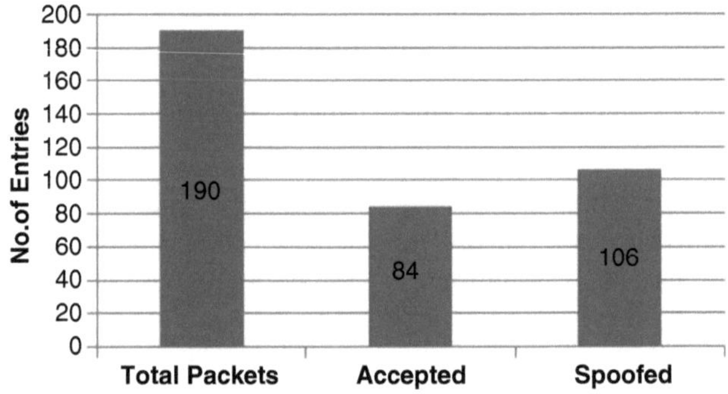

Fig. 4.8 Packet Status with IP Random Spoofing

'spoofed'. HCF can identify nearly 90% of spoofed IP addresses. To save space, the IP2HC table is clustered based on its 24-bit prefix [6], which both minimises retrieval times and economises on storage.

4.5.2 *Implementation on the Testbed*

The traffic traces are collected via the DDoS testbed. Packets coming from several nodes are routed through the MPLS-VPN cloud and then collected using Wireshark. The source IP address and TTL value are extracted and the IP-address-to-hop-count table is constructed at the victim end. Random IP spoofing is performed on generated traffic to check the effectiveness of the learning process.

The majority of spoofed traffic was detected, as shown in Fig. 4.8. The incoming traffic needs to be checked again in case legitimate users misbehave after passing through the filtering mechanism. Though the HCF algorithm is very effective against spoofed traffic, it doesn't attempt to protect the victim from flooding-based DDoS attacks from legitimate IP addresses. So even after passing the HCF filter, incoming traffic needs further treatment.

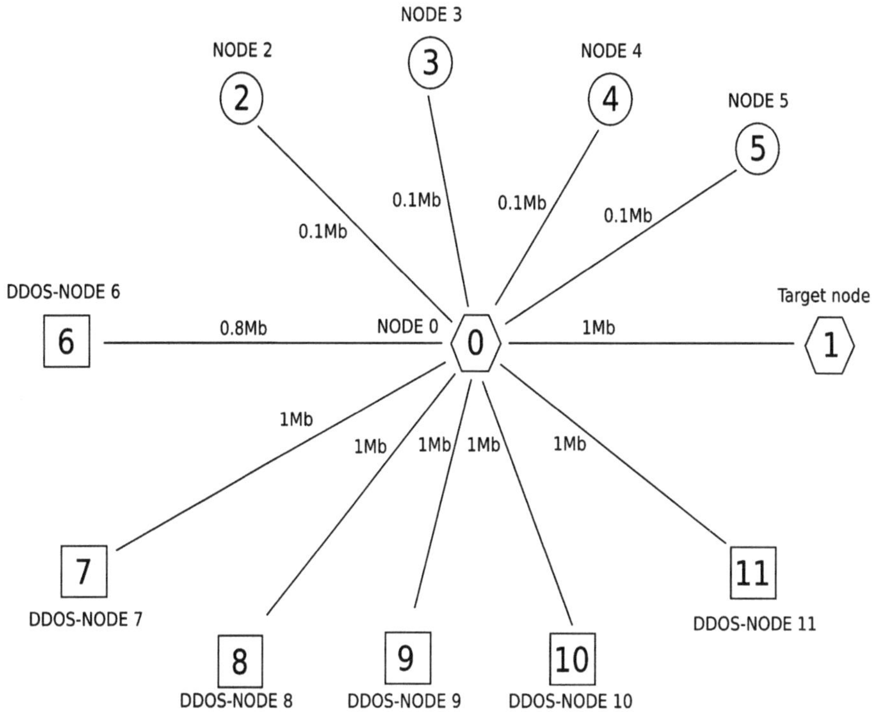

Fig. 4.9 Network topology under DDoS attack

4.5.3 Rate Limiting

Attackers are more aggressive than legitimate users, and hence their intention is to overwhelm the victim, without affecting other services. However, the end result is that requests from legitimate users are denied. Hence, filtered output from the HCF algorithm is passed through a rate limiter, which takes into account the aggregate traffic at the router. If this is too high, then the available bandwidth is reduced in steps of 10%. Although this also reduces the service to legitimate flows, it does avoid denial of service.

This has been tested in NS-2 [2], using a discrete event simulator with a simple topology comprising six attack sources and four legitimate sources with one target. The bottleneck link capacity is 1 Mbps as shown in Fig. 4.9.

4.6 Conclusion

Although the Indian and Australian testbeds differ in their physical topology, they share a design based on live experimentation. This approach yields more realistic results than either emulation- or simulation-based testbeds, at some cost in flexibility. Both testbeds use SNMP as a monitoring protocol to directly measure the responsiveness of services under DDoS attack, as well as SCP, SFTP and SSH for experimental control and configuration. The joint development of a customisable workload generator to generate both benign and attack traffic also shows promise as a way of generating realistic traffic on a small scale testbed as well as extending the number of agents by a factor of 10,000 using interface aliasing. In this way genuinely large-scale DDoS experiments can be carried out using the relatively modest physical resources of both testbeds. Experimental results from the Indian testbed show that a combination of hop-count filtering and rate limiting may be an effective means of mitigating high rate flooding attacks from spoofed and non-spoofed IP addresses.

References

1. The Swiss Education and Research Network. 2001. Default TTL values in TCP/IP. http://www.map.meteoswiss.ch/map-doc/ftp-probleme.htm. Accessed 1 Feb 2011.
2. The Network Simulator – ns-2. 2011. http://www.isi.edu/nsnam/ns/: Accessed 1 Feb 2011.
3. Apache. 2011. Apache flood. http://httpd.apache.org/test/flood/.
4. Benzel, T., B. Braden, T. Faber, J. Mircovic, S. Schwab, K. Sollins, and J. Wroclawski. 2009. Current developments in DETER cybersecurity testbed technology. In *Proceedings of the Cybersecurity Applications and Technology Conference For Homeland Security (CATCH 2009)*, Washington, Mar 2009.
5. Benzel, T., R. Braden, D. Kim, C. Neuman, A. Joseph, K. Sklower, R. Ostrenga, and S. Schwab. 2007. Design, deployment, and use of the DETER testbed. In *DETER Community Workshop on Cyber-Security and Test*, Berkeley, Aug 2007.
6. Beverly, R. and K. Sollins. 2008. An internet protocol address clustering algorithm. In *Proceedings of USENIX Tackling Computer Systems Problems with Machine Learning Techniques*, San Diego, Dec 2008.
7. Blackert, W.J., D.M. Gregg, A.K. Castner, R.L. Hom, R.M. Jokerst, and E.M. Kyle. 2003. Distributed denial of service defense attack tradeoff analysis (DDOS-DATA) demonstration overview. In *Proceedings of the DARPA Information Survivability Conference and Exposition (DISCEX'03)*, vol. II, 66–67, Apr 2003.
8. Botta, A., A. Dainotti, and A. Pescape. 2007. Multi-protocol and multi-platform traffic generation and measurement. In *INFOCOM 2007 DEMO Session*, Alaska, May 2007.
9. Calvet, J., J.M. Fernandez, P.-M. Bureau, and J.-Y. Marion. 2010. Large-scale malware experiments why, how, and so what? In *Proceedings of Virus Bulletin Conference*, 241–247, Sept 2010.
10. Gelenbe, E., M. Gellman, and G. Loukas. 2005. An autonomic approach to denial of service defence. In *Proceedings of the Sixth IEEE International Symposium on a World of Wireless Mobile and Multimedia Networks*, 537–541, 2005.
11. Huang, Y.I., J.D. Tygar, H.Y. Lin, L.Y. Yeh, H.Y. Tsai, K. Sklower, S.P. Shieh, C.C. Wu, P.H. Lu, S.Y. Chien, Z.S. Lin, L.W. Hsu, C.W. Hsu, C.T. Hsu, Y.C. Wu, and M.S. Leong. 2008. SWOON: A testbed for secure wireless overlay networks. In *CSNET '08 Workshop on Cyber Security Experimentation and Test*, Berkeley, July 2008.
12. Iakobashvili, R. and M. Moser. 2007. Curl-loader. http://curl-loader.sourceforge.net/. Accessed 11 Jan 2011.
13. Jin, C., H. Wang, and K.G. Shin. 2003. Hop-count filtering: An effective defense against spoofed traffic. In *Proceedings of the 10th ACM Conference on Computer and Communications Security*, Washington, 30–41, Oct 2003.

14. Kotenko, I.V. and A.V. Ulanov. 2006. Software testbed and experiments for exploring counteraction of attack and defense agents in the internet. In *Proceedings of the International Security and Counteracting Terrorism Conference*, 80–93, Lomonosov Moscow State University Intellectual Center, 2006.
15. Mirkovic, J., S. Dietrich, D. Dittrich, and P. Reiher. 2005. *Internet denial of service attack and defense mechanisms*. Upper Saddle River: Prentice Hall.
16. Mirkovic, J., S. Fahmy, P. Reiher, and R.K. Thomas. 2009. How to test DoS defenses. In *Cybersecurity Applications and Technology Conference for Homeland Security*, 103–117, Washington, 2009.
17. Mirkovic, J., A. Hussain, B. Wilson, S. Fahmy, P. Reiher, R. Thomas, W.-M. Yao, and S. Schwab. 2007. Towards user-centric metrics for denial-of-service measurement. In *Proceedings of the 2007 Workshop on Experimental Computer Science*, San Diego, Jun 2007.
18. Mirkovic, J., B. Wilson, A. Hussain, S. Fahmy, P. Reiher, R. Thomas, and S. Schwab. 2007. Automating DDoS experimentation. In *Deter Community Workshop on Cyber Security Experimentation and Testing*, Jul 2007.
19. Mosberger, D. and T. Jin. 1998. httperf: A tool for measuring web server performance. *Performance Evaluation Review* 26(3): 31–37.
20. Rajab, M.A., J. Zarfoss, F. Monrose, and A. Terzis. 2007. My botnet is bigger than yours (maybe, better than yours): Why size estimates remain challenging. In *HotBots'07 Proceedings of the First Conference on First Workshop on Hot Topics in Understanding Botnets*, Berkeley, Apr 2007.
21. Salah, K., K. El-Badawi, and F. Haidari. 2007. Performance analysis and comparison of interrupt-handling schemes in gigabit networks. *Computer Communications* 30: 3425–3441.
22. Schmidt, D., S. Suriadi, A. Tickle, A. Clark, G. Mohay, E. Ahmed, and J. Mackie. 2010. A distributed denial of service testbed. In *What Kind of Information Society? Governance, Virtuality, Surveillance, Sustainability, Resilience. Proceedings of 1st IFIP TC 11 International Conference, CIP 2010 Held as Part of WCC 2010*, eds. Australia, J. Berleur, M.D. Hercheui, and L.M. Hilty, 338–349, Sept 2010.
23. Schwab, S., B. Wilson, C. Ko, and A. Hussain. 2007. SEER: A security experimentation enviRonment for DETER. http://www.usenix.org/event/deter07/tech/full_papers/schwab/schwab.pdf. Accessed 16 Feb 2011.
24. Sommers, J. and P. Barford. 2004. Self-configuring network traffic generation. In *Proceedings of ACM Internet Measurement Conference*, Sicily, Oct 2004.
25. Stone-Gross, B. 2009. Your botnet is my botnet: Analysis of a botnet takeover. In *Proceedings of the ACM CCS*, 635–647, Chicago, 9–13 Nov 2009.
26. Swain, B.R. and B. Sahoo. 2009. Mitigating DDOS attack and saving computational time using a probabilistic approach and HCF method. In *IEEE International Advanced Computing Conference (IACC2009)*, 1170–1172, 6–7 Mar 2009.
27. Ting, Y.A., D. Ma, and K. Levitt. 2005. NTGC: A tool for network traffic generation control and coordination. http://wwwcsif.cs.ucdavis.edu/~tingy/NTGC.pdf. Accessed 16 Feb 2011.
28. Wang, H., C. Jin, and K.G. Shin. 2007. Defense against spoofed IP traffic using hop-count filtering. *IEEE/ACM Transactions on Networking* 15(1): 40–53.
29. White, B., J. Lepreau, L. Stoller, R. Ricci, S. Guruprasad, M. Newbold, M. Hibler, C. Barb, and A. Joglekar. 2002. An integrated experimental environment for distributed systems and networks. In *Proceedings of the 5th Symposium on Operating Systems Design and Implementation*, 255–270, New York, Dec 2002.
30. Wu, Z. and Z. Chen. 2006. A three-layer defense mechanism based on web servers against distributed denial of service attacks. In *First International Conference on Communications and Networking in China*, 1–5, 2006.

Chapter 5
Detection and Mitigation of High-Rate Flooding Attacks

G. Mohay, E. Ahmed, S. Bhatia, A. Nadarajan, B. Ravindran, A.B. Tickle, and R. Vijayasarathy

5.1 Introduction

Because high-rate flooding attacks constitute such a potent threat to the delivery of Internet-based services, the early and reliable detection of the onset of such an attack together with the formulation and implementation of an effective mitigation strategy are key security goals. However, the continuously evolving nature of such attacks means that they remain an area of active research and investigation. This chapter focuses largely on our research into attack detection, with some discussion of mitigation through IP address filtering. The chapter outlines leading-edge work on developing detection techniques that have the potential to identify a high-rate flooding attack reliably and in real time or, at least, in near real time. In addition, it formulates an architecture for a DoS Mitigation Module (DMM) to provide a vehicle for integrating the elements of the solution.

Section 5.2 reviews related work on network anomaly detection, while the Sect. 5.3 addresses the key issue of feature extraction and the allied problem of determining which features of Internet traffic can best be used to characterise

G. Mohay (✉) • E. Ahmed • S. Bhatia • A.B. Tickle
Information Security Institute, Queensland University of Technology, Brisbane, Australia
e-mail: g.mohay@qut.edu.au; e.ahmed@qut.edu.au; s.bhatia@qut.edu.au; ab.tickle@qut.edu.au

A. Nadarajan
Maths and Computer Application Department, PSG College of Technology, Coimbatore, India
e-mail: anitha_nadarajan@mca.psgtech.ac.in

B. Ravindran
Department of Computer Science and Engineering, Indian Institute of Technology Madras, Chennai, India
e-mail: ravib@iitm.ac.in

R. Vijayasarathy
Network Security Research Group, Society for Electronic Transactions and Security, Chennai, India
e-mail: vijayasarathy@setsindia.net

S.V. Raghavan and E. Dawson (eds.), *An Investigation into the Detection and Mitigation of Denial of Service (DoS) Attacks: Critical Information Infrastructure Protection*, DOI 10.1007/978-81-322-0277-6_5, © Springer India Pvt. Ltd. 2011

a DDoS attack. It also examines the problem of ranking the efficacy of the various machine-learning techniques that have been utilised in the detection process. The following three sections describe three different approaches to the detection problem. Section 5.4 shows how the CPA technique is used in combination with other techniques such as bit vectors and bloom filters to monitor the rate of arrival of new IP addresses which is then used as a basis to detect a high-rate flooding event. This is followed in Sect. 5.5 by a discussion of an approach to DDoS detection based on Naive Bayesian (NB) techniques that exploit characteristic features of the transmission control protocol (TCP) and User Datagram Protocol (UDP) to distinguish a high-rate flooding attack from normal traffic. It then shows how NB techniques could be used to detect a high-rate flooding attack in near real time. Section 5.6 discusses the Cumulative-sum-based Adaptive Neuro-Fuzzy Interface System (CANFIS) approach which comprises a blend of an artificial neural network (ANN), a Fuzzy Inference System (FIS) and the cumulative sum (CUSUM) algorithm. In this approach, different attacks are first modelled using CUSUM techniques, and the modelled traffic is then classified as being an attack or normal based on the Adaptive Neuro-Fuzzy Inference System (ANFIS). This is followed by Sect. 5.7 'Conclusion and Further Work'.

In addition to the approaches described in this chapter, readers should also note that Chap. 4 includes an account of some recent experiments on the Indian testbed using hop-count filtering (HCF) to mitigate flood-type attacks.

5.2 Review of Traffic Anomaly Detection Techniques

The very nature of high-rate flooding attacks suggests that the primary point of focus of corresponding attack detection techniques should be on detecting anomalies in network traffic. The intention here is to identify traffic that is likely to constitute high-rate flooding attacks. While signature detection techniques are designed to detect attacks based on signatures of attacks already learnt by them, anomaly detection techniques learn network traffic from a baseline profile and detect traffic anomalies which deviate significantly from the baseline profile. Signature detection techniques are effective against known attacks while anomaly detection has the ability to detect unknown (zero-day) attacks.

Mitigation approaches involve dropping traffic completely (and running the risk of dropping 'good' traffic), or throttling traffic (with its attendant negative effect on response times), or provisioning extra resources to cope with the high-rate traffic. Of these, the most intensively researched approach is the first, and the key requirements here necessarily are accuracy and timeliness [44]. Regarding accuracy, an alert should be raised and traffic dropped if and only if an anomalous traffic event occurs on the network [44]; furthermore, the number of false positives and false negatives must be kept to a minimum, the former so as not to disrupt the normal provision of services, the latter for obvious reasons. Regarding timeliness, the detection technique must be capable of detecting the anomalous situation in real time or in near real-time so that appropriate response/mitigation processes may be invoked as rapidly as practicable to limit the impact of the attack. There are significant challenges in satisfying these two requirements. For example, as Kline et al. [44] observe, there is no exact definition of what constitutes anomalous traffic patterns. Normal (or so-called non-anomalous) network traffic can sometimes exhibit characteristics that, at face value, would be indicators of 'anomalous' behaviour. This includes, for example, phenomena such as so-called flash events which are sudden and extreme bursts of (legitimate) traffic [20,44,78]. Moreover, the detection problem is compounded by the fact that not only is the actual composition of normal network traffic both diverse and continuously evolving [44], the pattern of traffic used in an actual attack can be programmed to mimic such behaviour [20].

For the purpose of this discussion, the detection process is considered to comprise three main elements:

- Recording and/or measurement of certain parameters of interest
- Data analysis
- Decision-making about whether or not the observed behaviour is classified as anomalous (and the subsequent triggering of a response such as generating an alert or dropping the anomalous traffic)

5.2.1 *Parameters of Interest and Approaches Used*

The analysis and decision-making approaches determine the number of separate parameters to be measured and the granularity of the measurements (e.g. the sampling interval). Commonly used measurements are those of the volume of traffic entering a network (also known as ingress), and sometimes the volume of traffic departing from the network (also known as egress). Network traffic volume (or traffic flow) is typically expressed in terms of both packets/second and bits (or bytes)/second [55, 61]. Other parameters of interest include, for example, source/destination IP addresses, port addresses and protocol type (e.g. to distinguish between TCP, UDP and Internet control message protocol (ICMP) packets) [20,57]. An alternative to focusing on measurements at the packet level and one that fits the inherent interlinking and sequencing of packets in the transmission control protocol/Internet protocol (TCP/IP) suite of protocols is to consider aggregates of packets as so-called traffic-flows [13, 14]. In this context, a traffic-flow could, for example, be the sequence of packets that comprises a complete TCP-level connection session. In this case, the packets within the designated traffic-flow are linked by common Internet protocol (IP) source address and destination address as well as TCP source and destination ports.

In 2006, Carl et al. [20] undertook a systematic survey of techniques for detecting various forms of flooding attacks. They grouped the detection techniques under the following categories:

(a) Activity profiling
(b) Sequential change-point detection
(c) Wavelet analysis

They then ranked the performance of individual techniques in each category on the basis of memory use and computational complexity. Obviously, a survey of this type and the corresponding rankings only represent a snap-shot of the techniques available at the time. However, they raise a number of concerns which are important in analysing subsequent work in this field. Basically, in their view, most of the results on which they based their comparison were of limited utility to the wider networking community because the results were not generated in an environment that replicated real-world network environments.

5.2.2 Detection Performance

Since the survey undertaken by Carl et al. [20], there has been continuous activity and refinement in this field [9, 10, 44, 63, 78]. For example, in the interests of achieving computation tractability in operational environments, Kline et al. [44] restrict their measurements (and subsequent analysis and decision-making) to traffic volume only, that is, their initial technique does not require packet header or payload inspection as used in other techniques. Their particular technique uses a wavelet-based method in the analysis phase and, in the interests of computational efficiency, a Bayesian network in the decision-making phase. Similarly, a recent technique proposed by Papadopoulos [63] uses cumulative packet rates as the primary parameter of interest. (This particular technique also uses a novel combination of change-point detection and discrete Fourier transforms to identify an attack.)

The focus on performance reinforces the point made earlier that one of the key challenges in implementing a workable solution to detecting a network flooding attack is to be able to operate the detection process in real time. In essence, this means executing all of the tasks described above (i.e. monitoring the traffic stream, the possible construction of feature vector based on a predetermined set of key characteristics, using the feature vector to compute a metric to be used as the basis for comparison and then deciding if a given segment of the network traffic stream is 'normal' or 'anomalous') at a speed commensurate with that of the underlying transmission links, that is, so-called wire speed. Such speeds are a moving target and are being driven inexorably upwards by improvements in technology. To be competitive the detection processes must be currently capable of operating at speeds of at least 10 Gbps [56, 61] and preferably higher. Fortunately the emergence of high-performance parallel-processing computer hardware and associated software development environments has made feasible the task of designing and implementing sophisticated detection processes for such hardware platforms and then operating them at speeds approaching those required. Commercial examples of network processing hardware specifically designed for this purpose (e.g. the Intel IXP2400, the AMCC np7510, the EZchip NP-1 and the Agere Fast Pattern Processor and Routing Switch Processor) are listed in [84]. In addition, discussion on using the high-performance parallelism of field-programmable gate arrays (FPGAs) to achieve the same goal can be found in [18, 23, 66].

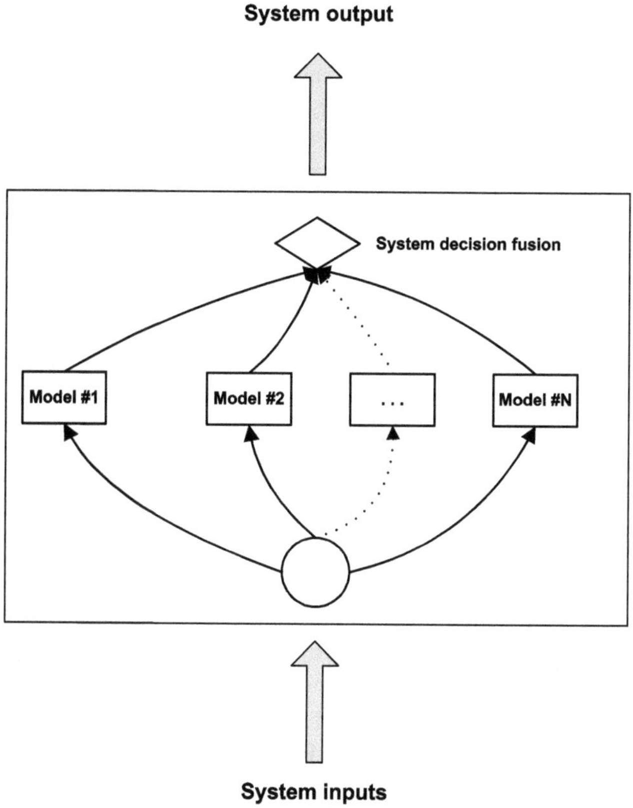

Fig. 5.1 Ensemble learning (Adapted from Wang et al. [83])

An alternative approach to using a single anomaly detection algorithm is that of Shanbhag and Wolf [73,74]. They exploit the potential of high-performance parallel processors to operate several anomaly detection algorithms in parallel. Under this regime, the output from each algorithm is normalised and then aggregated to produce a single anomaly metric. (This notion of using ensembles (Fig. 5.1) has a long history in the domain of machine learning [24, 83] and ensembles have been used successfully in a wide range of problems, e.g. see [76]).

While focusing on a simple metric such as sudden increases in traffic volume assists in maximising performance in order to meet the objective of timely detection, some approaches to detecting high-rate flooding attacks involve the inspection and analysis of both the header and content (payload) of packets. Notwithstanding their utility, there are two constraints on deploying this type of detection technique. The first, which applies, for example, at the ISP level, is that of privacy. The second is, of course, performance. Analysing both packet headers and packet contents is time- and resource-intensive. Consequently, systems that analyse both the header and content of each packet have the potential to become bottlenecks and, in the extreme case, an actual point of failure during a high-rate flooding attack.

In the interests of performance, header analysis has historically focused simply on source and destination IP addresses and packet type [11, 12, 52, 60]. In addition, a typical low-level packet analysis may involve verifying the TCP port number in the transport-layer header [17]. However, as mentioned previously, the advent of high-performance hardware (including FPGAs) and specialised network processors has created an environment in which it is now feasible to perform deeper and more complex levels of analysis in real time [18, 19, 23, 35, 66, 73, 74, 84]. For example, the capability now exists to perform 'deep-packet' inspection, including the ability to parse the application-level content of each packet [35]. At a commercial-product level, this capability appears in the form of so-called application-aware firewalls. There is now also the capability to exploit the inherent parallelism in network processing by disassembling traffic into individual connection streams and then analysing the sequence of packets in each such connection stream concurrently [66, 84]. As mentioned previously in the brief discussion on 'ensemble machine learning', the availability of high-performance parallel-processing hardware also affords the opportunity to operate a set of decision-making/filtering algorithms concurrently.

It is evident that in some circumstances, the successful identification of attack activity can be considerably assisted by having individual network devices or computers share information regarding the network traffic activity they see. Consequently a number of architectures have been proposed for allowing (primarily) sets of routers or devices of similar capability to act in concert [11, 12, 52, 77]. The focal point is often on the routers that operate on the network borders and where ingress and egress packet filtering can be performed efficiently. One final comment is that, as indicated previously, an important consideration in deploying such filters is their ability to operate at so-called wire speed lest the filters themselves adversely affect packet-forwarding performance [12]. The availability of high-speed parallel-processing hardware is an important step in mitigating this problem.

5.2.3 *Decision-Making and Mitigation*

The decision to accept or discard a packet can be made on a binary white-list/black-list basis. Such lists may take a variety of forms, for example, source or destination-based access control lists, real-time black-hole lists and DNS black lists. In addition, the information in such lists may be drawn from (trusted) external parties, for example, Spamhaus XBL (http://www.spamhaus.org/xbl) and CBL (http://cbl.abuseat.org). This builds upon the link between the notions underlying the creation and utilisation of such lists and the broader concept of building and deploying a reputation scheme as part of the decision-making process.

A considerable body of ideas and experience have now emerged as to how to design and execute the required mitigation strategy once an attack commences and is detected [12,52,55]. Whilst these defence mechanisms have become increasingly sophisticated, central to the strategy for mitigating the impact is to reduce the traffic volume to manageable levels as quickly as possible. Necessarily, this requires that individual IP packets or packet streams be either discarded (or blocked) or delayed. The inherent difficulty is in processing the packets (and streams) in real time in order to decide which packets to filter, which packets to allow, which packets to proceed unmolested, and where (i.e. in which networking device or devices) the actual decision is to be made.

In this context, it is worth noting the existence of a degree of similarity between the current problem of identifying and filtering anomalous packet streams in real-time and earlier work in real time network congestion management [51]. For example, algorithms such as random early detection (RED) [30] and fair random early detection (FRED) [43] were developed to decide probabilistically which TCP packets should be dropped when a given network device reached a point where the volume of traffic exceeded the available buffer space. A key difference between this early work on congestion management and the current approaches is that the capability now exists to involve a greater number of information elements in the decision-making process. This opens up a number of interesting research possibilities. For example, one potential avenue of investigation could be to use a reputation scheme to assign network traffic to a particular 'class' that can then be selectively controlled using Class-Based Queuing techniques [31] from congestion management.

5.3 Evaluating Machine-Learning Algorithms for Detecting DoS Attacks

5.3.1 *Related Work*

Two important and challenging research problems in detecting distributed denial of service (DDoS) attacks are:

1. Extracting a valid and sufficient subset of features that can be used to build efficient models to identify a DDoS attack
2. Ranking the efficacy of the various machine-learning techniques that have been utilised in the detection process

For most problem domains, the process of feature reduction which involves extracting the most significant and relevant attributes or features prior to applying modelling techniques (such as machine-learning and statistical techniques) can lead to a major improvement in the time required in training and testing the model. However, in comparison with other problem domains, extracting a set of features that characterise Internet traffic to the point of being able to distinguish normal traffic from anomalous traffic is particularly difficult. One problem, for example, is that nodes in the Internet experience widely differing traffic flux densities caused by the large variations in the number of users seen at each node. This makes it difficult to decide as to what constitutes 'normal' traffic on the Internet. Another problem, and one that can be seen from the discussion on the detection techniques already presented, is that there are potentially a large number of variables that can be used to characterise network traffic patterns. Nevertheless, extracting the important and relevant attributes from network traffic is crucial for modelling network behaviours so that attack behaviours can be differentiated clearly from normal behaviour. This feature-extraction problem has been studied by a number of groups. For example, Xu et al. [87] selected eight relative values as features that are independent from the network flow. Zargar et al. [91] proposed and investigated the identification of effective network features for probing attack detection using the Principal Component Analysis (PCA) method to determine an optimal feature set. Jin et al. [37] discussed the application of multivariate correlation analysis to DDoS detection and proposed a covariance analysis model for detecting flooding attacks. They used all of the flag-bits in the flag field of the TCP header as features in the covariance analysis model. The authors have demonstrated the successful use of the proposed method in detecting SYN flooding attacks which is an important form of DDoS attacks. However, the method has the major limitation that there is no guarantee that the six flags are valid or sufficient features to detect all forms of DDoS attack with consistent accuracy.

As has been discussed previously in Sect. 5.2, a widely diverse range of statistical methods and machine-learning techniques could be used to detect abnormal changes in the resource usage that are indicative of a DDoS attack. However, both approaches have their limitations. For example, one identifiable problem with statistics-based detection is that it is not possible to find out the normal network packet distribution. Rather, it can only be simulated as a uniform distribution. Some research papers suggest that this problem may be resolved by using clustering methodologies to formulate the normal patterns since one of the advantages of clustering methods over statistical methods is that they do not rely on any prior known data distribution. While machine-learning techniques, typically drawn from the allied field of data mining, have been shown to produce a high degree of accuracy in detecting DDoS attacks, they also have their limitations. For example, these techniques typically require a lengthy learning period and hence, currently, these methods typically cannot operate in real time.

Despite these current limitations, a solution to the problem of reliable DDoS detection will come from either or both these domains and considerable research effort continues to be directed to this end. For example, Seo et al. [72] have used a multi-class SVM classification model to detect DDoS attack as have Xu et al. [87]. In the work of Xu et al. [87], a group of new features was also introduced, including the composition of relative values as part of an expanded set of detection information. They also proposed a new approach of using attack intensity to detect a DDoS event. In [65], Paruchuri et al. proposed a new Probabilistic Packet Marking (PPM) scheme called TTL-based PPM scheme, where each packet is marked with a probability inversely proportional to the distance traversed by the packet so far, enabling a victim source to trace back the attack source. In [21], Cheng et al. proposed a novel algorithm to detect DDoS attacks using IP address feature values using support vector machine (SVM) classification. Nguyen et al. [62] have developed an anti-DDoS framework for detecting DDoS attack proactively utilising K-NN Classifier. They used the k-nearest neighbour method to classify the network status into each phase of DDoS attack. However, while the K-NN approach is excellent in attack detection, the detector is computationally expensive for real-time implementation when the number of processes simultaneously increases. As has been indicated previously, the problem of computational intensity is critical in the DDoS problem as it is in other applications of data mining where large databases are analysed.

One of the key resources used to evaluate the performance of DDoS detection techniques is the KDD dataset. The set contains 14 attacks which is used for testing and model creation. Several methods have been proposed to extract useful features from this dataset, and a wide range of classifiers drawn from areas such as statistics, machine learning and pattern recognition have been evaluated against this dataset. For example, in Kim et al. [42], the 1999 KDD dataset was pre-processed followed by a learning and testing process. In the learning process, they used polynomial, kernel functions linear and Radial Basis Function (RBF). A classification accuracy of 93.56% was achieved. An SVM-based one-class classifier is also used to perform anomaly detection in [25]. The training data in the feature space was mapped into a new feature space. Yuan et al. [90] used the cross-correlation analysis to capture the traffic patterns and then to decide where and when a DDoS attack may possibly arise.

The following study discusses the extraction of a feature set from two different sources of datasets of Internet traffic. These are the public-domain CAIDA dataset [5] and traffic collected on the Smart and Secure Environment (SSE) Network. Various types of DDoS attacks were studied to select the packets and traffic parameters that change unusually during such attacks. Approximately 23 features were collected. Ranking these features is done with Information Gain and chi-square statistic [82] which enables the number of features to be reduced to eight. All the features used in this paper are calculated at an interval of 1 s. Since these classes are well divided as attack and normal, it is possible to apply various machine-learning algorithms for the detection. The approach taken is to use the feature selection mechanism discussed previously and build the classifier using various machine-learning algorithms such as SVM, K-NN, Naive Bayesian, Decision Tree, K-means and Fuzzy c-means clustering. This phase of the study shows an evaluation of the performance of the selected set of machine-learning algorithms in detecting DDoS attacks. The performance measures were the Receiver Operating Characteristic (ROC) curve and F-measure. An important result from this work is that, of the various methods used, Fuzzy c-means clustering is a useful and very efficient way to detect a DDoS attack.

5.3.2 Feature Selection and Evaluation

A list of 23 features, namely,

1. One-Way Connection Density (OWCD)
2. Average length of IP flow
3. The ratio between incoming and outgoing packets
4. Entropy of IP flow length
5. Entropy of the packet ratios of the three protocols TCP, UDP and ICMP
6. Ratio of TCP protocol
7. Ratio of UDP protocol
8. Ratio of ICMP protocol
9. Number of data bytes from source to destination
10. Number of data bytes from destination to source
11. Number of packets in which destination port is mapped to a particular service
12. Type of the protocol, e.g. TCP, UDP, ICMP
13. The number of packets having the same source IP address and destination IP address
14. Number of wrong fragments
15. Number of connections that have SYN errors
16. Number of connections to the same source IP
17. Number of connections having the same destination host
18. Number of packets where URG flag is set
19. Number of packets where SYN flag is set
20. Number of packets where FIN flag is set
21. Number of packets where ACK flag is set
22. Number of packets where PSH flag is set
23. Number of packets where RST flag is set

was selected for evaluation. After ranking these 23 features with information gain and chi-square statistics [82], the following 8 features [87] were selected:

(a) One-Way Connection Density (OWCD):
An IP packet without a corresponding reverting packet composes a One-Way Connection (OWC). In a sampling interval T, the ratio of OWC packets to all packets is called One-Way Connection Density (OWCD):

$$OWCD = \frac{\sum OWC\ Packets}{\sum IP\ Packets} \times 100 \tag{5.1}$$

(b) Average Length of IP Flow (L_{ave_flow}):
IP flow, a concept which is used widely in network analysis area, means that a packet set has a same five-element-group (source IP address, source port, destination IP address, destination port and protocol). Length of IP flow means that the number of packets belong to a certain IP flow:

$$L_{ave\_flow} = \frac{\sum IP\ Packets}{\sum IP\ Flows} \tag{5.2}$$

(c) Incoming and Outgoing Ratio of IP packets (R_{io}):
Normally the ratio between incoming and outgoing packets is steady. But in a DDoS attack, R_{io} increases quickly:

$$R_{io} = \frac{\sum incoming\ IP\ Packets}{\sum outgoing\ IP\ Packets} \tag{5.3}$$

(d) Ratio of TCP Protocol (R_t):

$$R_t = \frac{\sum TCP\ Packets}{\sum IP\ Packets} \tag{5.4}$$

(e) Ratio of UDP Protocol (R_u):

$$R_t = \frac{\sum UDP\ Packets}{\sum IP\ Packets} \tag{5.5}$$

(f) Ratio of ICMP Protocol (R_i):

$$R_t = \frac{\sum ICMP\ Packets}{\sum IP\ Packets} \tag{5.6}$$

(g) Land: The number of packets having the same source IP address and destination IP address.
(h) Protocol-type: Type of the Protocol, e.g. TCP, UDP, ICMP, etc.

These eight features, which have been selected based on the principles mentioned in Xu et al. [87], are used to classify the network status. Each variable is normalised to eliminate the effect of difference between the scales of the variables, as proposed by Lee et al. [47]. With normalisation, variables become

$$z = \frac{x - \overline{x}}{\sigma} \tag{5.7}$$

where x, $\overline{x}$, and σ denote the value of each feature, the mean of the sample dataset and the standard deviation, respectively.

Additionally, chi-square statistic and information gain is applied to measure the rank of each feature. The Information gain of a given attribute X with respect to the class Y is the reduction in uncertainty about the value of Y, after observing values of X. This is denoted as $IG(Y|X)$. The uncertainty about the value of Y is measured by its entropy defined as

$$H(Y) = -\sum_i P(y_i) \log_2(P(y_i)) \tag{5.8}$$

Table 5.1 Feature ranking

Features	Chi-squared rank	Information gain rank
Ratio ICMP	1	1
Land	2	2
Ratio UDP	3	3
One-way ratio	4	4
Ratio TCP	5	5
Protocol type	6	6
Average length TCP/IP flow	7	8
Ratio of in/out packets	8	7

where $P(y_i)$ is the prior probabilities for all values of Y. The uncertainty about the value of Y after observing values of X is given by the conditional entropy of Y given X defined as

$$H(Y|X) = -\sum_j P(x_j) \sum_i P(y_i|x_j) \log_2(P(y_i|x_j) \log_2(P(y_i|x_j)) \quad (5.9)$$

where $P(y_i|x_j)$ is the posterior probabilities of Y given the values of X. The information gain is thus defined as

$$IG(Y|X) = H(Y) - H(Y|X) \quad (5.10)$$

According to this measure, an attribute X is regarded as being more correlated (i.e. makes more contributions) to class Y than attribute Z if $IG(Y|X) > IG(Y|Z)$. By calculating information gain, the correlations of each attribute can be ranked to the class. The most important attributes can then be selected based on the ranking.

The chi-square statistic [82] measures the lack of independence between a feature X and a cluster Y. It can be compared to the chi-square distribution with one degree of freedom to judge extremeness:

$$x^2 = \sum_{i=1}^{r} \sum_{j=1}^{k} \frac{(A_{ij} - E_{ij})^2}{E_{ij}} \quad (5.11)$$

where r is the number of features and k is the number of clusters, A_{ij} is the number of instances for which the value of a feature is i and the value of the cluster is j, E_{ij} is the expected number of instances of A_{ij}. The larger the x^2 value, the more important the feature is to the cluster. Thus, ranking the importance of each feature with respect to the clusters based on the value of x^2 for the proposed work is considered. Based on the x^2 value, eight features, as shown in Table 5.1, are considered as important features.

Table 5.2 Samples collected

Network Data	Data type	Total number of packets
Trained	Attack (CAIDA)	9,45,372
	Normal	1,10,535
Unseen test data	Attack (CAIDA)	3,24,098
	Normal	36,485

Table 5.3 Classification results

Method used	Correct classification %
Fuzzy c-means	98.7
Naive Bayesian	97.2
SVM	96.4
KNN	96.6
Decision tree	95.6
K-means	96.7

Table 5.4 F-Measure details of classifiers

Method	TP	FP	TN	FN	F-measure
Fuzzy c-means	298	2	270	3	0.987
Naive Bayesian	290	10	256	17	0.972
KNN	280	20	243	30	0.969
SVM	282	18	253	20	0.964
K-means	285	15	273	0	0.9669
Decision tree	278	22	218	55	0.956

The first step is to extract these eight features from the dataset consisting of both normal and attack data patterns. In the experiments, a sampling frequency of 1 s was used. The next step is to train the machine-learning techniques with these datasets. In the detection phase, the same set of eight features were computed for the given network traffic, and the traffic is labelled as attack or normal based on the majority of the values computed by the machine-learning classifiers.

5.3.3 *Experimental Results*

The CAIDA dataset [5] was used in the experiments as the attack component. Data collected on the SSE network provided the normal traffic component. Classification of attack and normal traffic was done using an open-source tool called KNIME (Konstanz Information Miner) version 3 [1]. Table 5.2 shows details of the CAIDA dataset and the normal traffic collected on the SSE network. Table 5.3 shows the correct classification, Table 5.4 shows the f-measure details and Fig. 5.2 shows the evaluation results using ROC curves for the selected machine-learning techniques. Based on the results of these experiments, the FCM-based classification gives the best results in detecting DDoS attacks.

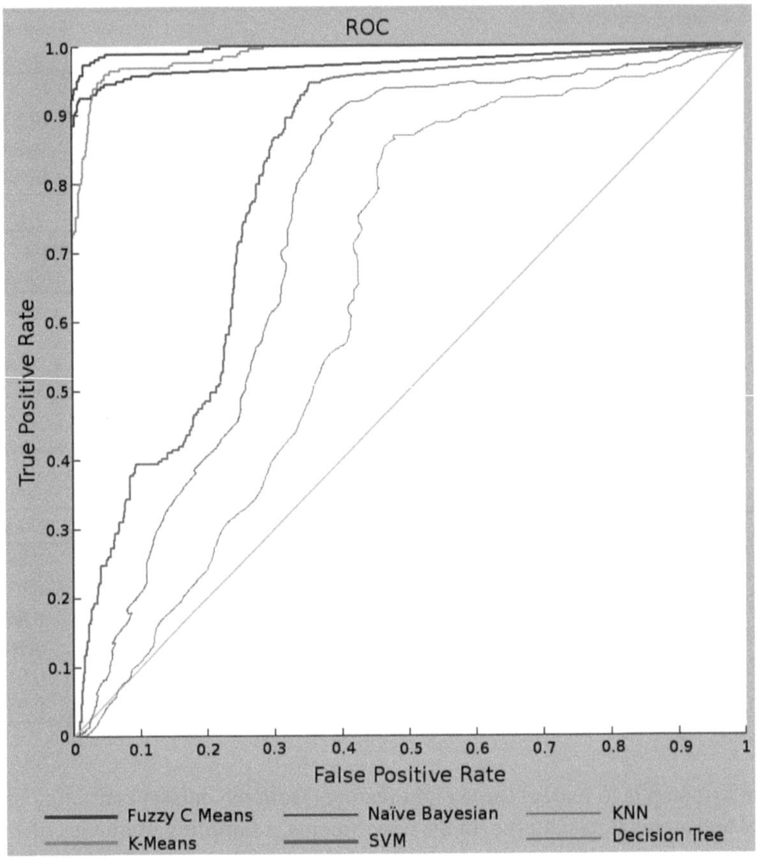

Fig. 5.2 False vs. true positive rate

5.4 DoS Detection Using Change Point Analysis (CPA) of Not Seen Previously (NSP) IP Addresses

5.4.1 Related Work

IP addresses are an integral part of communication over TCP/IP networks and are a valuable piece of information for uniquely identifying communicating entities. Given their significance, there is now a distinct body of knowledge surrounding the use of source IP address for detecting various network activities. This includes Denial of Service (DoS) and high-rate flooding attacks/DDoS [15, 21, 33, 39, 45, 67, 70, 78]. It also enables discrimination between network traffic sharing common characteristics such as anomalous network traffic and flash event [15, 39].

Central to the functioning of the Internet is its IP protocol within layer 3 of the TCP/IP suite. The IP protocol was designed for interconnection of hosts to networks without any support for the verification/authentication of IP header fields [22]. This allows attackers to inject falsified information into the IP header (such as source address spoofing), enabling them to conceal their identity and yet traverse the IP packet to the intended target. Historically, this has been a primary vector of attack. Hence, considerable amount of research has been carried out to detect such attacks with the assumption that, source address of the attack is being spoofed [29,46,50]. More recently, attackers have been able to gain control of a large number of computing resources in a so-called botnet herd. These do not need to spoof the source IP addresses. This allows them to imitate the behaviour of a legitimate user making it difficult to distinguish between normal network traffic and attack traffic [21,69,70]. An additional problem is that the traffic generated by individual bots can be varied dynamically so that the bot does not exhibit an easily predictable pattern of behaviour.

For the detection of attacks with such adaptive behaviour, it is important to use features that are difficult or impossible for an attacker to alter without being detected [69]. In this regard, considerable research has been conducted in utilising source IP address – related features as the primary means of detecting such attacks [21,39,67, 68,70]. One of the key issues associated with using the source IP address features to identify DDoS attacks is the difficulty of storing statistical data for all 2^{32} IP addresses. This can further be complicated due to a sharp increase in rate of arrival of new IP addresses during an attack [69].

To address this scalability issue, Gil and Poletto [33] proposed a scheme called MULTIOPS, consisting of a dynamic 4-byte 256-ary tree. The proposed solution was later found to be vulnerable to memory exhaustion attack by Peng et al. [69]. The memory issue can be addressed by storing information about only a subset of source IP addresses in a set of 2^{32} available addresses such as those completing the 3-way TCP handshake or sending more than a predetermined number of packets [67, 69,78]. A somewhat similar solution based on aggregating the source IP addresses was proposed by Peng et al. [70]. These alternatives solve the memory exhaustion (also known as scalability) issue identified previously but can still be vulnerable to similar problems under the next-generation of IP address (IPv6), where the number of addresses is many orders of magnitude larger than the IPv4 address space.

As discussed in [8], the key challenges in developing an adaptive solution for detecting DDoS attacks are:

- Operation of detection process at high speed
- Activation of response in real time to mitigate the impact of the attack

Ahmed et al. [8] provide the proof-of-concept implementation of the DMM architecture which integrates the DDoS detection and mitigation capability into a single architecture. This architecture is discussed briefly in Sect. 5.4.2, following which we present a detailed description of our detection approach.

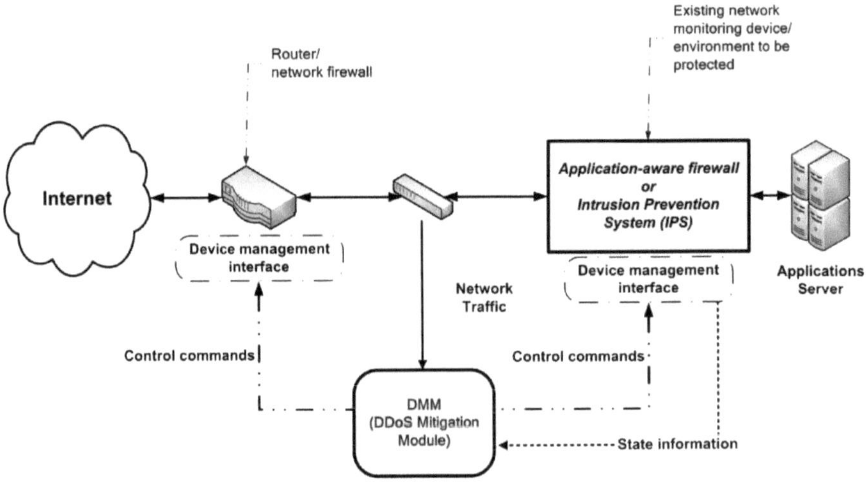

Fig. 5.3 A conceptual architecture of a DDoS Mitigation Module (DMM)

5.4.2 DMM Architecture

As has been indicated previously, reliable and timely attack detection is the crucial first step in successfully managing DDoS attacks. However, once the onset of an attack has been detected, the next step is to protect the system as a whole from failure by protecting the individual components. For example, a typical host application environment comprises not only of the actual application server but also devices whose role is to protect the application from attack. These include, for example, application-aware firewalls and intrusion prevention system (IPS). Typically, the role of the application-aware firewall is to filter packets using a set of rules based on incoming IP addresses, ports, payload, etc. An IPS performs a deep-packet analysis of network packets to detect any malicious payload targeted at the application-layer services (such as HTTP payload) again based on a set of rules. In both cases, the rule sets need to be updated dynamically so that the devices can respond to the continuously evolving threat environment. Whilst the primary focus is on protecting the applications host/server, it is equally important to protect these security devices which may themselves be at risk. In order to protect the set of components at risk, a security architecture, also known as the DDoS mitigation module (DMM) – is proposed in this chapter.

Figure 5.3 shows a schematic of a version of the proposed DMM. At a conceptual level, the design goal is for the DMM to have the capability of developing a continuous profile of the network traffic. This information is then used with a predictive model to form real-time estimates of the impact of the traffic on the processing capacity of the devices to be protected, viz., an application-aware firewall or a network-based intrusion detection system. In the event that DMM determines that the devices to be protected are at risk of failing, it would then initiate a traffic-management strategy to mitigate the situation. For example this could take the form of using a white list to allow access only to legitimate sources whilst the network is experiencing a period of heavy load and potentially when an attack is actually in progress. This capability helps the DMM to predict and then react to imminent failure of the network security/monitoring devices to be protected. Interested readers may refer to [8] for further details. Section 5.4.3 provides a detailed description of a potential high-rate flooding attack detection algorithm used in the implementation of DMM.

5.4.3 Detection Approach

The proposed detection algorithm to operate within the DMM, consists of two functions:

- Function *ipac* for classifying IP addresses
- A *ddos* function to identify an attack

The *ipac* function extracts the source IP address of each incoming packet and determines if it is a new IP address (not seen previously (NSP)). The resulting time series is then analysed using the *ddos* function to identify if the system is under attack. The *ddos* function maintains two states: NA (not under attack) and A (under attack).

In order to learn the behaviour of network under normal operating conditions, the *ipac* classification function is first applied on normal network traffic traces without the *ddos* function being activated. After training, the *ddos* function is invoked periodically (intervals of the order of 1 to 10 s), and based on the rate of arrival of new IP addresses calculated by the *ipac* function, the *ddos* function determines a transition between two states NA and A. A description of detection algorithm used in the *ddos* function is provided in Sect. 5.4.5.

Once a transition from state NA to A signals that an attack has been detected, the *ddos* function generates a white list which is then used to define a mitigation strategy, that is, only allow traffic from IP addresses within a white list. A transition from state A to NA signals the end of an attack and will result in the relaxation of mitigation strategy i.e. stop using white list as a filtering policy. It is to be noted that in the current implementation, the white list will contain all the IP addresses observed on the targeted network during normal network conditions. The *ddos* function does as follows:

```
if (in state NA) then
    if NOT (StateChange(NA)) then //no state change
    Update White-list (Add IP address/es to white-list)
else //state change to A
     state = A
```

```
    communicate White-list to the protected security
     device
if ((in state A) then
    if (StateChange(A)) then    //state change to NA
    state = NA
    communicate to the protected security device to
      stop
    using the white-list
```

There are also two obvious limitations in this approach. The first is that when the system is deemed to be under attack, new IP addresses are treated as malicious. This can give rise to false positives. Use of other features of IP addresses including IP address distribution can be used to limit the false positives and is part of the future work. The second limitation concerns the malicious IP addresses observed during normal network operations. For example, an attacker performing reconnaissance of the target network by sending a small number of packets (such as ICMP echo requests) would result in these IP addresses being added to the white-list addresses, giving rise to false negatives, before actually flooding the target with large amounts of packets. This can be addressed not only by using the other IP address features described previously but also by using smaller historical time periods, for example, only using IP addresses observed during the last 24 h as being trusted IP addresses.

5.4.4 IP Address Classification

The implementation of the IP address classification function (*ipac*) necessitates the use of a compact and efficient data structure. Using the rate of change of new IP addresses as a primary feature to detect the onset of an attack requires keeping track of source IP addresses already observed over the network to be protected. During normal network conditions, the rate of arrival of new IP addresses is relatively low, that is, only a small number of new IP addresses appear at the user site. A substantial increase in the rate of arrival of a number of new IP addresses is usually observed during a high-rate flooding attack [69]. Keeping in view the scalability of *ipac* function during normal and attack conditions, two different implementations of the IP address classification function (*ipac*) are provided:

- Bit vector
- Bloom filter

The classification of 32-bit IPv4 addresses using a bit vector would require an array of length 2^{29} with each array element of size 1 byte. This results in 0.5 GB storage space for representing the entire set of IPv4 addresses. In the *ipac* function, whenever a packet is received, a bit at the specific position in the bit vector is marked to indicate the presence of a given IP address. Using a separate counter, the number of new IP addresses observed during an interval is calculated resulting in a time series of the rate of change of IP addresses. Further details of *ipac* classification function using the bit-vector approach can be found in [8].

In contrast to IPv4, classification of 128-bit IPv6 addresses using a bit vector requires an array of length of at least 2^{125} with each array element of size 1 byte. The increase in address size from 32 bits to 128 bits results in a significantly larger storage requirement for representing the entire set of IPv6 addresses. One solution to the problem of scaling the *ipac* function to accommodate the storage requirement for the IPv6 address space is to use efficient data structures like bloom filters [16].

Bloom filters are data structures that offer inexact but compact representation of the elements within a set. In contrast to bit vectors, the benefits of using bloom filters are twofold. Firstly, they offer a compact representation of the elements within a given set whereas similar elements would have sparse representation if a bit vector were used. Secondly, for similar network traffic, a bloom filter can reduce the storage requirement significantly as compared to a bit vector. Given their advantages, bloom filters can be used to efficiently identify the presence or absence of elements within a set. For example, in case of IPv6 using bloom filters for membership query (in *ipac* function) of a given IPv6 address requires identification of an IP address as being seen (old) or not seen (new) previously. For a given IP address, the corresponding positions (array index) within a bloom filter are marked as set. The array indexes are identified by applying a hash function on the IP address.

The inexact and compact representation offered by the bloom filters comes at the cost of an increased number of false alarms as compared to bit vectors which have no false alarms. The bloom filter can identify an element as being within a set when it is not (collision). The false alarm rate can be reduced as a function of bloom filter size, the number of elements within a given set and the number of hash functions used. In order to identify the relationship between these three parameters, a theoretical analysis presented by Ripeanu et al. [71] has been conducted. The theoretical analysis helps analyse the relationship between these three parameters and helps in selecting the optimal size of a bloom filter for a given number of hash functions and false alarm rates. For the purpose of this discussion, different sizes of bloom filters with different numbers of hash functions for a given false alarm rate have been analysed.

Figure 5.4 shows the relationship between false positive rate, number of hash functions and bloom filter size (in terms of number of bits per set element). In this figure, the number of hash functions is plotted along the horizontal axis and the false positive rate (in log scale) along the vertical axis. Each graph represents a bloom filter of different size ranging from 10 to 40 bits per set element. In the graph legend, B represents the number of bits per set element. For example, B10 means bloom filter with 10 bits per set element. As shown in Fig. 5.4, the false positive rate can be reduced to a specific limit by increasing the number of hash functions, after which an increase in the number of hash functions as no significant effect on the false positive rate. On the basis of this analysis, a bloom filter of size 30 bits per set element with 10 hash functions is selected for experimentation (described in Sect. 5.4.6) which gives a false alarm rate of less than .00001%.

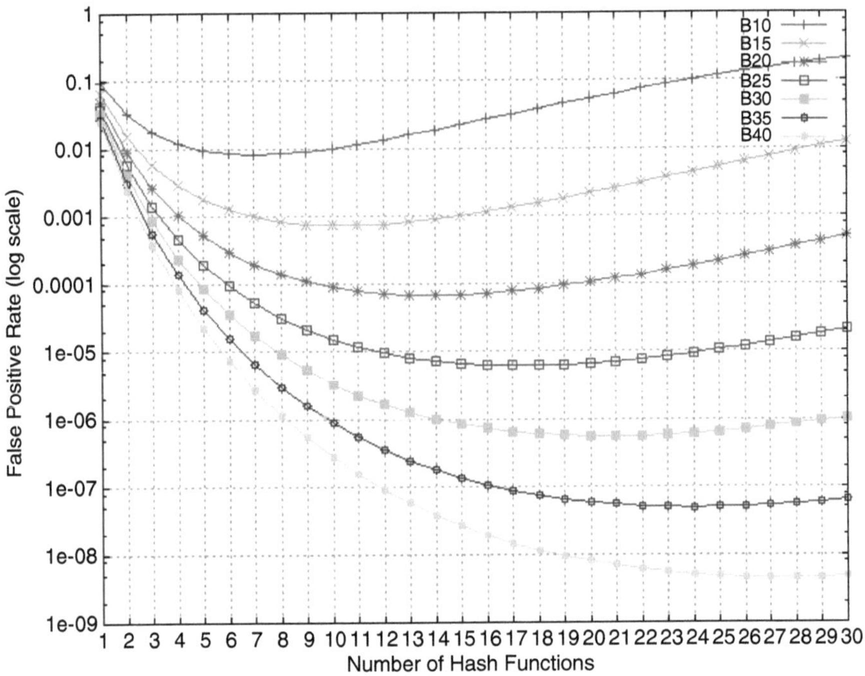

Fig. 5.4 Bloom filter: false positive rate as a function of the number of hash functions used. For example, in the case of a bloom filter with 30 bits per entry (graph third from the bottom), the false positive rate is at the minimum using 21 hash functions. Adding further hash functions does not significantly decrease the false positive rate

For the purpose of identifying the group of 10 hash functions (to be used with bloom filters) having small collision rates, an experimental analysis has been performed with 14 general purpose hash functions. A description of some of these hash functions can be found [64]. In this regard, two experiments – one without specifying the size of a bloom filter and the other with a specific size bloom filter – were conducted. In the latter case, the size was selected based on the results of the theoretical analysis performed previously; see Fig. 5.4 (a bloom filter of size 30 array indexes per set element is selected which gives less than .00001% false alarm rate). For the experimental analysis, different set sizes (number of source IP addresses) have been used with the minimum being 10 and the maximum being 400,000. Table 5.5 shows results of the number of collisions observed for different numbers of IP addresses.

To identify the maximum collision size for a given number of IP addresses, the maximum number of IP addresses that have been hashed to a same location (array index) in a bloom filter has been calculated (see Table 5.6). It was observed that almost all 14 hash functions performed extremely well with, on the average, 2–4 IP addresses being hashed to the same array index within a bloom filter. It should

Table 5.5 Number of collisions for bloom filter size 30 ranked in ascending order based on the last column

Hash functions	Number of IP addresses to be stored in the bloom filter											
	10	50	100	500	1,000	5,000	10,000	50,000	100,000	200,000	300,000	400,000
Fowler Noll Vo (FNV) algorithm	0	0	0	0	0	0	6	202	832	3,143	7,271	12,785
Arash Partow (AP) algorithm	0	0	0	0	0	2	12	238	828	3,225	7,250	12,958
Brian Kernighan and Dennis Ritchie (BKDR) algorithm	0	0	0	0	0	2	12	204	817	3,289	7,282	13,094
Robert Sedgwick (RS) algorithm	0	0	0	0	0	0	4	168	793	3,342	7,449	13,188
Rotation (ROT13) algorithm	0	0	0	0	0	0	10	232	891	3,267	7,386	13,201
Cyclic Redundancy Check (CRC32) algorithm	0	0	0	0	0	0	6	212	898	3,499	7,498	13,216
Justin Sobel (JS) algorithm	0	0	0	0	0	4	12	215	828	3,389	7,597	13,236
Bob Jenkins one-at-a-time algorithm	0	0	0	0	0	2	10	214	886	3,257	7,498	13,245
Leonid Yuriev (LY) algorithm	0	0	0	0	0	2	8	244	821	3,309	7,584	13,287
Daniel J. Bernstein (DJB) algorithm	0	0	0	0	0	2	10	216	891	3,587	7,952	14,272
Donald E. Knuth (DEK) algorithm	0	0	0	0	0	2	30	334	1,279	5,023	9,807	16,409
Substitute DBM (SDBM) algorithm	0	0	0	0	0	0	14	288	1,520	6,335	14,487	25,416
Executable and Linkable Format (ELF) algorithm	0	0	0	0	0	0	10	378	1,625	6,752	14,652	26,468
BP algorithm	0	0	0	6	14	329	1,333	25,377	73,059	182,172	290,372	395,012

Table 5.6 Maximum collision size for bloom filter size 30: the maximum number of IP addresses that have been hashed to a same location

Hash Functions	Number of IP addresses											
	10	50	100	500	1,000	5,000	10,000	50,000	100,000	200,000	300,000	400,000
Fowler Noll Vo (FNV) algorithm	1	1	1	1	1	1	2	2	2	3	3	4
Arash Partow (AP) algorithm	1	1	1	1	1	2	2	2	2	3	3	3
Brian Kernighan and Dennis Ritchie (BKDR) algorithm	1	1	1	1	1	2	2	2	3	3	4	4
Robert Sedgwick (RS) algorithm	1	1	1	1	1	1	2	2	3	3	4	4
Rotation (ROT13) algorithm	1	1	1	1	1	1	2	2	3	3	3	3
Cyclic Redundancy Check (CRC32) algorithm	1	1	1	1	1	1	2	2	2	3	4	4
Justin Sobel (JS) algorithm	1	1	1	1	1	2	2	3	3	3	3	4
Bob Jenkins one-at-a-time algorithm	1	1	1	1	1	2	2	2	3	3	3	3
Leonid Yuriev (LY) algorithm	1	1	1	1	1	2	2	2	3	3	3	3
Daniel J. Bernstein (DJB) algorithm	1	1	1	1	1	2	2	2	3	4	4	4
Donald E. Knuth (DEK) algorithm	1	1	1	1	1	2	2	2	3	4	4	4
Substitute DBM (SDBM) algorithm	1	1	1	1	1	1	2	2	3	4	4	4
Executable and Linkable Format (ELF) algorithm	1	1	1	1	1	1	2	3	3	4	4	5
BP algorithm	1	1	1	2	2	3	4	9	11	15	21	26

be noted that each hash function has been applied independently. Applying all hash functions on each source IP address and then setting the corresponding bits in a bloom filter has not been done and is a part of the future work.

Based on the experimental analysis, the first 10 hash functions have been used in the implementation of the DMM. In the current implementation, the user has been given the choice of selecting a number of hash functions (from 1 to 10) and the size of the bloom filter. The program captures the network traffic in real time and performs the analysis of the network traffic for a given measurement frequency (duration specified in terms of seconds).

The current implementation makes use of a counting bloom filter [26], which uses an array of n counters instead of bits. The counters track the number of elements currently hashed to that location. In counting bloom filters, the counters should be large enough to avoid overflow. Although it has been observed that a 4-bit-long counter is enough for the majority of the applications [26], an 8-bit counter has been selected in the current implementation of the DMM. Currently, the counting bloom filter is being used as a traditional bloom filter and can be enhanced in the future to minimise the number of collisions. This can be achieved by using linked lists together with a counting bloom filter to exactly identify the source IP addresses being hashed to a given array index.

5.4.5 DDoS Detection

An outbreak of anomalous activities including DDoS/high-rate flooding attack results in an increase in the rate of unsolicited packets arriving at the victim machine. This increase in the number of unsolicited packets usually results from a large number of compromised hosts sending voluminous traffic to the target. This results in not only an increase in traffic volume but also an increase in the number of sources sending such traffic. The problem of identifying malicious network activity in the presence of DDoS/high-rate flooding attacks can thus be formulated as a change detection problem. This involves identifying the change in the statistical properties of the network traffic parameter under investigation, that is, the rate of arrival of new IP addresses.

This abrupt increase in the rate of arrival of new IP addresses is being used in the DMM as a key observation in identifying and investigating potential malicious network activity. For change detection, a sliding-window-based non-parametric CUSUM proposed by Ahmed et al. [1, 7] has been used. The *ddos* function in DMM takes input from the IP classification function (*ipac*). The bloom filters in *ipac* identify the number of new source IP addresses during a given measurement frequency which then calls the *ddos* function (change detection algorithm) to identify an attack.

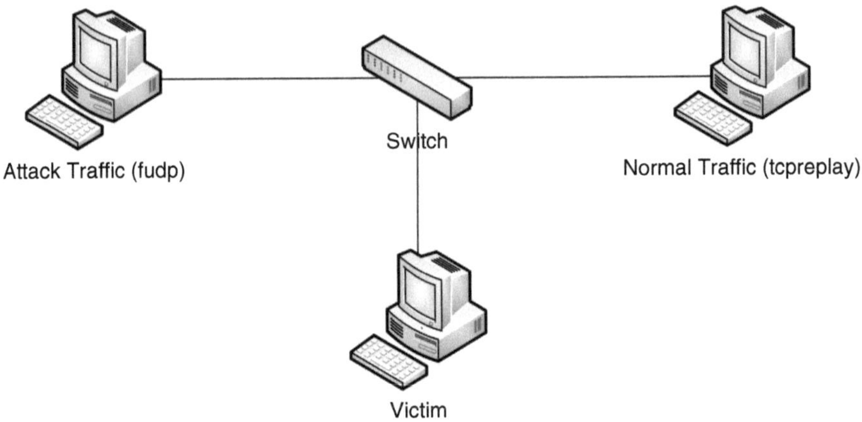

Fig. 5.5 The testbed architecture

5.4.6 *Experimental Results*

A proof-of-concept evaluation of the proposed Not Seen Previously (NSP) and CPA algorithms has been performed with both bit vector and bloom filter classification techniques. The testbed shown in Fig. 5.5 was used to evaluate the NSP algorithm using bit vector and bloom filter classification techniques. A detailed description of the testbed is provided in Chap. 4.

For the purpose of the experiments, it is necessary to construct a dataset with two distinct components, viz., background/normal traffic and the anomalous traffic. The background network traffic component is drawn from the data collected by the University of Auckland (Auckland VIII) [3]. It comprises 20 h of continuous bidirectional real-world network traffic collected on 12-03-2006. The IP addresses in the traffic trace have been mapped into 10.*.*.* using one-to-one hash mapping for privacy. Before using the data, it is required to be pre-processed. The dataset is first analysed to select the traffic destined for the busiest web server (destination IP address 10.0.0.63). The selected traffic is then processed to remove TCP flows with less than three packets (SYN attacks) and also flows with no data from the web server. The processed data is then converted into a one-way incoming traffic to destination web server 10.0.0.63.

For the analysis of NSP algorithm using bit-vector classification technique, the processed data is reproduced over the testbed using TCPREPLAY utility. The traffic is replayed at the rate of around 300 packets per second. The attack traffic component was constructed using the FUDP utility. It was used to flood the victim machine (see Fig. 5.5) with UDP packets having varying number of spoofed source IP addresses ranging from 35 to 150 sources. The rationale for using a different number of source IP addresses for the attack traffic was to enable an analysis of the behaviour of the proposed algorithm under different attack conditions. It should be noted that attacks with a large number of source IP addresses (e.g. UDP flooding attack with 150 sources per second) are trivial to detect as compared to attacks

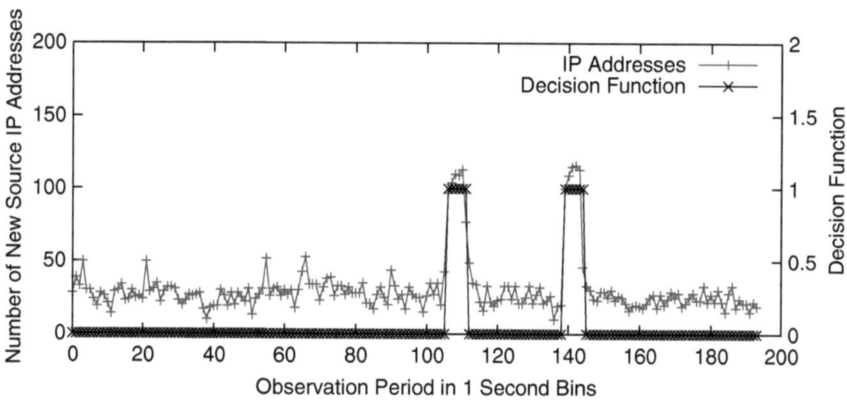

Fig. 5.6 Successful detection of two UDP flooding attacks using CPA (right y-axis). Each attack uses 75 source IP addresses vs. ca. 25–35 source IP addresses in the background traffic (left y-axis)

with a small number of sources. The average attack traffic rate is set to around 3000 packets per second (10 times the normal background traffic). The algorithm successfully detected all the attacks and generated a list of IP addresses observed under normal network conditions. A detailed analysis and evaluation of bit-vector-based NSP algorithm can be found in [8].

For the analysis of NSP algorithm using bloom filter classification technique, the processed data described above is replayed using TCPREPLAY utility at the rate of 4 Mbps, that is, nearly 6,500 packets per second. The attack traffic component was constructed using the FUDP utility with varying number of source IP addresses, including 75, 100 and 300 sources per second, being generated. Based on the theoretical analysis of the bloom filters described previously, for these experiments 10 hash functions with 30 bits per set element were used. This would give an expected false positive rate of less than .00001%. Unless otherwise specified, a measurement frequency of 1 s is used in all the experiments. Figure 5.6 shows the result of UDP flood using 75 unique source IP addresses.

In Fig. 5.6, the horizontal axis represents the observation period in 1 s bins, the left vertical axis is the total number of new source IP addresses in the measurement interval and the right vertical axis represents the CPA decision function with 1 being attack and 0 being no attack. Two instances of UDP flooding attacks were generated, as shown in the figure, and both attacks were successfully detected. The NSP algorithm was able to generate a list of legitimate IP addresses also known as the white list (IP addresses appearing under normal network operations). It is to be noted that the detection delay is bound by the measurement interval which in this case is 1 s and can further be reduced using smaller intervals.

Figure 5.7 shows the result of flooding attack with 100 source IP addresses. Similarly, Fig. 5.8 shows the result of UDP flooding attack with 300 source IP addresses. For UDP floods with 100 and 300 source IP addresses, only one attack instance was generated. The proposed algorithm was successful in identifying attacks in both cases.

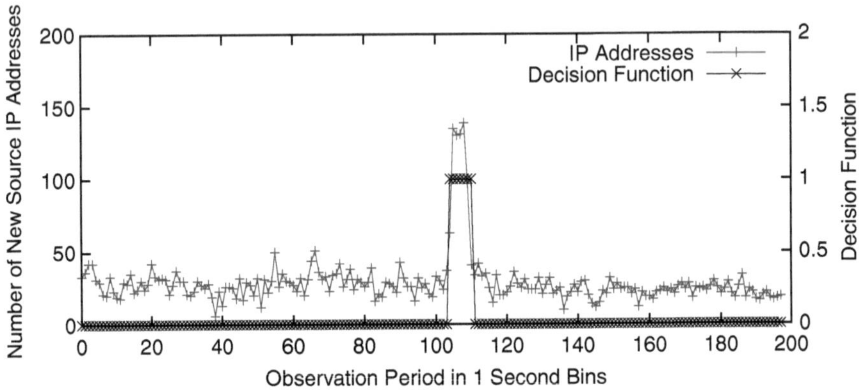

Fig. 5.7 Successful detection of UDP flooding attack using CPA (right y-axis). The UDP attack uses 100 source IP addresses vs. ca. 25–35 source IP addresses in the background traffic (left y-axis)

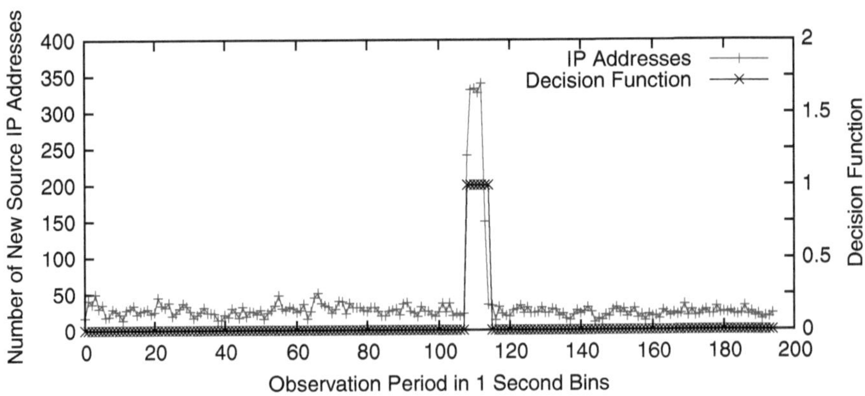

Fig. 5.8 Successful detection of UDP flooding attack using CPA. The UDP attack uses 300 source IP addresses vs. ca. 25–35 source IP addresses in the background traffic (left y-axis)

5.5 DoS Detection Using Naïve Bayesian Classifiers

5.5.1 Related Work

Recently, a substantial amount of work has been done on using various machine-learning techniques and other general statistical approaches in the detection of DoS attacks. In these works, DoS attacks have been classified into the broad category of intrusion attacks. Hence, solutions have been oriented more towards formal intrusions, including root escalation, scripting attacks, etc. A major factor that is often missed by classifying DoS attacks into intrusion attacks is the enormous volume that DoS detection solutions have to handle in comparison with intrusion attacks. While computationally intensive models could be used to detect intrusions, this extreme processing workload renders learning models such as hidden Markov models (HMMs) and ANNs impractical when it comes to real-time attack detection. These models would either be too space intensive or too time intensive. The detection mechanism for DoS attacks is required to be ultralightweight to support 'wire speeds'.

Another practical drawback of supervised/unsupervised learning models for DoS detection is the accuracy of (supposedly normal) traffic used in the building of the models. While it is common to assume that all traffic used is absolutely normal, for data collected at actual user sites, this may not necessarily be true and the training data may itself contain abnormalities. This may cause false negatives when the model is applied to other real-world data.

Another common problem for the entire research community on DoS attacks is the lack of data in all forms and, in particular, the unavailability of training data. Currently, the DoS research community is heavily dependent on two standard datasets, that is, the KDD dataset and the DARPA dataset [4, 34], for the purpose of learning and analysis. However, these have a number of inherent limitations and are better suited to the analysis of intrusion attacks [79].

The objective of any practical system to detect DoS attacks-including high rate flooding DDoS attacks, should be to ensure that the resultant detection system is (a) light weight, (b) accommodative of practical difficulties in learning, and (c) operable at close to line speeds.

Proposed DoS/DDoS detection techniques differ from each other in terms of the objectives to which they cater, the strategy used, the features chosen and the (reported) performance. These detection techniques include statistical approaches like [28] which proposes a chi-square-test on the entropy values of the packet headers. Jin and Yeung [37] discusses the effects of multivariate correlation analysis on DDoS detection and presents an example of detecting a SYN flooding attack. Jin and Yeung [38] uses a covariance matrix to present the relationship between each pair of network features in order to identify an attack. In [92] the CUSUM algorithm is used to detect a change in the number of packets directed towards the destination to identify the attack. Other techniques taken from pattern analysis and machine learning have also been proposed in the literature. For example, Xie et al. [86] consider application layer DDoS attacks and use hidden semi-Markov models to detect these types of attacks. Other classification algorithms such as support vector machines [72], genetic algorithms [75], artificial neural networks (ANN) [32, 85] and Bayesian learning [54] have also been applied. Hybrid modelling techniques such as [75] also provide interesting results. Hidden Markov model–based DoS attack solutions have also been proposed in [40, 88]. A taxonomy of DDoS attacks and defence mechanisms has been documented in [58]. Recent works [27, 41] have discussed the use of Bayesian classifiers towards intrusion detection, in general, which includes DoS attacks. Some especially relevant works are as follows:

- In [88], the authors model the rate of change of new and old IP addresses at nodes in the network using an hidden Markov model (HMM) and determine attacks based on the above characteristic. They also explain how the results appear to improve with attack information exchange between distributed nodes using Cooperative Reinforcement Learning techniques. The solution is a target end solution.
- Seo et al. [72] use the traffic rate analyser (TRA) to model the TCP flag rate which is a measure of the proportion of packets with chosen flag set to the total number of TCP packets. In addition, they use the protocol rate which is a measure of the proportion of packets from a selected protocol to the total number of incoming packets. Ten features, each representative of either of the above, are monitored and separately modelled for both normal packets and attack packets as a series of SVMs. Attacks are labelled as DoS, DDoS or DrDoS attacks based on the class to which the pattern belongs. The solution is a source end solution.
- Jin et al. [37] present a generalised view of multivariate correlation analysis for DDoS detection using SYN flooding attacks as an example. The idea projected in the paper is to differentiate between normal and attack traffic by considering the correlation between possible pairs of selected features among a set of features $f_1 \ldots f_p$ at different intervals, and comparing the distance between this correlation matrix and the average of correlation matrices seen during training. The authors consider the six flags in the TCP field as separate features and show the effectiveness of the method under these conditions. The solution is a target end solution.
- Kang et al. [40] proposed a solution which is deployed at the source end and uses three features, viz., a three-tuple {*source IP, destination IP, destination port*}, a 16-bit packet identifier and the TCP header flags on the same dataset. Independent HMMs is trained for each of these three features and then the HMMs are coupled to form the multi-stream fused HMM. If the probability of the traffic pattern is smaller than the threshold probability, the pattern is flagged as anomalous, and a suitable thresholding on the proportion of anomalous patterns to total patterns determines an attack. The solution is a source end solution.

5.5.2 Detection Approach

The detection approach described here is likewise to be situated and to operate within the DMM architecture described in Sect. 5.4.2, as is the case with the CPA NSP approach presented in Sect. 5.4.

While signature-based detection techniques use signatures to identify attacks, anomaly-based detection models build models of normal traffic and determine attacks as traffic which deviates significantly from these models. The latter is more suited to detecting flooding attacks and hence was chosen.

General design elements

- **(Traffic separation)**. Network traffic needs to be separated into streams in order to facilitate applying DoS handling techniques in the case of an attack. We define a stream by the two-tuple value *destination ip, destination port.*
- **(Windowing)**. Windowing essentially means splitting the input traffic into traffic subsets which fit into logical entities called windows. Windows may either be time windows or packet windows. Packet windows were chosen in order to decrease the reaction time and also to have control on the number of events that have to be modelled.
- **(Flagging an attack)**. Flagging an abnormal situation as an attack is a function of a series of abnormal windows, not necessarily consecutive. For example, if an attack were to be flagged after observing five consecutive abnormal windows, the attacker can cleverly escape attack detection by keeping the attack traffic just below 5. The target would still be reasonably safe from this action by the attacker. However, the detection mechanism will miss such attack attempts in the build-up of an attack. In order to overcome this problem, a step-based mechanism is used. At any point (until a reasonable timeout period) in the course of deployment, if the number of abnormal windows goes beyond a particular number (quantified as a parameter called abnormal window count (AWC)), then the system flags an attack. This will ensure that such attacks will be caught.

5.5.3 Modelling TCP Traffic

For attack detection, the modelling is done based on the protocol headers. The choice of parameter is driven by the fact that it should be able to differentiate between a normal and attack pattern. Among a set of various other header parameters, TCP flags appear to be best suited. At the time of selecting parameters, it should also be noted that keeping down the number of such parameters will reduce the computation load on the detection system.

In [80], the authors present a detailed account of the proposed methodology for TCP traffic modelling. Central to modelling TCP traffic is the TCP flags field, available as a field in the TCP header. The TCP flags field is a collection of 8-bits, each bit representing one flag. Individual flags or combinations of flags symbolise specific actions in TCP – for example, connection establishment, connection closure, requesting data, etc. The different TCP flags are: SYN, ACK, PSH, FIN, URG, RST (standard TCP flags) and ECE, CWR. The last two flags, it was observed, are seldom used in contemporary Internet traffic, and so we are concerned only about events raised from six flags. Hence, there are 2^6 different observables. A packet window is technically defined as a collection of flag sets observed with every packet in the window. The probability of a given packet window is a function of individual probabilities of different flag sets observed inside the window. Even though some of these flag sets may be dependent on each other (TCP is a causal system), inside small windows (in comparison to the total training traffic) the dependency between individual flags may not exist and are assumed to be totally independent.

However, TCP traffic is observed to be highly skewed in nature. This is evidenced by the presence (and prescription) of only very few flag combinations in inbound traffic. This gives scope to modelling traffic with fewer than 2^6 events as mentioned above. Also inducing some form of valid grouping will reduce the number of events and will be more space efficient. After experiments on various datasets to determine a reasonable grouping of these flags, the following groups were arrived at:

1. T_1: Packets with RST bit set (irrespective of other bits) – 32 packet types
2. T_2: SYN packets – 1 packet type
3. T_3: ACK packets – 1 packet type
4. T_4: FIN/ACK packets – 1 packet type
5. T_5: PSH/ACK packets – 1 packet type
6. T_6: Rest of the packets – 28 packet types. Includes seldom used packets and invalid packets

For a packet window of size N, there are $(N + 1)$ instances to be monitored, thereby making the total number of instances in probability space (and hence the total number of floating points in memory) $6 * (N + 1)$. However, as mentioned above, many instances are very unlikely to occur in normal traffic because TCP traffic is highly skewed in terms of its use of TCP flags. Hence, a fewer number of events could be sufficient in modelling the traffic, and this offers a reason for choosing NB classifiers with simplistic assumptions. However, the existence of instances that are unlikely to occur may result in events with zero probability. To avoid zero probability events, a simple technique like Laplacian smoothing (used for speedy smoothing) can be used. However, this may produce inaccurate results when applied to a large number of instances, particularly given that we started with the assumption of working with smaller amounts of training data.

We reduce the $N + 1$ number of possible events per observable group (N is the packet window size) by using a constant number of bands K, where each band groups together events of similar probability. This reduces the number of zero probability events and also improves smoothing. To achieve this, the following procedure is used:

1. Observe occurrences for each event and consider jumps between event occurrences in descending order. Arranging events in descending order of their occurrences brings like events closer. This will help grouping like events together. The jump between two observations of events in the array is denoted as $J_i = O_i - O_{i+1}$.
2. Consider top $K - 1$ jumps in J_i s and form K bands bordering them.
   To illustrate:
   Band grouping – An Example

   - Let the set representing number of windows during learning against each event (from $n = 0$ to $n = 10$) be $A = \{500, 291, 271, 36, 222, 111, 1211, 3, 1, 31, 1\}$, corresponding to the packet type T_i. Let the fixed number of bands be $K = 5$.
   - Sorted array $A_s = \{1211, 500, 291, 271, 222, 111, 36, 31, 3, 1, 1\}$, which brings like events closer.

- Jump array $J = \{711, 209, 20, 49, 111, 75, 5, 28, 2, 0\}$.
- Top 4 jumps $J_s[0] \ldots J_s[3] = \{711, 209, 111, 75\}$.
- Index array $I = \{0, 1, 4, 5\}$.
- Sorted index array $I_s = \{0, 1, 4, 5\}$.
- Band groups: $Bg_1 : A_s[0]$ (1 value), $Bg_2 : A_s[1]$ (1 value), $Bg_3 : A_s[2] - A_s[4]$ (3 values), $Bg_4 : A_s[5]$ (1 value), $Bg_5 : A_s[6] - A_s[10]$ (5 values).
- Bands[indices]: B_1: $\{6\}$, B_2: $\{0\}$, B_3: $\{1, 2, 4\}$, B_4: $\{5\}$, B_5: $\{3, 9, 7, 8, 10\}$.

After clustering the $N + 1$ instances per observable group T_i into K clusters, we have reduced the storage requirements from 6*($N + 1$) to 6*K floating points for the detection system. Interested readers may refer to [80] for further details.

5.5.3.1 Training Goals for TCP

In essence, the objectives of the training phase for TCP traffic are as follows:

- To group the flag instances into bands for each observable.
- To learn events from input traffic based on the above set of events and determine probabilities for each event.
- To determine an *appropriate* threshold probability, a probability below which a window will be classified as abnormal.

5.5.3.2 Setting Threshold Probabilities

The important consideration for setting threshold probabilities is the assumption that abnormal traffic may be present, although in small quantities, within the training traffic. Based on this consideration, a technique called cross-validation is used, along with sieving of the lower t percent of probabilities during threshold determination. The input parameter t is the sensitivity factor or error proportion which denotes the proportion of abnormal traffic which may be present in the training traffic.

5.5.4 *Modelling UDP Traffic*

The objective of monitoring protocol headers is to evolve a distinguishing pattern in them, which will subsequently help detecting DoS attacks against the victim. A characteristic feature of UDP is that it is connection-less, and hence ensures speedy communication. However, unlike TCP, the smaller UDP header does not contain fields like flags, which will determine the state of the communication. Since UDP is connection-less, most DoS attacks performed using UDP are only bandwidth-based attacks, essentially trying to exhaust bandwidth resources available to the server. Consequently, UDP header information cannot be used as a critical parameter for detecting DoS attacks using UDP.

Hence, the parameter considered here is the window arrival time (WAT) of a packet window. WAT of a packet window is the duration in which a packet window has arrived. Technically, it is the difference in time between packets P_1 and P_N of a window, where N is the size of the window. Due to various constraints, only non-overlapping windows are considered. During the training phase, the WATs of windows are monitored, and a model evolved to accommodate these WAT events. During the deployment phase, the model probabilities are used to determine the probability of an incoming window. If the probability is less than a threshold probability (determined from of the learning input itself), then the window is classified as abnormal.

As is the case with TCP, the events are so sparse in nature that there appears to be a lot of scope to group these events in order to reduce the number of probabilities handled by the model. This requires that WATs be further grouped into a constant number of bands defined by time bounds. Hence, each band is a collection of a contiguous range of WATs. The probability of a band will be common to all WATs falling under the band. During deployment, the WAT of each window is computed, and the probability of the window is the probability of the band inside which the WAT falls. If this probability were to be smaller than a threshold probability, then the window is considered abnormal. The threshold probability is determined by adopting cross-validation methods similar to that done in TCP.

5.5.4.1 Training Goals for UDP

The primary goals of the training phase with respect to UDP are as follows:

1. To compute per UDP stream, the upper and lower intervals of each band of intervals of WATs, given that a constant number of bands(K) exist. For example, if the window size N = 100, then the different bands could be $B_1 = (0 - 10)$ s,[1] $B_2 = (11 - 235)$ s, $B_3 = (236 - 499)$ s and so on. It should be noted that the band intervals span the entire time.
2. To compute per stream, the probabilities of each of these bands (e.g. probability of WAT in B_2, etc).
3. To compute per stream, the dynamic threshold probability associated with the stream. This threshold probability is the probability below which the window will be considered abnormal.

The definition of a band is fundamentally different for UDP due to the way UDP traffic is modelled. The variables *upperbound* and *lowerbound* are the time bounds for the band of WATs, and the variable probability denotes the probability of WAT falling into the time band.

[1]Unit of time is assumed to be seconds.

The problems in the case of UDP are very similar to that of TCP, and similar mechanisms are adopted.

5.5.4.2 UDP Bands

Given an array of numbers WATs with size of array equivalent to the total number of windows during learning, $A = \{a_0, a_1, a_2, a_3, \ldots a_{TW}\}$, where TW = The total number of windows seen during learning divide the array of WATs into B time bands.

The traffic is modelled in terms of the WAT, that is, the time taken for a (non-overlapping) window of packets to arrive. An approach similar to the one used in TCP is utilised. In this case, only contiguous time ranges are considered for grouping since the learning traffic may not have captured all time variations. This is different from TCP where there is only a finite set of events to model. However, the approach of making contiguous bands has a drawback – inaccuracy of estimation of probabilities is possible if events with like probabilities are distributed non-contiguously. This was the case in TCP where equally likely events were put into different bands and were estimated by different probabilities. However, it is assumed that in the case of UDP, consecutive WAT events (e.g. WAT = 500, WAT = 501, WAT = 502, etc.) are most likely to be equally probable, and hence this method of dividing along the X-axis can be adopted.

In summary, a simple clustering method is used to solve the problem, as follows:

1. Observe occurrences for each event and consider jumps between event occurrences in descending order. Arranging events in descending order of their occurrences brings like events closer, which will help grouping like events together. The jump between two observations of events in the array is denoted as $J_i = O_i - O_{i+1}$.
2. Consider top $B - 2$ jumps in J_i s and form B bands bordering them.

5.5.4.3 Band Expansion

The above algorithm describes the method of clustering and grouping UDP traffic based on WATs of incoming windows. However, it has to be noted that the exact boundaries of these bands may not be accurately known from the WATs observed during learning, since the learning traffic may not have captured all possible values of WATs. Hence, there should be a mechanism for meaningful expansion of bands in order to accommodate events which have not occurred before. Only after expansion can the entire time be spanned. A bad choice of expansion strategy can result in inaccurate probabilities.

A simple strategy to expand bands is followed: The difference between the upper bound of a band and the lower bound of its neighbouring band (the conflict zone) is equally shared between the consecutive bands. Hence, every dynamic band (bands except the first and the last band, which are already expanded to accommodate unseen events) can be expanded on both its bounds to accommodate the entire time.

5.5.4.4 Updating of Learning Probabilities

Updating of learning probabilities for UDP streams is straightforward, after time bands have been frozen. It involves re-counting the number of windows in the learning traffic falling under each band (according to the WAT of the window) and estimating the probability of each band as the number of windows seen in the band by the total number of windows seen during learning. It is to be seen that the window counters against each band have to be smoothed before estimating probabilities.

5.5.4.5 UDP Threshold Probability Determination

Given learning traffic statistics for a site, the objective here is to determine the threshold probability. The same strategy of 10-fold cross validation is employed to arrive at a threshold probability for UDP streams based on the learning traffic. Vijaysarathy et al. [80] describe in detail the cross-validation technique for TCP. The probabilities, in this case, would be the UDP window probabilities.

5.5.5 Experimental Results

Experiments were conducted primarily on the following datasets:

1. DARPA dataset for TCP – Consists of many streams at various levels of traffic. The stream with the most number of packets was chosen. The stream was a telnet stream with roughly 70,600 inbound packets.
2. SETS dataset for TCP – This represents the user dataset which contains traffic addressed to the SETS web server on port 80. The stream consists of roughly 240,900 inbound packets.
3. SETS UDP dataset for UDP – This represents the user dataset for UDP which contains traffic addressed to the SETS internal DNS server on port 53. The stream consists of roughly 60,400 packets.

Critical system parameters, namely, accuracy (Acc), false alarm rate (FAR), and miss rate (MR) were computed based on the number of true and false predictions. To understand the effect of window size on the problem, experiments were conducted with different window sizes. The results are tabulated in Tables 5.7 and 5.8.

Table 5.7 Experimental results for TCP

	DARPA at t =1%			SETS at t =5%		
WS	Acc	FAR	MR	Acc	FAR	MR
50	98.3%	2.3%	0	97%	7%	0
100	98.6%	2%	0	97.4%	6.2%	0.06%
200	98.7%	1.8%	0	97.1%	5%	1.1%

Table 5.8 Experimental results for UDP

	SETS at t =1%		
WS	Acc	FAR	MR
100	99.5%	0.4%	0.1%
200	99.2%	0.5%	0.6%

For the tagged DARPA dataset and at an error proportion of 1%, the accuracy (Acc) improves with increasing window size. There is a corresponding decrease in the false alarm rate (FAR) and the miss rate (MR) is zero, that is, no attacks were missed. For the untagged SETS dataset, it is safe to assume a slightly bigger error proportion of 5%. In this case, the false alarm rate falls with increasing window size but the miss rate increases marginally. Hence, although untagged datasets show a slightly different pattern in performance, it can be concluded that larger window sizes are bound to improve system performance.

5.6 DoS Detection Using CUSUM and Adaptive Neuro-Fuzzy Inference System

Our approach in this work has been to apply the CUSUM algorithm to track variations of the attack characteristic variable $X(n)$ from the observed traffic (specific to different kinds of attacks) and raise an alarm when the cumulative sum gets too large (based on a threshold value). But often this threshold mechanism produces many false alarms. A fuzzy inference system (FIS) can be employed instead of a threshold-based system as it removes the abrupt separation between normality and abnormality and thereby reduces the number of false alarms comparatively. The output of a FIS depends on the membership function chosen and its parameters. Hence, for a better output it is important to appropriately select the membership function parameters; ANFIS serves as a suitable technique for this purpose. In ANFIS, the membership function parameters of FIS are fine-tuned based on training data collected in a real-time environment.

5.6.1 Related Work

DoS/DDoS defence mechanisms can be broadly categorised based on the attack detection strategy into pattern detection and anomaly detection. Further, in anomaly detection we have two Normal Behaviour Specifications: Standard and Trained [4]. In pattern detection mechanism, the signatures of known attacks are stored in a database, and each communication is monitored for the presence of these patterns. The drawback here is that only known attacks can be detected, whereas attacks with slight variations of old attacks go unobserved. On the contrary, known attacks are easily and unfailingly detected, and no false positives come across. Snort [2] provides some examples of a DDoS detection system that uses pattern attack detection. An identical approach has been supportive in controlling computer viruses also. Similar to virus detection software, signature databases must be frequently updated to account for new attacks.

Methods that set up anomaly detection have a representation of normal system behaviour, such as normal traffic dynamics or anticipated system performance. The current state of the system is periodically compared with the models to detect anomalies. Mirkovic et al. [58] proposed a technique called D_WARD (DDoS Network Attack Recognition and Defence). It is a source end solution whose goal is to autonomously detect and stop outgoing attacks from the deploying network. It provides a dynamic response that is self-adjusting. By carefully choosing the criteria for adjustment, D_WARD is able to promptly react to network conditions while being resilient to an attacker's attempts. It is efficient in autonomous mode. However, the authors note that their approach sometimes results in wrong actions.

Mahajan et al. [53] proposed the aggregate congestion control (ACC) system, a mechanism which reacts as soon as attacks are detected, but does not give a mechanism to detect ongoing attacks. For both traffic monitoring and attack detection, it may suffice to focus on large flows. XenoServices [89] is an infrastructure for a distributed network of web hosts that responds to an attack on any website by replicating the website rapidly and widely among the XenoService servers, thereby allowing the attacked site to acquire more network connectivity to absorb a packet flood. Although such infrastructure can ensure QoS during DDoS attacks, it is doubtful that a large number of ISPs will adopt such an infrastructure quickly. Multops [33] proposes a heuristic and a data structure that network devices maintain a multi-level tree, monitoring certain traffic characteristics and storing data in nodes corresponding to subnet prefixes. The tree expands and contracts within a fixed budget. The attack is detected by abnormal packet ratio values and offending flows are rate-limited. The system is designed so that it can operate as either a source-end or victim-end DDoS defence system.

Methods presented in [33, 53, 59, 89] provide examples of anomaly detection approaches. The advantage of anomaly detection over pattern detection is that previously unknown attacks can also be discovered in anomaly detection.

Based on the specification of a normal behaviour, we can segregate anomaly detection mechanisms into standard and trained. Mechanisms that employ standard specifications of normal behaviour depend on some protocol standard or a set of rules. For example, the TCP protocol specification describes a three-way handshake that has to be performed for TCP connection establishment. An attack detection mechanism can make use of this specification to detect half-open TCP connections and discard them from the queue. Some protocol stack enhancement approaches like SYN cookies and SYN cache are proposed, as described in [48] and [14, 15] respectively, to mitigate attacks like the SYN flood which uses TCP protocol. The advantage of a standard-based specification is that it produces no false positives; all legitimate traffic must meet the terms of the specified behaviour. The disadvantage is that attackers can still perform complicated attacks which, on the surface, seem amenable to the standard and hence unnoticed.

Mechanisms that use trained specifications of normal behaviour observe network traffic and system behaviour and produce threshold values for different parameters. All connections exceeding one or more (depending on the approach) of these values are considered as anomalous. One such widely used threshold-based approach is the CUSUM algorithm. Wang et al. [81] proposed a detection mechanism very specifically for SYN flood attack; Peng et al. [67] proposed a detection mechanism by monitoring Source IP address; Zhou et al. [92] proposed a detection based on CUSUM and space similarity at each host in a P2P network; Leu et al. [49] proposed an intrusion prevention system named Cumulative Sum–based Intrusion Prevention System (CSIPS). Though all the above-mentioned work have their own novel ideas, one common drawback is the selection of threshold, which is important since selection of low thresholds lead to a lot of false positives, whereas high thresholds reduce the sensitivity of the detection mechanism. For this reason, in our proposed approach, more intelligent ANFIS engines are used.

5.6.2 Detection Approach

The detection approach described here is likewise to be situated and to operate within the DMM architecture described in Sect. 5.4.2, as is the case with the CPA NSP approach presented in Sect. 5.4.

The proposed model involves the CUSUM technique and ANFIS. The CUSUM measure for each attack is found and passed on to their respective ANFIS. The classification of the analysed traffic as an attack or normal is decided based on the output of ANFIS. Figure 5.9 gives the block diagram of the proposed model. As shown in Fig. 5.9, the real-time traffic data is collected and the characteristics needed for modelling the attack using CUSUM is extracted (pre-processing) which is then passed on to the corresponding ANFIS engine.

The characteristics differ for different types of attacks which is explained elaborately in this section. Considering the fact that it may not be possible to model the traffic in dynamic and complex systems like the Internet using simple parametric description, we chose the non-parametric version of the CUSUM algorithm. The CUSUM algorithm dynamically checks if the observed time series is statistically homogenous and, if not, it finds the point at which the change happens and responds accordingly. Consider $X(n)$, which denotes the value of the attack characteristic variable during the nth time interval and its corresponding weights $W(n)$, which is a derived value from $X(n)$.

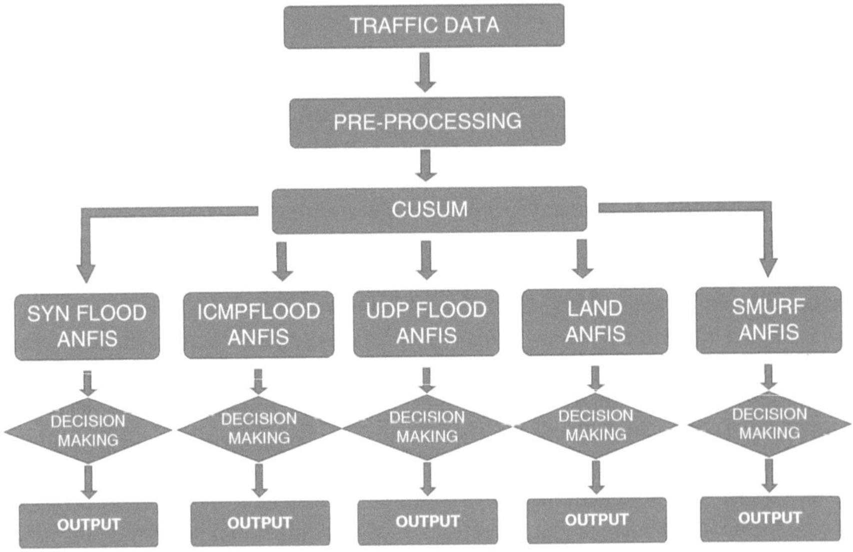

Fig. 5.9 Proposed model

We have modelled $X(n)$ for five different attacks where $XSYN(n)$, $XLAND(n)$, $XSMURF(n)$, $XUDP(n)$ and $XICMP(n)$ represent SYN Flood, Land, Smurf, UDP Flood and ICMP Flood attacks, respectively, and are collectively termed as attack characteristic variables. These variables are premeditated in such a way that it shows a sudden increase in value only during an attack, thus allowing the CUSUM algorithm to detect sudden variations in traffic more accurately and restricting false alarms to a large extent. Some of the existing DDoS detection techniques using CUSUM keep record of only the packet counts observed in a sample interval [49]. But in the proposed technique, CUSUM modelling is done in such a way that it reflects exactly the attack behaviour. The variable $X(n)$ for a specific attack category keeps track of a unique quantity corresponding to the same. Its value will be large when there is a high attack.

We model the above five different attacks as follows:

- SYN flood attack where we are taking into consideration the counts of RST, SYN and SYN/ACK packets:

$$X_{SYN}(n) = (N_{RST} + N_{SYN}) - N_{SYN/ACK} \tag{5.12}$$

- Land attack where $N_{[(SRC\_IP=DST\_IP)\&(SYNset)]}$ represents the number of incoming packets having the same source IP address and destination IP address with its SYN FLAG set:

$$X_{LAND}(n) = N_{[(SRC\_IP=DST\_IP)\&(SYNset)]} \tag{5.13}$$

- Smurf attack where $N_{(DEST\_ADDR=BADDR)}$ denotes the number of ICMP requests made to the broadcast address, exploits the vulnerability in the ICMP protocol:

$$X_{SMURF}(n) = N_{(DEST\_ADDR=BADDR)} \quad (5.14)$$

- UDP flooding attack where $N_{(DEST\_ADDR=HOST\_IP)}$ denotes the number of incoming UDP packets, $N_{(SRC\_ADDR=HOST\_IP)}$ denotes the number of outgoing UDP Packets and N_{ICMP_error} denotes the number of ICMP Destination Port Unreachable Error packets:

$$X_{UDP}(n) = (N_{(DEST\_ADDR=HOST\_IP)} - N_{(SRC\_ADDR=HOST\_IP)}) + N_{ICMP\_error} \quad (5.15)$$

- ICMP flood attack:

$$X_{ICMP}(n) = \text{Total payload size of the ICMP request packets} \quad (5.16)$$

(The above characteristic variable keeps a record of the summation of the sizes of incoming ICMP request packets since the time of observation.)

After modelling the attack characteristic variables, the values that are collected from the real-time traffic are passed on to their corresponding trained ANFIS engines which is explained in the next section.

5.6.3 ANFIS Engines

Fuzzy logic and neural networks help to embark upon issues such as vagueness and unknown variations in parameters more efficiently, and hence improve the robustness of the overall defence mechanism. Neuro-fuzzy techniques have been developed by a blend of the Artificial Neural Network (ANN) and the Fuzzy Inference System (FIS) and is termed 'ANFIS'. Using ANFIS with the CUSUM algorithm provides twofold advantage. First 3 it helps in removing the crisp threshold-based alarm-raising mechanism of CUSUM by a more comprehensible fuzzy logic–based mechanism, and second it fine-tunes the membership function parameters involved in FIS using the neural network–based learning technique. An adaptive network (Fig. 5.10) is a five-layer feed-forward network in which each node performs a particular function (*node function*) on incoming signals as well as having a set of fuzzy membership parameters pertaining to this node [36]. Following are the fuzzy rules base of Mamdani type which we have modelled for each type of attacks discussed above.

RULE 1: If (X(n) is HIGH) then attack is HIGH
RULE 2: If (X(n) is MEDIUM) then attack is MEDIUM
RULE 3: If (X(n) is LOW) then attack is LOW

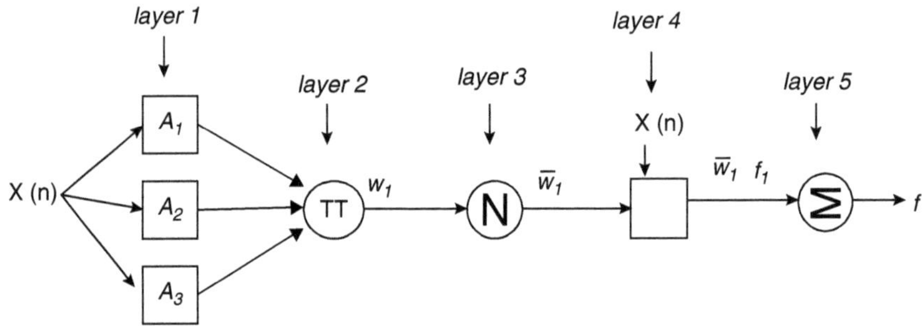

Fig. 5.10 Feed-forward networks

where $X(n)$ can either be $X_{SYN}(n)$, or $X_{LAND}(n)$, or $X_{SMURF}(n)$, or $X_{UDP}(n)$, or $X_{ICMP}(n)$ for SYN Flood, Land, Smurf, UDP Flood and ICMP Flood attacks, respectively.

Each node i in layer 1 is a square node with node function $O_i^1 = \mu_{A_i}(X(n))$, where A_i is the linguistic label (LOW, MEDIUM, HIGH) associated with this node function. In other words, O_i^1 is the membership function of A_i, and it specifies the degree to which the given *X(n)* satisfies the quantifier A_i . In our case all the membership functions are Gaussian,

$$O_i^1 = \mu_{A_i}(X(n)) = \exp\left(-\frac{(c_i - X(n))^2}{2\sigma_i^2}\right) \tag{5.17}$$

where $\{c_i, \sigma_i\}$ is the parameter set. As the values of these parameters change, the Gaussian function varies accordingly, thus exhibiting various forms of membership functions on the linguistic label A_i. Every node in layer 2 is a circle node labelled II which multiplies the incoming signal and sends the product out, but since in our model there is only one incoming signal, we get $w_i = \mu_{A_i}(X(n)), i = 1$. Each node output represents the firing strength of a rule. In layer 3, every node is a circle node labelled N. The ith node calculates the ratio of the ith rules̆ firing strength to the sum of all rules̆ firing strength:

$$\bar{w_i} = \frac{w_i}{w_1} \tag{5.18}$$

The output of this layer is called normalised firing strengths. Every node i in layer 2 is a square node with a node function

$$O_i^4 = \bar{w_i} f_i = \bar{w_i}(p_i X(n) + q_i) \tag{5.19}$$

where $\bar{w_i}$ is the output of layer 3, and $\{p_i, q_i\}$ is the parameter set known as consequent parameters. Variable f_i is not explained here. The single node in layer 5 is a circle node labelled $\sum$ that computes the overall output as the summation of all incoming signals, that is,

$$O_i^5 = overall\_output = \sum \bar{w_i} f_i = \frac{\sum_i w_i f_i}{\sum_i w_i} \tag{5.20}$$

All the membership functions used are Gaussian functions whose parameters are trained using the above-explained network with a hybrid learning algorithm. More specifically, in the forward pass of the hybrid learning algorithm, functional signal goes forward until layer 4 and the consequent parameters are identified by the least square estimate. In the backward pass, the error rates propagate backward and the premise parameters are updated by the gradient descent. Thus, it enables the FIS to adjust its parameters from the given training data. Also, it affords best trade-off between fuzzy and neural network by providing both smoothness and adaptability. In the proposed model, a separate adaptive network, as explained above, is formed for each of the five different attacks. Since the proposed system is intended to provide real-time detection, the training phase of ANFIS is carried out in offline mode over and over again to come out with a perfect training model, and during the deployment of this system the testing is done. Thus, the five ANFIS engines would not turn out to be a heavy computation because it is only the FIS that will be evaluated during the detection phase and the number of ANFIS engines can be further increased if required.

5.6.4 Decision-Making

After the evaluation of all the five ANFIS engines with CUSUM measures, the output (defuzzified value) of each ANFIS engine is collected and a decision is made based on those defuzzified values. Depending on the intensity of the attack, the final decision is given as LOW, MEDIUM or HIGH, which denotes the risk level of the network being monitored. The intensity of each attack can be known from the output of its corresponding ANFIS engine. The risk level 'LOW' means that the observed traffic is a NORMAL traffic. The risk level 'MEDIUM' warns the administrator of a possible attack which might occur. However, such an attack might not occur sometimes. Also, since the attack level is MEDIUM, the system may not be affected much. In this case, the administrator may not be sure if he can take any action because he would not know if it is going to be a serious attack. Therefore, the administrator can wait for some time and can take necessary action after he receives the next alarm. Thus, in these situations, it is solely the decision of the administrator to take necessary countermeasures. The risk level 'HIGH', alarms the administrator to immediately take necessary actions.

5.6.5 Experimental Results

All the experiments were carried out in the eight node SSE (Smart and Secure Environment) network laboratories distributed across different geographic locations and connected through an MPLS VPN cloud with each lab consisting of around nine workstations and one server. A detailed description of the testbed is provided in Chap. 4. During a period of 10 days, 1,000 sample sets S_i, $1 \leq i \leq 1000$ with each set consisting of 16 different attack characteristic variables (i.e. features) $s_{ij}, 1 \leq i \leq 1000, 1 \leq j \leq 16$ are collected from a network traffic consisting of a mixture of attack and normal traffic. Each attack characteristic variable in a sample set is a count of the number of packets collected in a 5 s interval. These features are related to a particular characteristic of traffic packets such as packets with SYN flag set, RST flag set, SYN/ACK flags set, UDP packets, ICMP packets and Smurf packets.

From the collected sample sets, the attack characteristic variable of each attack is then used for training its corresponding ANFIS engine. The training process is done many times with different sample sets (S_i) in order to fine-tune the corresponding ANFIS engine, as mentioned in Sect. 5.6.3. The training process is carried out in offline mode, thus not adding any extra computation cost during the deployment in real-time traffic. Since the SSE network is a closed network, normally there will not be any attacks, and the training datasets are assumed to be clean.

Once the ANFIS is trained, many trials of testing were done in real-time traffic against the trained model, and comparisons are made with one of the well-known open source network Intrusion Detection System, (IDS) called Snort. For testing, 1,100 sample sets are collected comprising 200 sample sets each of SYN flood, LAND, SMURF, ICMP flood and UDP flood attack traffic along with 100 samples of normal traffic. The detection delay, DP, which is the time difference between the attack sample given as an input to the proposed scheme and its detection for each attack sample, is observed. Similarly DS, the detection delay using Snort for the same samples, is also observed. After observing the detection delays whenever the attack is detected, the maximum detection delay in each category is considered for DS as well as DP.

Table 5.9 compares the detection time of CANFIS and Snort by testing them using online samples. The specification of each sample regarding its average incoming packets per second and the type of attack traffic injected along is tabulated. A sampling interval of 0.5 s is fixed for analysing the samples.

As shown in Table 5.9, the detection time of the proposed technique is less than that of Snort. Table 5.10 compares the detection time of CANFIS and Snort by testing them offline. For this purpose, a well-known DDoS attack dataset is used [5]. The specification of each sample regarding its average packets per second, the attack duration and the type of attacks involved are tabulated above. The datasets are of varying attack levels – most of the datasets correspond to high-level DoS attacks. In most of the above-mentioned cases, the detection time of the proposed technique is less than that of the Snort. Thus, it is reasonable to infer from these facts that the proposed technique is faster when compared to Snort. In Table 5.11, the accuracy of the proposed technique is shown. The values of various evaluation parameters for each attack, namely True Positive (TP), True Negative (TN), False Positive (FP), False Negative (FN), True Positive Rate (TPR $=$ TP/(TP $+$ FN)), False Positive Rate (FPR $=$ FP/(FP $+$ TN)) and the Accuracy (ACC $=$ (TP $+$ TN)/(FP $+$ TP $+$ FN $+$ TN)) are shown.

Table 5.9 Comparison of detection time of CANFIS and Snort when tested online

S. no.	Sample name	Average packets/s	Attacks injected	Detection time CANFIS (s)	Detection time SNORT (s)
1	Sample_17	1,955	SYN flood, ICMP flood, Smurf	2.1	3.1
2	Sample_238	1,404	ICMP flood, SYN flood, Land, UDP flood	1.8	2.8
3	Sample_394	1,897	Land, Smurf	2.6	Not detected
4	Sample_431	1,540	UDP flood, ICMP flood, Land	1.8	3.1
5	Sample_479	1,478	SYN flood, UDP flood, Smurf	2.1	3.1
6	Sample_563	1,562	ICMP flood, SYN flood, UDP flood	2.5	2.8
7	Sample_704	1,457	Smurf, Land, ICMP flood	2.7	3.4
8	Sample_859	1,784	Smurf, Land, SYN flood	2.3	2.6
9	Sample_947	1,609	Smurf, SYN flood, Land, ICMP flood, UDP flood	2.8	3.8

Table 5.10 Comparison of detection time of CANFIS and Snort – using CAIDA dataset (offline)

S. no.	Sample name	Average packets/s	Attack duration (s)	Attacks injected	Detection time CANFIS (s)	Detection time SNORT (s)
1	ddos.2007 0804_134936	555	300	ICMP flood	4.6	5.1
2	ddostrace.2007 0804_135436	362	300	ICMP flood	5.1	5.5
3	ddostrace.2007 0804_143936	15,2619	300	ICMP flood, SYN flood	4.2	5.6
4	ddostrace.2007 0804_142936	153,180	300	ICMP flood, SYN flood	4.5	5.1
5	ddostrace.2007 0804_143436	147,740	300	ICMP flood, SYN flood, UDP flood	4.2	4.4
6	ddostrace.2007 0804_144436	165,226	300	ICMP flood, SYN flood	4.8	4.8
7	ddostrace.2007 0804_145436	172,566	55	ICMP flood, SYN flood	4.4	4.5
8	ddos.2007 0804_142436	164,824	300	ICMP flood, SYN flood, UDP flood	5.5	5.6

Table 5.11 Accuracy of CANFIS and Snort

Attack		FP	FN	TP	TN	TPR (%)	FPR (%)	Accuracy (%)
SYN	CANFIS	2	5	98	95	95.2	2.1	96.5
flood	SNORT	9	20	81	90	80.2	9.1	85.5
Land	CANFIS	1	3	97	99	97.0	1.0	98.0
	SNORT	10	25	75	90	75.0	10.0	82.5
Smurf	CANFIS	3	6	92	99	93.9	2.9	95.5
	SNORT	8	26	82	84	75.9	8.7	83.0
ICMP	CANFIS	4	1	98	97	99.0	3.9	97.5
flood	SNORT	11	17	92	80	84.4	12.1	86.0
UDP	CANFIS	2	1	99	98	99.0	2.0	98.5
flood	SNORT	7	23	80	90	77.7	7.2	85.0

For calculating the various evaluation parameters as listed in Table 5.11, a test sample containing 100 sample sets of attack traffic of each type and 100 sample sets of normal traffic is created. For example, for a SYN flood attack, 100 sample sets each containing traffic related to SYN flood and 100 sample sets of normal traffic are taken, and the various evaluation parameters were calculated. For all the attacks, the accuracy of CANFIS is quite better than Snort, and the false alarms also are also comparatively less when compared to it. It can be inferred from Tables 5.9 to 5.11 that our proposed scheme is computationally faster and has less false alarms when compared to Snort IDS.

5.7 Conclusion and Further Work

As has been discussed elsewhere, high-rate flooding attacks remain a potent threat to the delivery of Internet-based services. Moreover, the early and reliable detection of the onset of such an attack remains a challenging problem as does the formulation and implementation of an effective mitigation strategy. The preceding discussion has covered both the detection and mitigation elements of the solution strategy and includes leading-edge work on developing detection techniques that can reliably identify a high-rate flooding attack in real time or, at least, in near real time.

The initial discussion was on the extraction and ranking of a feature set that characterises Internet traffic in a way that distinguishes normal traffic from anomalous traffic. These features were then used to evaluate the performance of the selected set of machine-learning algorithms in detecting DDoS attacks. An important result from this work is that, of the various methods used, Fuzzy c-means clustering is a useful and very efficient way to detect a DDoS attack. The next technique discussed was based on Change Point Analysis. The specific focus was on exploiting a characteristic of high-rate flooding attacks, viz., a marked increase in the number of new users (i.e. IP addresses) attempting to interact with the target site. The discussion showed how the Change Point Analysis technique could be used in combination with other techniques such as bit vectors and bloom filters to monitor the rate of arrival of new IP addresses and then use this as a basis to detect a high-rate flooding event. This was followed by discussion on an approach that exploited characteristic features of the TCP and UDP protocols which potentially distinguish a high-rate flooding attack from normal traffic. This approach also established that Naive Bayesian techniques could be used to detect a high-rate flooding attack in near real time. The final approach is to apply the CUSUM algorithm to track variations in the characteristic variable from the observed traffic that is specific to different kinds of attacks. Instead of using a threshold system to signify an attack, the discussion shows how a FIS can be employed which removes the abrupt separation between normality and abnormality and thereby reduces the number of false alarms.

The other key element of the discussion was on mitigation techniques. Central to this was the formulation of an architecture for a DDoS Mitigation Module (DMM). In particular, the discussion showed how the creation and maintenance of a white list could, in principle, meet one of the design goals for an effective DMM. Having established a proof of concept for the key elements of a DMM architecture, the next phase is to investigate the performance of both bit vector and bloom filter classification techniques under high-speed network flooding attacks in both IPv4 and IPv6 networks. Moreover, the analysis of detection algorithms under different and diverse flooding attacks needs to be investigated. We also expect to extend the basis for rejection of packets beyond the simple property of IPs Not Seen Previously. Further work will attempt to identify malicious traffic using other features to be used in tandem with NSP – features like subnet source addresses, IP distribution, traffic volume, traffic volume per IP address, packet inter-arrival time, etc. In addition, another line of investigation is the realisation of key detection algorithms in hardware so that they are genuinely capable of detecting the onset of high-rate flooding attacks in realtime.

References

1. KNIME. 2011. http://www.knime.org. Accessed 7 Feb 2011.
2. Snort: The open source network intrusion detection systems. http://www.snort.org/. Accessed 31 Aug 2011.
3. Waikato Applied Network Dynamic Research Group. http://wand.cs.waikato.ac.nz/. Accessed 1st Oct 2010.
4. DARPA Intrusion Detection Datasets, 1991. http://www.ll.mit.edu/mission/communications/ist/corpora/ideval/data/. Accessed 31 Aug 2011.
5. UCSD Network Telescope – Code-Red Worms Dataset, 2001. The Cooperative Association for Internet Data Analysis http://www.caida.org/data/passive/codered_worms_dataset.xml. Accessed 7 Feb 2009.
6. Ahmed, E., A. Clark, and G. Mohay. 2008. A novel sliding window based change detection algorithm for asymmetric traffic. In *Proceedings of the IFIP International Conference on Network and Parallel Computing*, 168–175, Oct 2008.
7. Ahmed, E., A. Clark, and G. Mohay. 2009. Effective change detection in large repositories of unsolicited traffic. In *Proceedings of the Fourth International Conference on Internet Monitoring and Protection*, May 2009.

8. Ahmed, E., G. Mohay, A. Tickle, and S. Bhatia. 2010. Use of IP addresses for high rate flooding attack detection. In *Security and Privacy Silver Linings in the Cloud*, vol. 330, 124–135. Boston: Springer.
9. Almotairi, S., A. Clark, G. Mohay, and J. Zimmermann. 2008. Characterization of attackers' activities in honeypot traffic using principal component analysis. In *Proceedings of the IFIP International Conference on Network and Parallel Computing*, 147–154, Washington, DC, 2008. IEEE Computer Society.
10. Almotairi, S., A. Clark, G. Mohay, and J. Zimmermann. 2009. A technique for detecting new attacks in low-interaction honeypot traffic. In *Proceedings of the Fourth International Conference on Internet Monitoring and Protection*, 7–13, Washington, DC, 2009. IEEE Computer Society.
11. Argyraki, K. and D.R. Cheriton. 2005. Active internet traffic filtering: Real-time response to denial-of-service attacks. In *Proceedings of the Annual Conference on USENIX Annual Technical Conference*, ATEC '05, 10–10, Berkeley, 2005 . USENIX Association.
12. Argyraki, K. and D.R. Cheriton. 2009. Scalable network-layer defense against internet bandwidth-flooding attacks. *IEEE/ACM Transactions on Networking* 17: 1284–1297.
13. Baldi, M., E. Baralis, and F. Risso. 2004. Data mining techniques for effective flow-based analysis of multi-gigabit network traffic. In *Proceedings of IEEE 12th International Conference on Software, Telecommunications and Computer Networks*, 330–334, Split, Croatia, 2004.
14. Baldi, M., E. Baralis, and F. Risso. 2005. Data mining techniques for effective and scalable traffic analysis. In *Proceedings of the Ninth IFIP/IEEE International Symposium on Integrated Network Management*, 105–118, Nice, France, 2005.
15. Barford, P. and D. Plonka. 2001. Characteristics of network traffic flow anomalies. In *Proceedings of ACM SIGCOMM Internet Measurement Workshop*, 2001.
16. Bloom, B. 1970. Space/time trade-offs in hash coding with allowable errors. *Communications of the ACM* 13: 422–426.
17. Bocan, V. 2004. Developments in DoS research and mitigating technologiess. *Transactions on AUTOMATIC CONTROL and COMPUTER SCIENCE* 49(63): 1–6.
18. Bos, H. and K. Huang. 2005. Towards software-based signature detection for intrusion prevention on the network card. In *Proceedings of Eighth International Symposium on Recent Advances in Intrusion Detection*, Seattle, WA, 2005.
19. Bruijn, W.D., A. Slowinska, K. Reeuwijk, T. Hruby, L. Xu, and H. Bos. 2006. Safecard: A gigabit IPS on the network card. In *Proceedings of Ninth International Symposium on Recent Advances in Intrusion Detection*, Hamburg, 2006.
20. Carl, G., G. Kesidis, R.R. Brooks, and S. Rai. 2006. Denial-of-service attack - detection techniques. *IEEE Internet Computing* 10(1): 82–89.
21. Cheng, J., J. Yin, Y. Liu, Z. Cai, and M. Li. 2009. DDoS attack detection algorithm using IP address features. In *Frontiers in Algorithmics*, eds. X. Deng, J. Hopcroft, and J. Xue, vol. 5598, *Lecture notes in computer science*, 207–215. Berlin: Springer.
22. Clark, D.D. 1995. The design philosophy of the darpa internet protocols. *SIGCOMM Computer Communication Review* 25: 102–111.
23. Deri, L. 2007. High-speed dynamic packet filtering. *Journal of Network and Systems Management* 15(3): 401–415.
24. Dietterich, T.G. 2000. Ensemble methods in machine learning. In *Proceedings of the First International Workshop on Multiple Classifier Systems, MCS '00*, London, 1–15. Springer-Verlag.
25. Erskin, E., A. Arnold, M. Prerau, and L. Portnoy. 2002. A geometric framework for unsupervised anomaly detection: Detecting intrusions in unlabeled data. In *Applications of Data Mining in Computer Security*, eds. D. Barbará and S. Jajodia, 77–102. Kluwer.
26. Fan, L., P. Cao, J. Almeida, and A.Z. Broder. 2000. Summary cache: A scalable wide-area web cache sharing protocol. *IEEE/ACM Transactions on Networking* 8: 281–293.
27. Farid, D.M., N. Harbi, and M.Z. Rahman. 2010. Combining naive bayes and decision tree for adaptive intrusion detection. *CoRR*, abs/1005.4496.

28. Feinstein, L., D. Schnackenberg, R. Balupari, and D. Kindred. 2003. Statistical approaches to ddos attack detection and response. In *Proceedings of the DARPA Information Survivability Conference and Exposition*, vol. 1, 303–314, 2003.
29. Ferguson, P. and D. Senie. 2000. Network ingress filtering: Defeating denial of service attacks which employ IP address spoofing, BCP 38, RFC 2827, May 2000.
30. Floyd, S. and V. Jacobson. 1993. Random early detection gateways for congestion avoidance. *IEEE/ACM Transactions on Networking* 1(4): 397–413.
31. Floyd, S. and V. Jacobson. 1995. Link-sharing and resource management models for packet networks. *IEEE/ACM Transactions on Networking* 3(4): 365–386.
32. Gavrilis, D. and E. Dermatas. 2005. Real-time detection of distributed denial-of-service attacks using rbf networks and statistical features. *Computer Networks* 48(2): 235 – 245.
33. Gil, T.M. and M. Poletto. 2001. Multops: A data-structure for bandwidth attack detection. In *Proceedings of the Tenth Conference on USENIX Security Symposium*, 3–3. USENIX Association.
34. Hettich, S. and S. D. Bay. 1999. The UCI KDD archive [http://kdd.ics.uci.edu]. University of California, Department of Information and Computer Science.
35. Hruby, T., K.V. Reeuwijk, and H. Bos. 2007. Ruler: high-speed packet matching and rewriting on npus. In *Proceedings of the Third ACM/IEEE Symposium on Architecture for Networking and Communications Systems, ANCS '07*, 1–10, New York, 2007. ACM.
36. Jang, J.S.R. 1993. ANFIS: adaptive-network-based fuzzy inference system. *IEEE Transactions on Systems, Man and Cybernetics* 23(3): 665–685.
37. Jin, S. and D. Yeung. 2004a. A covariance analysis model for DDOS attack detection. In *Proceedings of IEEE International Conference on Communications*, vol. 4, 1882–1886,20–24 June 2004.
38. Jin, S.Y. and D.S. Yeung. 2004b. DDoS detection based on feature space modeling. In *Proceedings of 2004 International Conference on Machine Learning and Cybernetics*, vol. 7, 4210–4215, 2004.
39. Jung, J., B. Krishnamurthy, and M. Rabinovich. 2002. Flash crowds and denial of service attacks: Characterization and implications for CDNs and web sites. In *Proceeding of 11th World Wide Web Conference*, 252–262, Honolulu, 2002.
40. Kang, J., Y. Zhang, and J.B. Jus. 2006. Detecting DDoS attacks based on multi-stream fused HMM in source-end network. In *Cryptology and Network Security*, vol. 4301, *Lecture Notes in Computer Science*, eds. D. Pointcheval, Y. Mu, and K. Chen, 342–353. Berlin: Springer.
41. Khor, K.C., C.T. Ting, and S.P. Amnuaisuk. 2009. From feature selection to building of bayesian classifiers: A network intrusion detection perspective. *American Journal of Applied Sciences* 6(11): 1949–1960.
42. Kim, D. and J. Park. 2003. *Network-based intrusion detection with support vector machines*, *Lecture Notes in Computer Science*, vol. 2662, 747–756. Springer, Berlin.
43. Kim, W.J. and B.G. Lee. 1998. Fred – fair random early detection algorithm for tcp over atm networks. *Electronic Letters* 34(2): 152–153.
44. Kline, J., S. Nam, P. Barford, D. Plonka, and A. Ron. 2008. Traffic anomaly detection at fine time scales with bayes nets. In *Proceedings of the Third International Conference on Internet Monitoring and Protection*, 37–46, Washington, DC 2008. IEEE Computer Society.
45. Le, Q., M. Zhanikeev, and Y. Tanaka. 2007. Methods of distinguishing flash crowds from spoofed dos attacks. In *Proceedings of the Third EuroNGI Conference on Next Generation Internet Networks*, 167–173, 2007.
46. Lee, H. and K. Park. 2001. On the effectiveness of probabilistic packet marking for ip traceback under denial of service attack. In *Proceedings of the IEEE INFOCOM*, 338–347, 2001.
47. Lee, K., J. Kim, K.H. Kwon, Y. Han, and S. Kim. 2008. DDoS attack detection method using cluster analysis. *Expert Systems with Applications* 34(3): 1659–1665.
48. Lemon, J. 2002. Resisting syn flood dos attacks with a syn cache. In *Proceedings of the BSD Conference, BSDC'02*, 10–10, Berkeley, 2002. USENIX Association.
49. Leu, F.Y. and Z.Y. Li. 2009. Detecting dos and ddos attacks by using an intrusion detection and remote prevention system. In *Proceedings of the Fifth International Conference on Information Assurance and Security*, vol. 2, 251–254.

50. Li, J., J. Mirkovic, M. Wang, P. Reiher, and L. Zhang. 2002. Save: Source address validity enforcement protocol. In *Proceedings of the IEEE INFOCOM*, 1557–1566, 2002.
51. Lin, D. and R. Morris. 1997. Dynamics of random early detection. *SIGCOMM Computer Communication Review* 27(4): 127–137
52. Liu, X., X. Yang, and Y. Lu. 2008. To filter or to authorize: Network-layer DoS defense against multimillion-node botnets. *SIGCOMM Computer Communication Review* 38(4): 195–206.
53. Mahajan, R., S.M. Bellovin, S. Floyd, J. Ioannidis, V. Paxson, and S. Shenker. 2002. Controlling high bandwidth aggregates in the network. *ACM Computer Communication Review* 32: 62–73.
54. Mahoney, M. and P. Chan. 2002. Learning nonstationary models of normal network traffic for detecting novel attacks. In *Proceedings of the Eighth ACM SIGKDD International Conference on Knowledge Discovery and Data Mining, KDD '02*, 376–385, New York, 2002. ACM.
55. McPherson, D., C. Labovitz, M. Hollyman, J. Nazario, and G.R. Malan. 2008. Worldwide infrastructure security report. Technical report, Arbor Networks.
56. Miercom. 2008. Enterprise firewall: Lab test summary report. Technical report.
57. Mirkovic, J., G. Prier, and P.L. Reiher. 2002. Attacking DDoS at the source. In *Proceedings of the Tenth IEEE International Conference on Network Protocols, ICNP '02*, 312–321, Washington, DC, 2002. IEEE Computer Society.
58. Mirkovic, J. and P. Reiher. 2004. A taxonomy of DDoS attack and DDoS defense mechanisms. *SIGCOMM Computer Communication Review* 34:39–53.
59. Mirkovic, J. and P. Reiher. 2005. D_WARD: A source-end defense against flooding denial-of-service attacks. *IEEE Transactions on Dependable and Secure Computing* 2: 216–232.
60. Molsa, J. 2005. Mitigating denial of service attacks: a tutorial. *Journal of Computer Security* 13(6): 807–837.
61. Nazario, J. 2008. Political ddos: Estonia and beyond (invited talk). In *Proceedings of the Seventeenth USENIX Security Symposium*, San Josa, 2008.
62. Nguyen, H.V. and Y. Choi. 2009. Proactive detection of DDoS attacks utilizing K-NN classifier in an anti-DDos framework. *International Journal of Electrical and Electronics Engineering* 4(4): 247–252.
63. Papadopoulos, C., A.G. Tartakovsky, and A.S. Polunchenko. 2008. A hybrid approach to efficient detection of distributed denial-of-service attacks. Technical Report, June 2008.
64. Partow, A. 2008. General purpose hash function algorithms. http://www.partow.net/programming/hashfunctions/. Accessed 25 Feb 2011.
65. Paruchuri, V., A. Durresi, and S. Chellappan. 2008. TTL based packet marking for IP traceback. In *Proceedings of the IEEE Global Telecommunications Conference*, 2552–2556, Los Angels, 30 Nov–4 Dec 2008. IEEE.
66. Paxson, V., K. Asanovic, S. Dharmapurikar, J. Lockwood, R. Pang, R. Sommer, and N. Weaver. 2006. Rethinking hardware support for network analysis and intrusion prevention. In *Proceedings of the First USENIX Workshop on Hot Topics in Security*, 63–68.
67. Peng, T., C. Leckie, and K. Ramamohanarao. 2004. Proactively detecting distributed denial of service attacks using source IP address monitoring. In *Networking Technologies, Services, and Protocols; Performance of Computer and Communication Networks; Mobile and Wireless Communications: NETWORKING 2004*, 771–782, 2004.
68. Peng, T., C. Leckie, and K. Ramamohanarao. 2007. Information sharing for distributed intrusion detection systems. *Journal of Network and Computer Applications* 30(3): 877–899. 1231771.
69. Peng, T., C. Leckie, and K. Ramamohanarao. 2007. Survey of network-based defense mechanisms countering the DoS and DDoS problems. *ACM Computing Surveys* 39(1): 3. 1216373.
70. Peng, T., C. Leckie, and K. Ramamohanarao. 2008. System and process for detecting anomalous network traffic. United States Patent Application 20100138919. http://www.freepatentsonline.com/y2010/0138919.html. Accessed 31 Aug 2011.
71. Ripeanu, M. and A. Iamnitchi. 2001. Bloom filters – Short tutorial. Technical report, Dept. of Computer Science, University of Chicago.

72. Seo, J., C. Lee, T. Shon, K.H. Cho, and J. Moon. 2005. A new DDoS detection model using multiple SVMs and TRA. *Lecture notes in computer science*, vol. 3823, 976–985. Berlin: Springer.
73. Shanbhag, S. and T. Wolf. 2008. Evaluation of an online parallel anomaly detection system. In *Proceedings of the IEEE Global Telecommunications Conference*, 1–6, 2008.
74. Shanbhag, S. and T. Wolf. 2008. Massively parallel anomaly detection in online network measurement. In *Proceedings of Seventeenth International Conference on Computer Communications and Networks*, 1–6.
75. Shon, T., Y. Kim, C. Lee, and J. Moon. 2005. A machine learning framework for network anomaly detection using svm and ga. In *Proceedings of the Sixth Annual IEEE Information Assurance Workshop*, 176–183, 2005.
76. Simmons, K., J. Kinney, A. Owens, D.A. Kleier, K. Bloch, D. Argentar, A. Walsh, and G. Vaidyanathan. 2008. Practical outcomes of applying ensemble machine learning classifiers to high-throughput screening (hts) data analysis and screening. *Journal of Chemical Information and Modeling* 48(11): 2196–2206.
77. Sterne, D.F., K. Djahandari, B. Wilson, B. Babsonl, D. Schnackenberg, H. Holliday, and T. Reid. 2001. Autonomic response to distributed denial of service attacks. In *Proceedings of the Fourth International Symposium on Recent Advances in Intrusion Detection, RAID '00*, 134–149, London, 2001. Springer-Verlag.
78. Takada, H.H. and A. Anzaloni. 2006. Protecting servers against DDoS attacks with improved source IP address monitoring scheme. In *Proceedings of the Second Conference on Next Generation Internet Design and Engineering*, p. 6, 2006.
79. Tavallaee, M., E. Bagheri, W. Lu, and A.A. Ghorbani. 2009. A detailed analysis of the KDD CUP 99 data set. In *Proceedings of the Second IEEE International Conference on Computational Intelligence for Security and Defense Applications, CISDA'09*, 53–58, Piscataway, 2009. IEEE Press.
80. Vijayasarathy, R., B. Ravindran, and S.V. Raghavan. 2011. A systems approach to network modeling for DDoS detection using naive Bayesian classifier. In *Proceedings of the Third International Conference on Communication and Networks*, 2011.
81. Wang, H., D. Zhang, and K.G. Shin. 2002. Detecting SYN flooding attacks. In *Proceedings of the IEEE Infocom*, 1530–1539, 2002. IEEE.
82. Wang, W. and S. Gombault. 2008. Efficient detection of DDoS attacks with important attributes. In *Proceedings of the Third International Conference on Risks and Security of Internet and Systems*, 61–67, Oct 2008.
83. Wang, W., G.R. Guile, J.A. Shaqsi, A.A. Aulamie, R. Harrison, and W. Zhang. 2007. Machine learning ensemble methodology, 2007. http://www.uea.ac.uk/cmp/research/mma/kdd/projects/ensemble-methods/Machine+Learning+Ensemble+Methodology. Accessed 31 Aug 2011.
84. Weng, N. and T. Wolf. 2009. Analytic modeling of network processors for parallel workload mapping. *ACM Transactions in Embedded Computing Systems* 8(3): 1–29.
85. Xiang, Y. and W. Zhou. 2005. Mark-aided distributed filtering by using neural network for DDoS defense. In *Proceedings of the IEEE Global Telecommunications Conference*, vol. 3, 5.
86. Xie, Y. and S. Yu. 2006. A novel model for detecting application layer DDoS attacks. In *Proceedings of the First International Multi-Symposiums on Computer and Computational Sciences, IMSCCS '06*, 56–63, Washington, DC, 2006. IEEE Computer Society.
87. Xu, T., D. He, and Y. Luo. 2007. DDoS attack detection based on RLT features. In *Proceedings of the International Conference on Computational Intelligence and Security*, 697–701, China, 15–19 Dec 2007.
88. Xu, X., Y. Sun, and Z. Huang. 2007. Defending DDoS attacks using hidden Markov models and cooperative reinforcement learning. In *Intelligence and Security Informatics, Lecture notes in computer science*, vol. 4430, 196–207, 2007. Springer, Berlin.
89. Yan, J., S. Early, and R. Anderson. 2000. The xenoservice – A distributed defeat for distributed denial of service. In *Proceedings of the Information Survivability Workshop*, Oct 2000.
90. Yuan, J. and K. Mills. 2005. Monitoring the macroscopic effect of DDoS flooding attacks. *IEEE Transactions on Dependable and Secure Computing* 2: 324–335.

91. Zargar, G.R. and P. Kabiri. 2009. Identification of effective network features for probing attack detection. In *Proceedings of the First International Conference on Networked Digital Technologies*, 392–397, July 2009.
92. Zhou, Z., D. Xie, and W. Xiong. 2009. Novel distributed detection scheme against DDoS attack. *Journal of Networks* 4: 921–928.

Chapter 6
Cryptographic Approaches to Denial-of-Service Resistance

C. Boyd, J. Gonzalez-Nieto, L. Kuppusamy, H. Narasimhan, C. Pandu Rangan, J. Rangasamy, J. Smith, D. Stebila, and V. Varadarajan

6.1 Introduction

Authentication is a promising way to treat denial-of-service (DoS) threats against nonpublic services because it allows servers to restrict connections only to authorised users. However, there is a catch with this argument since authentication itself is typically a computationally intensive rocess that is necessarily exposed to unauthenticated entities. This means that the authentication protocol can become a source of denial-of-service vulnerability itself, thereby causing the same problem it is aimed at solving.

In order to overcome this dilemma, Meadows [43] suggested the principle of gradual authentication which allows servers to commit only limited resources *before* obtaining some level of assurance regarding a client's identity. Aura et al. [5] proposed the notion of client puzzles which can be used during the gradual authentication process to show that the client is an entity willing to commit computational resources in order to obtain a connection. The puzzle is generated by the server for a specific client and the client must return a solution before the server will continue with the authentication process.

C. Boyd (✉) • J. Gonzalez-Nieto • L. Kuppusamy • J. Rangasamy • J. Smith • D. Stebila
Information Security Institute, Queensland University of Technology, Brisbane, Australia
e-mail: c.boyd@qut.edu.au; j.gonzaleznieto@qut.edu.au; l.kuppusamy@qut.edu.au; j.rangasamy@qut.edu.au; j4.smith@qut.edu.au; stebila@qut.edu.au

C.P. Rangan
Department of Computer Science and Engineering, Indian Institute of Technology Madras, Chennai, India
e-mail: prangan@iitm.ac.in

H. Narasimhan • V. Varadarajan
Department of Computer Science and Engineering, College of Engineering Guindy, Anna University, Chennai, India
e-mail: nhari88@gmail.com; venk1989@gmail.com

S.V. Raghavan and E. Dawson (eds.), *An Investigation into the Detection and Mitigation of Denial of Service (DoS) Attacks: Critical Information Infrastructure Protection*, DOI 10.1007/978-81-322-0277-6_6, © Springer India Pvt. Ltd. 2011

This chapter explores cryptographic approaches to DoS attacks, focusing on the design and analysis of DoS-resistant authenticated key establishment protocols. Key establishment protocols are the natural choice of authentication protocol to consider for DoS mitigation. They enable two parties to authenticate each other and establish cryptographic key material using an adversarially controlled network. Key establishment is usually executed as a precursor to accessing online services. The cryptographic keys derived from the key establishment protocol are typically used to secure the communications channel by providing confidentiality and integrity services using, e.g. IPsec [33]. In a key establishment protocol, the party that initiates the key establishment protocol by transmitting the first message is the initiator. The party that receives and responds to the first message is the responder.

Sections 6.2 and 6.3 survey strategies and techniques for resisting denial-of-service attacks in key establishment protocols. The main strategies employed to strengthen authentication protocols against DoS attacks entail counterbalancing memory expenditure, counterbalancing computational expenditure, and gradually authenticating requests. A common way of counterbalancing memory expenditure is via the use of cookies, which allow servers to remain stateless during the execution of the protocol. Counterbalancing of computational expenditure is mainly achieved by using client puzzles, which have the effect of increasing the amount of computational resources needed to mount a DoS attack. Client puzzles are the principal key cryptographic tool used in strengthening protocols against DoS attack.

In order to obtain confidence in their security, protocols for authenticated key exchange are today expected to come with a formal proof of security. It is a natural requirement that resistance to denial-of-service should be part of a formal analysis. Section 6.4 is devoted to the formalisation and analysis of client puzzles and DoS resistance protocols using two complementary approaches, namely, provable security and game theory. Finally, Sect. 6.5 applies the formal security model to a range of existing authentication protocols that incorporate puzzle-based denial of service protection to establish/verify their security properties.

Parts of this chapter have already been published in the proceedings of Topics in Cryptology (CT-RSA) 2011 [60] and the proceedings of Information Security, Practice and Experience (ISPEC) 2010 [49].

6.2 Denial-of-Service Resistance Strategies

To successfully disrupt the execution of a protocol between an initiator and a responder will require an attacker to possess: (1) prerequisite skills or knowledge, relating to the protocol in use and its implementation; (2) computational, storage, and network resources (bandwidth) in order to generate the messages that will interfere with the protocol execution; and (3) one or more attack sources. These are indicated as the x, y, and z axes of the graph in Fig. 6.1. Where the three axes intersect (shown in Fig. 6.1a) represents the point of highest vulnerability for

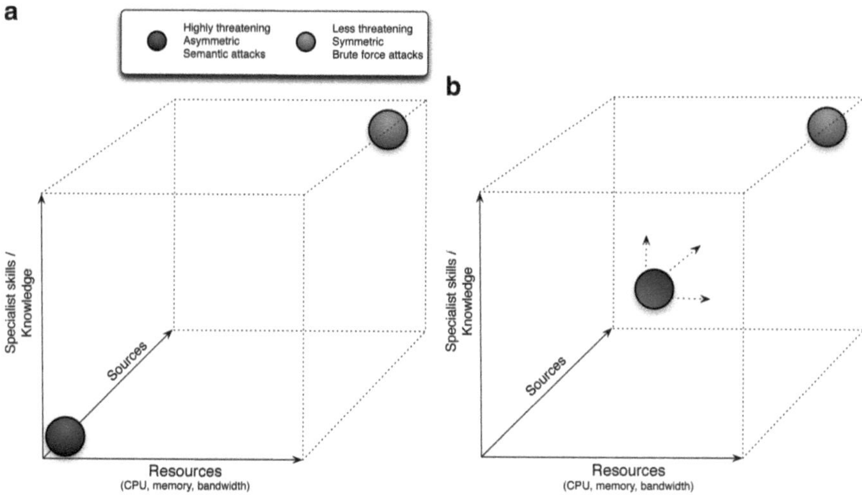

Fig. 6.1 (**a**) Denial-of-service threats versus resources. (**b**) Improving denial-of-service resistance by increasing the knowledge, resources, or attack sources needed

a protocol. Such a protocol could be successfully attacked by an adversary with limited knowledge, resources, and attack sources. The goal of protocol designers in increasing the denial-of-service resistance of protocols then is to implement strategies and techniques that increase the resources required by an attacker to successfully attack the protocol. Ideally increased resources should be required in all three dimensions (as shown in Fig. 6.1b). In practice, however, no assessment can be made of the adversary's knowledge of the system so it must be assumed that they have complete knowledge (this is consistent with Kerckhoffs' principle which is also known as Shannon's maxim). Therefore, the protocol designer must focus on dictating the level of resources or attack sources that must be required to attack the protocol.

The ability of a protocol to withstand attempts to exhaust responder resources via denial-of-service attack is termed the *denial-of-service resistance* of the protocol. The goal of denial-of-service resistance in key establishment protocols is to ensure that attackers cannot prevent a legitimate initiator and responder deriving cryptographic keys without expending resources beyond a responder determined threshold. In essence, this allows the responder to define the set of attackers with resources capable of interfering with the service they provide.

While either the initiator or responder may be the target of a denial-of-service attack, this work restricts itself to consideration of malicious initiators and the techniques used to defend responders against attack. The reason for focusing on responders is that responders typically handle requests for service from many initiators, so the impact of denial-of-service attacks against a responder is significant. The effect of a denial-of-service attack targeted at an initiator is far more localised, only impacting the specific target of the attack.

A responder in a key establishment protocol will have access to finite processing, storage, and network resources in order to complete its functions. Unless these resources are committed diligently, they may be exhausted by malicious initiators and the responder will have insufficient resources remaining to process legitimate incoming requests.

The requirement for key establishment protocols to exhibit denial-of-service resistance is well recognised by the protocol engineering community, and a number of design strategies have emerged that promote the judicious allocation of resources when processing initiator requests [37, 44]. The proposed strategies fall within the attack prevention, protocol security goal and the denial-of-service prevention, and resource accounting method of the Mirkovic and Reiher taxonomy [46]. The strategies can be broadly classified as approaches:

1. **Counterbalancing memory expenditure**: Ensuring that initiators must commit their memory resources to maintaining protocol state until the responder has some assurance that a denial-of-service attack is not underway reduces the vulnerability of the responder to state- or memory-based denial service attacks and increases the memory resources an attacker will need to attack the responder.
2. **Counterbalancing computational expenditure**: By counterbalancing computational expenditure at the responder, the protocol designer can ensure that the computational resources of an initiator will be exhausted before those of the responder. Achieving this goal may require artificially increasing the computational expenditure of the initiator to ensure the survivability of the responder [37], or having the initiator perform computations on behalf of the responder, thereby reducing the relative cost of computation to the responder.
3. **Gradual authentication**: While initiators must be authenticated at some point during the protocol execution, immediate and strong authentication of requests merely aggravates the denial-of-service problem. The suggested strategy for balancing the need for authentication and computational expenditure is to use weak and computationally cheap authentication when the protocol is initiated and gradually increase the strength of authentication as the protocol proceeds [44]. The strategy of gradual authentication can be used to detect attacks (e.g. based on IP spoofing for example), verify the computational and memory commitments of the initiator, and link messages from the same source (even though the exact identity of that source may be unknown).

   The protocols discussed in Sect. 6.5 provide examples of how gradual authentication can be implemented in key establishment protocols.

Combining the strategies of counterbalancing computational expenditure, counterbalancing memory expenditure, and gradually authenticating requests ensures that malicious initiators are unable to prevent the establishment of cryptographic keys between legitimate initiators and responders, unless they are prepared to expend significant resources of their own.

6.3 Denial-of-Service Resistance Techniques

Having identified the strategies employed to make responders in a key establishment protocol more resistant to denial-of-service attacks in Sect. 6.2, we now describe the specific techniques used. The techniques described may be considered primitives, some of which are capable of implementing more than one strategy and some of which can be combined to meet more complex goals such as gradual authentication. For each technique identified, we discuss its construction, the DoS resistance strategies it is capable of supporting, and how it might be combined with other techniques.

6.3.1 *Cookies*

When engaging in a protocol with a peer, a responder may be required to maintain information about the particular session currently being executed. In storing state, the memory resources available to a responder in a protocol can become a bottleneck, placing a limit on the maximum number of simultaneous connections the responder can accommodate. Such a limitation can be exploited by an attacker in a connection depletion attack, where all available memory resources are preemptively consumed to prevent legitimate connection requests completing. Aura and Nikander [4, 7] demonstrated, however, that instead of maintaining state in local memory, a protocol responder can append the required state information to the messages exchanged in a protocol as shown in Fig. 6.2.

Further, when these messages are exchanged between untrusting peers or using an untrusted network, the state contained in messages may be subject to eavesdropping, manipulation, or replay attacks.

Where the state information requires confidentiality protection, it can be encrypted by the responder prior to transmission. Integrity protection of state transferred between responder and initiator is achieved via the use of a cryptographic

a

	I		R
1.		$\xrightarrow{Msg_1}$	$Store(State_{R_{Msg_1}})$
2.		$\xleftarrow{Msg_2}$	
3.		$\xrightarrow{Msg_3}$	$Store(State_{R_{Msg_3}})$
4.		$\xleftarrow{Msg_4}$	
		...	

b

	I		R
1.		$\xrightarrow{Msg_1}$	
2.		$\xleftarrow{Msg_2,\, State_{R_{Msg_1}}}$	
3.		$\xrightarrow{Msg_3,\, State_{R_{Msg_1}}}$	
4.		$\xleftarrow{Msg_4,\, State_{R_{Msg_3}}}$	
		...	

Fig. 6.2 (**a**) Stateful protocol responder. (**b**) Stateless protocol equivalent

	I		R
1.		$\xrightarrow{Msg_1}$	
			$MAC = H_{K_R}(State_{R_{Msg_1}})$
2.		$\xleftarrow{Msg_2,\, MAC,\, State_{R_{Msg_1}}}$	
3.		$\xrightarrow{Msg_3,\, MAC,\, State_{R_{Msg_1}}}$	
			$MAC \stackrel{?}{=} H_{K_R}(State_{R_{Msg_1}})$
		...	

Fig. 6.3 Stateless protocol responder with integrity and replay protection

message authentication code (MAC). Replay protection can be achieved in many ways, but is commonly provided by including a timestamp with the state prior to generation of a MAC, or by periodically updating the cryptographic key used by the responder in generating MACs. A time variant responder cryptographic key and no requirement for confidentiality protection is assumed for the protocol shown in Fig. 6.3.

A common way to implement stateless connections in a protocol is via the use of cookies. Cookies are time variant, unpredictable data issued by the responder on receipt of a request for service that allow the responder to remain stateless and initiate gradual authentication of the initiator. First introduced in Photuris [30] and subsequently extended for resisting SYN flooding DoS attacks [38], cookies are now widely used (as will be seen in the protocols reviewed in Sect. 6.5).

Typically a cookie is constructed by taking some connection-specific parameters and transforming them with a time-variant local secret; a keyed hash [45, Section 9.5] of the initiator IP address and nonce for example. It is vitally important that the responder store no state when constructing cookies or that any state created is not a function parameterised by the number of initiator requests. In order to remain stateless and thereby prevent memory exhaustion, any relevant state required by the responder can also be encoded in the cookie and returned with the next message from the initiator.

On receipt of a valid cookie, the responder is able to reconstruct and validate any state encoded in the cookie and has weak assurance that it is in round-trip communication with the initiator. Round-trip communication implies that the initiator is not using a spoofed address. This assurance can only be considered weak, as an adversary with control of an intermediary link, between a claimed address and the responder, would be able to receive cookies for any address they wished to claim.

Unless cookies are carefully constructed, however, the responder may remain vulnerable to attack. Simpson [57] identified a state exhaustion attack, called a 'cookie crumb' attack, in the Internet Security Association and Key Management

Protocol (ISAKMP) implementation of cookies. In contrast to remaining stateless when constructing cookies, ISAKMP cookies required the storage of a small amount of state on each connection request. Even though the state information stored per request is very small (a 'crumb'), it is easy for an attacker to initiate a large number of requests, exhausting available memory resources.

In addition to ensuring that no state is stored on the construction of a cookie, Karn and Simpson [30] identified that the technique used for generating cookies must also satisfy the following three requirements:

- The cookie must depend on the participating entities.
- It must not be possible for anyone other than the issuing entity to generate a cookie that will be accepted by that entity.
- The cookie generation and verification methods must be computationally efficient.

The first requirement prevents an attacker from obtaining valid cookies, intended for other initiators, and using those cookies to generate a large number of requests with spoofed IP addresses. The second requirement secures the cookie generating process. The use of a secret value in generating the cookie prevents others from forging cookies and making this value time variant ensures that cookies must be used within a predetermined time frame, preventing the hoarding of valid cookies. Finally, the third requirement prevents denial-of-service attacks directed at the cookie mechanism itself.

6.3.2 *Proofs of Work*

Proofs of work, or puzzles, are hard but tractable problems that allow an initiator to prove to a responder that a verifiable level of computational effort has been expended. They permit the responder to gain some assurance of an initiator's willingness to commit resources to the protocol and provide a mechanism for counterbalancing computational expenditure in the event that the responder is exposed to a denial-of-service attack.

The concept was first proposed by Dwork and Naor [20] to control junk email by having recipients only accept emails if they were accompanied by a correct puzzle solution. It has since been extended to protect authentication protocols [5, 29] and permit clients to bid for limited service resources [64] using the difficulty of the puzzle as currency. Jakobsson and Juels [28] formalised the notion of reusable proofs of work, where the computational effort expended by the prover in generating the puzzle solution can be reused for some useful function, and described how a reusable proof of work could be used for minting electronic coins.

Puzzles serving as proofs of work can be constructed from a number of underlying problems which introduce a minimal and configurable overhead for

	I		*R*
1.		$\xrightarrow{\text{Puzzle?}}$	
			$X = h(t, M, s)$ $Y = h(X)$ $X' = X \;\&\; 0_0, 0_1, ..., 0_{k-1}1_k1_{k+1}1_{n-1}$
2.		$\xleftarrow{X', Y, t}$	
	Find p such that,		
3.	$h(p) = Y$	$\xrightarrow{p, t}$	
			Check that t is recent $X = h(t, M, s)$ $h(p) \stackrel{?}{=} h(X)$

Fig. 6.4 Juels and Brainard client puzzle protocol (simplified)

legitimate initiators but result in a significant computational burden for attackers who wish to send large numbers of requests to a responder.

Properties of a good puzzle include [5, 29]:

- Generation and verification is inexpensive for the responder.
- Level of difficulty can easily be adjusted from trivial to practically impossible.
- Solutions should not require specialised client hardware.
- Solutions cannot be precomputed.
- Issuing a puzzle does not require the responder to store any state.
- Knowledge of the solution to one client's puzzle is of no benefit in solving other puzzles, so that the same puzzle may be provided to numerous clients.
- Initiators can reuse a puzzle by creating new instances of it.
- The server must not be able to forward puzzle solutions it receives to other servers for which it is a client.[1]

6.3.2.1 Hash-Based Puzzles

Juels and Brainard [29] describe the construction of client puzzles to protect TCP and SSL against connection depletion (SYN flooding) attacks. In their proposal, when a server becomes heavily loaded (inferred from buffer occupancy), connections are only accepted if they are accompanied by a proof of work. The puzzle protocol exchange is depicted in Fig. 6.4 and message components

[1]This requirement addresses a flaw in the use of client puzzles proposed by Aura et al. [6] that was identified by Price [51].

a

$$
\begin{aligned}
X &= h(t, M, s)\\
Y &= h(X)\\
X' &= X \ \& \ 0_0, 0_1, ..., 0_{k-1} 1_k 1_{k+1} 1_{n-1}\\
puzzle &= (X', Y)\\
solution\ (p)\ &:\ p\ such\ that\ h(p) = Y\\
verification\ &:\ X = h(t, M, s)\\
& \quad h(p) \stackrel{?}{=} h(X)
\end{aligned}
$$

b

$$h(N_R || X) = \underbrace{000...000}_{k-bits} ||, ...$$

Fig. 6.5 Hash-based puzzle constructions. (**a**) Juels and Brainard construction. (**b**) Aura et al. construction

summarised in Fig. 6.5a. The scheme employs a cryptographic hash function (h). In this scheme the puzzle is constructed by first hashing session parameters (M), the current time (t), and a responder secret (s). The inclusion of session-specific parameters allows the responder to remain stateless. The incorporation of a timestamp restricts the time solutions will be accepted for a given puzzle. The n-bit output of this hash operation (X) becomes the preimage to another application of a hash function, whose output (Y) forms part of the puzzle. The initiator is provided the partial preimage X' (X with k-bits masked out), and the hash digest value Y.

In order to solve the puzzle, the initiator must test up to 2^k possible preimages until the correct output is achieved. On average this will take 2^{k-1} hash operations. To verify the solution (p), the responder checks the time value t is recent (i.e. the puzzle has not yet expired), and then confirms, with three hash operations, that the proposed solution (p) is correct:

$$X = h(t, M, s)$$

$$h(p) \stackrel{?}{=} h(X)$$

This construction allows the responder to remain stateless and ensures that initiators will on average perform 2^{k-1}, but no more than 2^k, hash operations to solve puzzles. This construction, however, does require the responder to expend computational effort of its own in constructing the puzzle.

An alternative construction is proposed by Aura et al. [5] and is shown in Fig. 6.5b. In this proposal the puzzle consists of a time variant responder nonce (N_R) and a difficulty parameter (k). To solve the puzzle the initiator must find the value X that, when hashed with the responder nonce (N_R), produces a digest output whose first k-bits are zeros.

The responder verifies the puzzle solution by checking that N_R is recent and that the solution (X) when hashed with the nonce produces an output with the first k-bits as zero.

Note, that a simplified description of the Aura et al. construction has been presented. In reality, the value X includes the initiator identity and an initiator nonce. Aura et al. acknowledge that a consequence of using a broadcast puzzle is that the verifier of a puzzle solution must store state ('keep book') related to puzzle solutions it has already seen for a given challenge. Of particular interest to this report, however, is the manner in which hash operations can be used as a proof of work, not the specific instantiation of the technique in the authentication protocol described by Aura et al.

The construction of Aura et al. [5] was designed specifically for use in authentication protocols. The puzzle construction does not include any initiator specific parameters, allowing the responder to sign a single puzzle and issue it to multiple initiators. Signing a responder selected value periodically prevents a responder from computationally based attacks, while simultaneously allowing initiators to authenticate the responder via a digital signature verification. Conversely, the inclusion of initiator specific parameters in the Juels and Brainard [29] construction would make the signing of puzzles prohibitively expensive and leave the responder vulnerable to computational denial-of-service attacks.

6.3.2.2 Granularity of Puzzle Difficulty

Both the construction of Juels and Brainard and Aura et al.'s require exponential solving time $\mathcal{O}(2^k)$, where the difficulty parameter k is the number of bits in the search space. Thus, the expected number of hashes needed to solve a puzzle of difficulty parameter $k + 1$ is twice that required to solve a puzzle of difficulty k. In order to allow responders to fix the difficulty of puzzle with finer granularity, Juels and Brainard [29] suggest the combination of *multiple puzzles*. For example, a responder could send a puzzle of difficulty $k + 1$ and a puzzle of difficulty k to an initiator. Solving both puzzles is 3 times harder than solving a single puzzle of difficulty k, whereas solving a puzzle of difficulty $k+2$ would be 4 times harder. The main disadvantage of combining puzzles is that the complexity of puzzle generation and bandwidth requirement increases linearly.

Another approach named *hint-based hash-reversal puzzles* was put forward by Feng et al. [23] as a variant of the puzzles of Juels and Brainard. Here the responder provides a 'hint' to the initiator to help him solve the puzzle in linear time. To generate a puzzle with linear complexity $\mathcal{O}(k)$, the responder chooses a random integer value x and computes $h(x)$. The responder also computes $a = x - b$ where

$b \in [0, \ldots, k]$ is chosen uniformly at random. The responder then sends $h(x)$ and the hint value a to the initiator. To solve the puzzle the initiator searches through all values $x' = a + i$ for $i = 0, \ldots, k$ until $h(x') = h(x)$.

6.3.2.3 Puzzles, Reachability, and Statelessness

To operate effectively as a replacement for cookies (providing weak authentication of initiator reachability and stateless connections) proofs of work must be constructed so that:

- The IP address of the initiator must be part of the puzzle construction
- Required state information is also encoded in the puzzle

If the puzzle construction is unable to meet these requirements, then the puzzle cannot replace the functions of a cookie in the protocol.

Reconsider the puzzle construction of Aura et al. [5], which does not include the IP address of the initiator in the puzzle construction. It would be possible for a malicious initiator to receive a puzzle challenge on one IP address and construct puzzle solutions for any number of spoofed IP addresses. Admittedly, as the puzzle solution must contain an initiator identity, the initiator would have to generate a unique solution for each identity (address) it wished to claim, but in spite of receiving a valid puzzle solution, the responder would have no assurance that the initiator was actually reachable at the claimed IP address. In order to gain assurance that an initiator is able to send and receive messages from a claimed IP address, protocols using the Aura et al. puzzle construction as a proof of work should also use a cookie, or modify the construction to meet the requirements identified. The latter has been done by Smith et al. [59]. They replace N_R in the original puzzle of Aura et al. (Fig. 6.5b) with a cookie, which is computed by the responder as a MAC of session state information including the initiator's address.

The Juels and Brainard [29] puzzle includes the IP address of the initiator and other related state in its construction, so protocols using this construction are able to use the puzzle as a replacement for a cookie.

6.3.2.4 Recycling CPU Cycles

Proofs of work can be viewed as a way for an initiator to make a payment to a responder for the services it will provide. Additionally, a proof of work when accompanying a request can serve as a useful discriminator between malicious and legitimate requests. The computational effort expended in generating the proof of work can be: (1) wasted, (2) reused by the initiator in completing the protocol (as is done in the modified Internet Key Exchange (IKE) protocol presented in Sect. 6.5), (3) reused by the initiator for some other purpose; (4) reused by the responder in completing the protocol, or (5) reused by the responder for some other purpose.

Client puzzles are typically nonreusable proofs of work, so the computational effort expended in generating the proof of work is wasted. The modified version of IKE proposed by Matsuura and Imai [42] adopts a reusable proof of work based on signature verification of the responder, an action that a legitimate initiator will have to perform in order to complete the protocol. Castellucia et al. [16] propose a puzzle construction implementing *client-aided RSA* computation within the SSL protocol proposed.

As identified earlier, Jakobsson and Juels [28] describe a proof of work in which the computational effort expended in generating the proof of work is reused by the responder for another application. There are currently no examples of key establishment protocols implementing this type of reusable proof of work and unless an initiator implicitly trusts the responder to delegate its computational resources, initiators must be aware that the computational effort expended in generating a proof of work may be reused for malicious purposes. In fact, Price describes how a number of protocols implementing proofs of work based on hash-based client puzzles are vulnerable to a form of chosen protocol attack, where an initiator of a protocol is able to convince third parties to generate puzzle solutions on their behalf [51].

6.3.2.5 Puzzle Freshness

It is a common requirement for puzzle schemes that responders be able to verify that a solution presented by an initiator corresponds to a fresh puzzle, that is it is the solution to a puzzle generated recently by the responder within some threshold value. For example, in Juels and Brainard's puzzles (Fig. 6.5a) this is achieved by including the time value t, while in Auras et al.'s (Fig. 6.5b) the time-varying nonce N_R assures the responder of the freshness of the solutions. In both cases the generation of a puzzle requires a random input which is unpredictable by the initiator and that consequently makes the puzzle also unpredictable. Otherwise, if the puzzle were predictable, the initiator would be able to precompute its solution ahead of time.

In most applications, puzzles are generated on the fly by responders using responder-generated random challenges (nonces) for freshness. A disadvantage of responder-generated puzzles is that the puzzle generation process can be costly and consequently become a cause of DoS vulnerability. Waters et al. [65] investigate numerous techniques for outsourcing the generation of puzzles in order to remove the computational burden of puzzle generation from the responder. One of their proposals is based on a Diffie-Hellman construction, in which given a generator g, and a random value a in the range r to $r + k$, the puzzle issued to the initiator contains the values g^a and r. The initiator searches for a solution by trying each candidate value in the range r to $r + k$ until it finds c, such that $g^c = g^a$. To bind the solution to a specific server, the initiator calculates y^a, where y is the responder's Diffie-Hellman public key. The responder verifies the solution with a single modular exponentiation, raising the challenge value to the exponent (x) of its private key, that is g^{ax}. The system described by Waters et al. [65] limits puzzles to a given timeslot.

The solutions for a given timeslot are all precomputed by the responder, so puzzle verification is performed by a table lookup, not an online modular exponentiation.

Chan et al. [17] provide another method for outsourcing puzzle generation that employs verifiable public broadcast nonces, which they called *rhythmic nonces*. In the proposal of Chen et al., a time server broadcasts a signed nonce t_i for period i. Assuming that the initiator knows the difficulty parameter k required by the responder, the initiator chooses a random value n and has to find x such that $h(t_i \| n \| x)$ has k leading zeroes. The initiator sends t_i, n, and x to the responder. The responder then verifies the freshness of t_i. This is done using a public verification function which takes as input t_i and the most recent broadcast nonce t_j. The responder proceeds by checking that $h(t_i \| n \| x)$ has k leading zeroes. Chan et al.'s scheme depends strongly on the efficiency of the public verification function. Their concrete proposal, however, incurs in expensive public-key cryptographic computations.

6.3.2.6 No-Parallelisable Puzzles

Hash-based constructions have the property that exhaustive preimage searching is a parallelisable task. Thus, an attacker with many processors available can drastically reduce the puzzle solution time. Despite this apparent disadvantage of hash-based constructions, non-parallelisability has been generally disregarded in the denial-of-service literature. This is because an attacker that tries to flood a responder with requests would seem to gain little by parallelising the solution of puzzles for two reasons. First, the attacker would incur in the overhead of managing the distribution of the task amongst processors. Second, even disregarding this overhead, the aggregate throughput of request (i.e. average request rate) will not be greater when the solution is parallelised.

It is conceivable nevertheless that as the use of puzzles evolves (and is used in puzzle auctions for example [64]), alternative puzzle constructions which are not parallelisable may be required. Rivest et al. [54] proposed *time-lock puzzles* which are based on the notion that a client has to spend a predetermined amount of computational time performing repeated squaring to exponentiate an integer modulo a large composite number in order to find a solution and unlock the puzzle. The puzzle issuing server calculates the number of squaring operations which a client can perform, and determines the amount of time it wants a client to spend solving the puzzle. Unlike hash-based puzzles, time-lock puzzles are inherently sequential and non-parallelisable. Tritilanunt et al. [62] also considered non-parallelisability in puzzles and proposed a new puzzle based on the *subset sum problem*.

6.3.2.7 Memory Bound Functions

Proofs of work based on processor-bound functions result in puzzle constructions that can be solved in negligible time on a modern desktop computer, but may take an inordinately long time on lightweight devices such as mobile phones or personal digital assistants. An effective strategy for authenticating the platform that a puzzle was being issued to could allow the responder to tune the difficulty of the puzzle. As the reliable authentication of a remote platform configuration is largely an open problem [32, 56], alternative constructions based on memory-bound functions are being investigated [1, 19]. Memory-bound functions should exhibit a more uniform response time across a range of devices, as the difference in memory access speeds between lightweight and more powerful devices is far less significant than the difference between processor speeds.

However, memory-bound puzzles can also be expensive to set up. For example, in [1], to create a memory-bound puzzle that took a client 2^2 s to solve required 2^{-7} s of setup time on a 2.4 GHz Pentium 4 server. However, a computer of similar speed could do a 1024-bit modular exponentiation in only 2^{-9} s: setting up a memory-bound puzzle is more expensive than doing RSA exponentiation. In order for memory-bound puzzles to be viable denial-of-service puzzles, they will need to be inexpensive to generate.

6.3.2.8 Puzzle Difficulty Management

The difficulty of puzzles must be adjusted dynamically by the responder in order to balance the quality of service provided to legitimate initiators against the effectiveness of the puzzles against denial-of-service. A basic approach employed for example by Feng et al. [23] adjusts the difficulty parameter proportionally to the current load of the responder. When an initiator makes a request to the responder, the initiator is asked to solve a puzzle of difficulty proportional to the load of the responder at the moment of the request. This approach has some weaknesses. First, it allows attackers to herd simple puzzles from the responder while it is at low load, solve the puzzles, and then flood the responder. Also this approach does not take into consideration the quality of service to legitimate requesters.

The interaction between a denial-of-service attacker and a responder can be viewed as a game in which the objective of the attacker is to maximise the resources committed by the responder while minimising its own commitment. The goal of the responder is to minimise resource expenditure, but also that of legitimate requesters. Game theoretic techniques have been applied to the problem of adjusting the difficulty of puzzles. Bencsath et al. [13] use the model of single-shot strategic-form game to find the requester's equilibrium strategy. Their analysis suggests that the optimal strategy for the requester is to issue a mixture of puzzle difficulties at any one moment. More recently, Fallah [22] has extended the use of game theory to client puzzles by considering history-dependent strategies as well as distributed attackers. In Fallah's technique the defender can choose his action based on the attacker's previous actions. Section 6.4.2 extends the game theoretic approach of Fallah to analyse a new type of client puzzles, where the difficulty of the puzzle is hidden from clients.

6.3.3 *Gradual Authentication*

While the expense of strongly authenticating initiators using digital signatures will be dependent on many parameters, the computational expense of a signature verification will not always be prohibitively expensive. Rabin signatures [52] with a public exponent of 2 or RSA signatures [53] with a public exponent of 3 can be verified with only one or two modular multiplications, respectively. While the cost of signature verification with these parameters is low, signature generation is somewhat more expensive, which may not be suitable for all deployment scenarios. Other signature schemes, RSA with larger public exponents for example, increase the cost of signature verification, requiring the responder to perform expensive modular exponentiations. While newly proposed key establishment protocols can be specified to accommodate cheap signature verification for responders, the requirement to improve resistance to denial-of-service attack remains for already deployed protocols, and protocols that for other reasons are restricted in the choice of signature schemes they must implement. Gradual authentication provides a mechanism for weakly authenticating an initiator, prior to performing stronger and more expensive cryptographic authentication.

The idea of combining weak and strong authentication was first introduced by Meadows [44] and is proposed as a technique to increase resistance to denial-of-service attacks by combining weak authentication when the protocol is initiated and moving to strong authentication as it completes.

Cookies and client puzzles can be considered forms of weak authenticators. Cookies provide some assurance that the initiator is able to send and receive packets from the claimed address – implying that the request is not part of a connection depletion attack, which typically relies on using random spoofed addresses. Receipt of a correct solution to a client puzzle provides some assurance to the responder that the initiator is willing to expend her own resources in order to get the protocol to proceed.

Other cryptographic techniques, such as the use of message authentication codes and release of hash digest preimages, that allow the responder to cheaply verify messages are being adopted by recently proposed protocols such as Just Fast Keying (JFK) as discussed in Sect. 6.5.

While the use of techniques such as cookies, client puzzles, and releasing hash preimages do not meet strong notions of authentication, when generated using cryptographically sound primitives they can be combined in ways which enable a responder to discount a range of denial-of-service attacks and present a number of hurdles that must be overcome by an attacker intent on disrupting the protocol execution. A key characteristic of the techniques used in gradual authentication is that they are all cheap for the responder to verify, while their fabrication is relatively expensive for an attacker. Even when signature schemes that minimise verification costs to a responder are adopted, the cost of verifying gradual authenticators such as client puzzles is still cheaper, costing only a single hash operation.

Key establishment protocols typically complete with strong authentication of the initiator. Strong authentication should be deferred, however, until the responder has gained adequate assurance that the initiator's requests are not part of a denial-of-service attack. Such assurance can be gained through gradual authentication. Gradual authentication is employed by the modified IKE and the JFK protocols, discussed in more detail in Sect. 6.5.

6.4 Analysing Denial-of-Service Resistance

In Sect. 6.3.2, we presented a variety of existing client puzzle schemes. Although there have been many puzzle constructions, only a few of these use any formal notion of security, and there has been little work in developing formal definitions of client puzzle difficulty or denial-of-service resistance. In this section, we examine two methods of formally analysing puzzle difficulty and denial-of-service resistance. The first approach is through the language of provable security, in which we aim to prove that a puzzle takes a certain amount of resources to solve. The second approach is through the language of game theory, in which we compare the resources of an adversary and a server, aiming to find an optimal solution to setting puzzle difficulty.

6.4.1 Provable Security

In cryptography, the provable security framework is used to analyse the security of proposed schemes by answering two questions: What does it mean for a scheme to be secure? How can we mathematically prove that it is secure? To define security, we construct a security experiment played between an adversary and a challenger that aims to capture the desired security functionality. To show that a scheme is secure, we aim to prove bounds on the ability of the adversary to win the experiment; generally a scheme is said to be *computationally secure* if the probability that a polynomial-time adversary can win the security experiment is negligible.

In this subsection, we will use the provable security framework to define what it means for a client puzzle to be difficult, and then what it means for a protocol to be denial-of-service resistant.

6.4.1.1 Previous Client Puzzle Difficulty Definitions

The first client puzzle difficulty definition was given by Jakobsson and Juels [27], and another by Canetti et al. [14]. Some memory-bound puzzles [19, 21] include proofs of amortised difficulty. A richer difficulty definition was given by Chen et al. [18], using two security experiments: unforgeability and puzzle difficulty. Importantly, the difficulty definition only addresses the ability of an adversary to solve a single puzzle. They describe a basic generic client puzzle protocol $\Pi(\mathsf{CPuz})$. Finally, they give a generic client puzzle construction from a pseudorandom function and a one-way function (essentially a MAC and a hash function).

We argue that these previous definitions are inadequate. In a public network setting, a server will be providing service to many clients at a time. A DoS countermeasure based on client puzzles should require appropriate work to be done for *each* client request: it should not be possible to solve many puzzles easily. While the existing models [18,27,61] describe the difficulty of DoS countermeasures when faced with an adversary trying to solve one puzzle, these models do not adequately defend against *powerful adversaries* who can expend more than the effort required to solve a single puzzle.

We now consider an example where a single instance cannot be solved easily by an attacker, satisfying existing difficulty definitions, but where an attacker can solve n puzzles more efficiently than just n times the cost of solving a single puzzle. This motivates our stronger definition of puzzle difficulty later in the section.

Chen et al. [18] proposed a generic client puzzle construction based on a pseudorandom function $\mathcal{F}$ and a one-way function ϕ. The challenger selects a secret $s \in \mathcal{K}$ with $|\mathcal{K}| = 2^k$ and public parameters (not relevant to our discussion here), denoted by $*$, to generate a puzzle. The challenger computes $x \leftarrow \mathcal{F}(s, *)$, where $x \in \mathcal{X}$ and $|\mathcal{X}| \geq |\mathcal{K}|$, and then sets $y \leftarrow \phi(x)$. The solver, given the challenge $(y, *)$, has to find a pre-image z such that $\phi(z) = y$.

This generic construction satisfies the puzzle unforgeability and puzzle difficulty security properties provided certain bounds are met: namely, $|\mathcal{X}| \geq |\mathcal{K}|$ and $\frac{|\phi^{-1}(y)|}{|\mathcal{X}|} \leq \frac{1}{2^k}$, for all y. Suppose we have that $|\phi^{-1}(y)| \leq 1$ and $|\mathcal{X}| = 2^k$. Then the bounds in the generic construction are satisfied and solving a single puzzle instance requires approximately 2^k searches in $\mathcal{X}$. But to solve n puzzles, the solver can find the value s with at most 2^k searches and then obtain a solution with one application of $\mathcal{F}$ for each puzzle. That is solving n puzzles would require $2^k + n$ operations rather than the desired $n \cdot 2^k$ computations.

6.4.1.2 Defining Puzzles

Our definition of puzzle difficulty starts from the Chen et al. [18] definition, but with a number of differences. First, we eliminated the unforgeability property. The unforgeability property is important for their protocol $\Pi(\mathsf{CPuz})$, but is not an essential feature of client puzzles. In fact, to define noninteractive puzzles, in which the client can generate the puzzle itself, we must remove unforgeability. Next, we strengthen the difficulty definition to consider an adversary who solves many puzzles.

We start by giving a formal definition of a client puzzle scheme:

Definition 6.1 (Client Puzzle). A *client puzzle* Puz is a tuple consisting of the following algorithms:

- $\mathsf{Setup}(1^k)$ (p.p.t. setup algorithm):
  1. Choose the long-term secret key space sSpace, puzzle difficulty space $\mathsf{diffSpace}$, string space $\mathsf{strSpace}$, puzzle space $\mathsf{puzSpace}$, and solution space $\mathsf{solnSpace}$.
  2. Set $s \leftarrow_R \mathsf{sSpace}$.
  3. Set $params \leftarrow (\mathsf{sSpace}, \mathsf{puzSpace}, \mathsf{solnSpace}, \mathsf{diffSpace}, \Pi)$, where Π is any additional public information, such as a description of puzzle algorithms, required for the client puzzle.
  4. Return $(params, s)$.
- $\mathsf{GenPuz}(s \in \mathsf{sSpace}, d \in \mathsf{diffSpace}, str \in \mathsf{strSpace})$ (p.p.t. puzzle generation algorithm): Return $puz \in \mathsf{puzSpace}$.
- $\mathsf{FindSoln}(str \in \mathsf{strSpace}, puz \in \mathsf{puzSpace}, t \in \mathbb{N})$ (probabilistic solution finding algorithm): Return a potential solution $soln \in \mathsf{solnSpace}$ after running time at most t[2].
- $\mathsf{VerSoln}(s \in \mathsf{sSpace}, str \in \mathsf{strSpace}, puz \in \mathsf{puzSpace}, soln \in \mathsf{solnSpace})$ (d.p.t. puzzle solution verification algorithm): Returns **true** or **false**.

For *correctness*, we require that if $(params, s) \leftarrow \mathsf{Setup}(1^k)$ and $puz \leftarrow \mathsf{GenPuz}(s, d, str)$, for $d \in \mathsf{diffSpace}$ and $str \in \mathsf{strSpace}$, then there exists $t \in \mathbb{N}$ with

$$\Pr\left(\mathsf{VerSoln}(s, str, puz, soln) = \textbf{true} : soln \leftarrow \mathsf{FindSoln}(str, puz, t)\right) = 1.$$

6.4.1.3 Strong Puzzle Difficulty

Next we consider what it means for a client puzzle to be difficult. Informally, a client puzzle is difficult if the success probability of an adversary is proportional to its runtime. We will define two formal notions of client puzzle difficulty, one for interactive settings and one for noninteractive settings.

A puzzle satisfies strong puzzle difficulty if the probability that a runtime-bounded-adversary can output a list of n fresh, valid puzzle solutions is upper-bounded by a function of the puzzle difficulty parameter and n. This is formalised in the following two experiments for the interactive and noninteractive settings.

---

[2] $\mathsf{FindSoln}$ runs in time *at most t* so that a client can stop searching for a puzzle after a specified amount of time; our difficulty definitions yield that a client must spend *at least* a certain amount of time to find a valid solution.

We first need to define additional helper oracles as follows:

- $\mathsf{GetPuz}(str)$: Set $puz \leftarrow \mathsf{GenPuz}(s, d, str)$ and record (str, puz) in a list. Return puz.
- $\mathsf{GetSoln}(str, puz)$: If (str, puz) was not recorded by GetPuz, then return $\perp$. Otherwise, find $soln$ such that $\mathsf{VerSoln}(s, str, puz, soln) = \textbf{true}$. Record $(str, puz, soln)$. Return $soln$[3].
- $\mathsf{V}(str, puz, soln)$: Return $\mathsf{VerSoln}(s, str, puz, soln)$.

6.4.1.4 Interactive Strong Puzzle Difficulty

In this setting, we imagine a solver interacting with a challenger: the solver submits a request for a puzzle, the challenger issues a puzzle, the solver sends a solution to the challenger, and the challenger checks the solution. The solver can only submit solutions to puzzles that were issued by the challenger: this immediately rules out puzzle forgery or generation of puzzles by the solver. The challenger also allows the solver, via queries, to see solutions to other puzzles.

Let k be a security parameter, let d be a difficulty parameter, let $n \geq 1$, and let $\mathcal{A}$ be an algorithm. The security experiment $\mathsf{Exec}^{\text{INT-STR-DIFF}}_{\mathcal{A},d,\mathsf{Puz}}(k)$ for interactive strong puzzle difficulty of a puzzle Puz is defined as follows:

- $\mathsf{Exec}^{\text{INT-STR-DIFF}}_{\mathcal{A},n,d,\mathsf{Puz}}(k)$:

  1. Set $(params, s) \leftarrow \mathsf{Setup}(1^k)$.
  2. Set $\{(str_i, puz_i, soln_i) : i = 1, \ldots, n\} \leftarrow \mathcal{A}^{\mathsf{GetPuz},\mathsf{GetSoln},\mathsf{V}}(params)$.
  3. If $\mathsf{VerSoln}(s, str_i, puz_i, soln_i) = \textbf{true}$, the tuple (str_i, puz_i) was recorded by GetPuz, and $(str_i, puz_i, soln_i)$ was not recorded by $\mathsf{GetSoln}$ for all $i = 1, \ldots, n$, then return **true**, otherwise return **false**.

Definition 6.2 (Interactive Strong Puzzle Difficulty). Let $\epsilon_{d,k,n}(t)$ be a family of functions monotonically increasing in t, where $\epsilon_{d,k,n}(t) \leq \epsilon_{d,k,1}(t/n)$ for all t, n such that $\epsilon_{d,k,n}(t) \leq 1$. Fix a security parameter k and difficulty parameter d. Let $n \geq 1$. Then Puz is an $\epsilon_{d,k,n}(\cdot)$*-strongly difficult interactive client puzzle* if, for all probabilistic algorithms $\mathcal{A}$ running in time at most t,

$$\Pr\left(\mathsf{Exec}^{\text{INT-STR-DIFF}}_{\mathcal{A},n,d,\mathsf{Puz}}(k) = \textbf{true}\right) \leq \epsilon_{d,k,n}(t).$$

[3]Note that $\mathsf{GetSoln}$ is only obligated to find a solution if puz was actually generated by the challenger. If $\mathcal{A}$ generated puz, then $\mathcal{A}$ may need to employ $\mathsf{FindSoln}$ to find a solution. Compared to $\mathsf{FindSoln}$, $\mathsf{GetSoln}$ has access to additional secret information that may allow it to find a solution more easily.

In the random oracle model[4], To our knowledge, this is the first formal justification for the security of Hashcash. we can define interactive and noninteractive strong puzzle difficulty in terms of the number of oracle queries made by the adversary instead of its running time.

Remark 6.3. The condition that $\epsilon_{d,k,n}(t) \leq \epsilon_{d,k,1}(t/n)$, for all t and n such that $\epsilon_{d,k,n}(t) \leq 1$, captures the property that solving n puzzles should cost n times the cost of solving one puzzle, at least until the adversary spends enough time t to solve n puzzles with probability 1.

Remark 6.4. This bound is quite abstract; let us consider a concrete function for $\epsilon_{d,k,n}(t)$. For example, suppose each Puz instance should take approximately 2^d steps to solve. Then we might aim for Puz to be a $\epsilon_{d,k,n}(\cdot)$-strongly difficult interactive client puzzle, where $\epsilon_{d,k,n}(t) \approx t/2^d n + \text{negl}(k)$.

Remark 6.5. In the security experiment, the adversary is allowed to request many more than n puzzles using GetPuz. The adversary can then pick which n puzzles it submits as its allegedly solved puzzles $\{(str_i, puz_i, soln_i) : i = 1, \ldots, n\}$. In other words, the adversary could request many puzzles and hope to find some easy-to-solve instances. This means, for example, that puzzles for which 1% of instances are trivially solved could not be proven secure (with a reasonable $\epsilon_{d,k,n}(t)$) according to this difficulty definition.

Remark 6.6. The Chen et al. generic puzzle construction earlier in this section does not satisfy our definition of strong puzzle difficulty. From Theorem 2 of [18], we have that the Chen et al. generic construction is $\epsilon_{d,k}(t)$-difficult, with $\epsilon_{d,k}(t) \lessapprox 2\nu_k(t) + (1 + t/(2^{k-t}))\gamma_d(t)$, where $\nu_k(t)$ is the probability of breaking the pseudorandom function family (with security parameter k) in time t and $\gamma_d(t)$ is the probability of breaking the one-way function (with security parameter d) in time t. Thus, there exists an adversary that can win the strongly-difficulty interactive puzzle game with probability at least $\epsilon'_{d,k,n}(t) \gtrapprox \nu_k(t) + \gamma_d(t)/n$, which does not satisfy $\epsilon'_{d,k,n}(t) \leq \epsilon'_{d,k,1}(t/n)$.

We describe a client puzzle based on hash function inversion, similar to the subpuzzle used by Juels and Brainard [29] or the partial inversion proof of work of Jakobsson and Juels [27]. Moreover, we show that this puzzle is an interactive strongly difficult puzzle, satisfying Definition 6.2.

Let $H : \{0,1\}^* \rightarrow \{0,1\}^k$ be a hash function. Define $\mathbf{SPuz}_H$ as the following tuple of algorithms:

- Setup(1^k): Set sSpace $\leftarrow \{\perp\}$, diffSpace $\leftarrow \{0, 1, \ldots, k\}$, strSpace $\leftarrow \{0,1\}^*$, puzSpace $\leftarrow \{0,1\}^* \times \{0,1\}^k$, solnSpace $\leftarrow \{0,1\}^*$, and $s \leftarrow \perp$.
- GenPuz($\perp, d, str$): Set $x \leftarrow_R \{0,1\}^k$; let x' be the first d bits of x and x'' be the remaining $k - d$ bits of x. Set $y \leftarrow H(x, d, str)$. Return $puz \leftarrow (x'', y)$.

[4] In the *random oracle model*, a hash function is modelled as an ideal random function accessible to the adversary solely as an oracle [12].

- FindSoln$(str, (x'', y), t)$: For z from 0 to $\max\{t, 2^d - 1\}$: set $soln \leftarrow z$ (in $\{0,1\}^d$); if $H(soln||x'', d, str) = y$ then return $soln$.
- VerSoln$(\perp, str, (x'', y), soln)$: If $H(soln||x'', d, str) = y$ then return **true**, otherwise return **false**.

Theorem 6.7. *Let H be a random oracle. Let $\epsilon_{d,k,n}(q) = \left(\frac{q+n}{n2^d}\right)^n$. Then $\mathbf{SPuz}_H$ is an $\epsilon_{d,k,n}(q)$-strongly difficult interactive client puzzle, where q is the number of distinct queries to H.*

Proof. For the $\mathsf{Exec}^{\text{INT-STR-DIFF}}$ experiment, we need to specify how the GetSoln oracle obtains a solution to a generated puzzle.

- GetSoln$(str, (x'', y))$: If (x'', y) was recorded by GetPuz, then return x', where $x = x'||x''$ was the random bit string chosen in GenPuz for this puzzle; otherwise, return $\perp$.

The proof proceeds using a counting argument. Fix d. Let $\mathcal{A}$ be a probabilistic algorithm. Clearly, there is a strategy for $\mathcal{A}$ to win the experiment with probability 1, by making at most $n2^d$ calls to H: for each of n puzzles $puz_i = (x_i'', y_i)$, try all strings z of length d until one is found such that $H(z||x_i'', d, str_i) = y_i$. In the random oracle model, this is essentially the optimal strategy.

Let $Z_i = \{z_{i,1}, \ldots, z_{i,q_i+1}\} \subseteq \{0,1\}^*$ with $|Z_i| = q_i + 1$ (i.e. the set contains no repetitions). Let $E_{i,j}$ be the event that $H(z_{i,j}) = y_i$ for $j = 1, \ldots, q_i$. Since the output of H is independent and uniformly random, and since x_i was chosen independently and uniformly at random, we have that $\Pr(E_{i,j}) \leq 2^{-d}$. Let F_i be the event that there exists $z_{i,j} \in Z_i$ such that $H(z_{i,j}) = y_i$; in other words, $F_i = \bigvee_{j=1}^{q_i+1} E_{i,j}$. Let $q_i \in \{0, \ldots, d\}$ be the number of queries issued to H for puzzle $i = 1, \ldots, n$, so that $q = \sum_{i=1}^n q_i$. Then,

$$\Pr\left(\bigwedge_{i=1}^n F_i\right) = \prod_{i=1}^n \Pr(F_i) = \prod_{i=1}^n \Pr\left(\bigvee_{j=1}^{q_i+1} E_{i,j}\right) \leq \prod_{i=1}^n \sum_{j=1}^{q_i+1} \Pr(E_{i,j})$$

$$= \prod_{i=1}^n \frac{q_i+1}{2^d} \leq \left(\frac{\sum_{i=1}^n (q_i+1)}{n2^d}\right)^n = \left(\frac{q+n}{n2^d}\right)^n$$

We note that any adversary making q_i queries to H has at best a probability of $(q_i+1)/2^d$ of returning a value z_i that satisfies $H(z_i) = y_i$: checking q_i values using H, and, if that fails then guessing at random one of the remaining values. Thus,

$$\Pr\left(\mathsf{Exec}_{\mathcal{A},n,d,\mathbf{SPuz}}^{\text{INT-STR-DIFF}}(k) = \mathbf{true}\right) \leq \Pr\left(\bigwedge_{i=1}^{n} F_i\right) \lessapprox \left(\frac{q+n}{n2^d}\right)^n = \epsilon_{d,k,n}(q).$$

Finally it is easy to see that $\epsilon_{2^d,k,n}(q) \leq \epsilon_{2^d,k,1}(q/n)$ for all n and q such that $\epsilon_{d^d,k,n}(q) \leq 1$.

Thus, **SPuz** is an $\epsilon_{2^d,k,n}(q)$-strongly difficult interactive client puzzle.

6.4.1.5 Non-interactive Strong Puzzle Difficulty

Noninteractive strong puzzle difficulty models the case of client-generated puzzles. Besides being useful in their originally proposed setting as an email spam countermeasure [8, 9], they can be useful in protocols that are inherently asynchronous, such as the Internet Protocol (IP), or have a fixed message flow, such as the Transport Layer Security (TLS) protocol.

The technical difference between interactive and noninteractive strongly difficult puzzles is whether the adversary can return solutions only to puzzles generated by the challenger (interactive) or can also return solutions to puzzles it generated itself (noninteractive).

The security experiment $\mathsf{Exec}^{\text{NINT-STR-DIFF}}_{\mathcal{A},n,d,\mathsf{Puz}}(k)$ for noninteractive strong puzzle difficulty is as in the interactive case with a change to line 3 of the experiment:

- $\mathsf{Exec}^{\text{NINT-STR-DIFF}}_{\mathcal{A},n,d,\mathsf{Puz}}(k)$:

  3. If $\mathsf{VerSoln}(s, str_i, puz_i, soln_i) = \textbf{true}$ and the tuple $(str_i, puz_i, soln_i)$ was not recorded by $\mathsf{GetSoln}$ for all $i = 1, \ldots, n$, then return **true**, otherwise return **false**.

The definition of $\epsilon_{d,k,n}(\cdot)$*-strongly difficult noninteractive client puzzles* follows analogously.

Remark 6.8. If Puz is an $\epsilon_{d,k,n}(\cdot)$-strongly difficult noninteractive puzzle, then it is also $\epsilon_{d,k,n}(\cdot)$-strongly-difficult interactive puzzle.

We can show that one of the earliest client puzzles, Hashcash [8, 9], satisfies the definition of a strongly difficult noninteractive client puzzle in the random oracle model.

While Hashcash was originally proposed to reduce email spam, the current specification (stamp format version 1 [9]) can be applied to any resource. Hashcash is noninteractive: the puzzle is generated by the same person who solves the puzzle. Hence, it should be difficult for a client to generate a puzzle that can be easily solved. Hashcash is based on the difficulty of finding a partial preimage of a string starting with a certain number of zeros in the SHA-1 hash function.

A Hashcash *stamp* is a string of the form `ver:bits:date:resource:[ext]:rand:counter`. The field `bits` denotes the 'value' of the stamp (the number of zeros at the start of the output) and `counter` is the solution to the puzzle. A stamp is *valid* if $H(stamp)_{[1\ldots\texttt{bits}]} = 0\ldots0$. In the context of real-world email applications, there may be additional restrictions on the validity of a stamp, such as whether `date` is within a reasonable range and whether the email address (`resource`) specified is acceptable.

Theorem 6.9. *Let* $H : \{0,1\}^* \rightarrow \{0,1\}^k$*, where* $k \geq d$*, be a random oracle. Let* $\epsilon_{d,k,n}(q) = \frac{q+n}{n2^d}$*. Then Hashcash is an* $\epsilon_{d,k,n}(q)$*-strongly difficult noninteractive puzzle, where* q *is the number of queries made by* $\mathcal{A}$ *to* H*.*

The proof of this theorem follows a counting argument similar to that of Theorem 6.7.

6.4.1.6 Defining Denial-of-Service Resistance of Protocols

Although we have defined what a good client puzzle is, it does not immediately follow that using a good client puzzle in a protocol yields DoS resistance. In this subsection, we describe what it means for a protocol to be DoS-resistant. We begin by reviewing previous approaches to analysing the DoS resistance of protocols.

Meadows [43] first presented a cost-based framework for identifying DoS attacks in network protocols (e.g. Smith et al.'s DoS attack [59] on the JFK key exchange protocol [3]), but can only be used to identify and quantify a DoS attack, not prove that a protocol is DoS-resistant.

Stebila and Ustaoglu [61] gave a provable security model for the DoS resistance of key agreement protocols based on the eCK model for key agreement security [36]. The model splits key exchange into two portions: a presession for the DoS countermeasure and a session for the key exchange. They give an example protocol using hash function inversions for the DoS countermeasure and building on CMQV [63] for the key exchange protocol. One of their main motivations was to avoid the DoS attack of Mao and Paterson [41] which derived from an authentication failure where messages could be redirected and accepted.

Our definition of DoS resistance for protocols shares some of these characteristics: it uses a presession for the DoS countermeasure and is suitable for a multiuser network setting. However, it can be used to analyse all protocols, not just key exchange protocols, and it uses a stronger notion of security, considering an adversary who solves many puzzles, not just one. By separating the definition of a puzzle from the definition of a DoS-resistant protocol, we can perform a modular analysis of each component separately and then combine them.

We now proceed by formalising a protocol and developing a definition of DoS resistance. As with client puzzles, we develop a formal definition involving a security experiment played between the adversary and a set of challengers, in this case the set of legitimate clients and servers.

Defining a Protocol

A *protocol* is a message-driven interaction, taking place amongst disjoint sets of clients $\mathsf{Clients}$ and servers $\mathsf{Servers}$, where each *party* is a probabilistic polynomial-time Turing machine. An execution of the protocol is called a *presession*. During execution, each party $\hat{U}$ may have multiple instances of the protocol running, with each instance indexed by a value $i \in \mathbb{Z}_+$; these instances are denoted by $\Pi_i^{\hat{U}}$. A protocol consists of the following algorithms:

- $\mathsf{GlobalSetup}(1^k)$ (p.p.t. protocol setup algorithm): Select the long-term secret key space $\rho\mathsf{Space}$. Choose global public parameters Π of the scheme and return $params \leftarrow (\rho\mathsf{Space}, \Pi)$; this is assumed to be an implicit input to all remaining algorithms.

- $\mathsf{ServerSetup}(\hat{S} \in \mathsf{Servers})$ (p.p.t. party setup algorithm): Select $\rho_{\hat{S}} \in \rho\mathsf{Space}$. Perform any additional setup required by *params*.
- $\mathsf{CAction}j(\hat{C} \in \mathsf{Clients}, i \in \mathbb{Z}_+, m_{j-1}, M'_{j-1})$, for $j = 1, \ldots$ (p.p.t. protocol client action algorithm): Instance i of party $\hat{C}$ produces its jth protocol message for the run of the protocol, based on the instance's previous private state m_{j-1} and the received message M'_{j-1}. The output (M_j, m_j) consists of its outgoing message M_j and its new private state m_j.
- $\mathsf{SAction}j(\hat{S} \in \mathsf{Servers}, i \in \mathbb{Z}_+, m'_{j-1}, M_j)$, for $j = 1, \ldots$ (p.p.t. protocol server action algorithm): Instance i of party $\hat{S}$ produces its jth protocol message for this instance, based on $\hat{S}$'s long-term secret, the previous private state m'_{j-1}, and the received message M_j. The output (M'_j, m'_j) consists of its outgoing message M'_j and its new private state m'_j.

The client is assumed to be the initiator. An instance records its current progress through the protocol with the value j of the last completed action.

After receiving some sequence of $\mathsf{SAction}j(\hat{S}, i, \ldots)$ calls, a server instance will either *accept* or *reject*; if it accepts, it outputs a *presession* identified by a tuple of the form $[\hat{C}, \hat{S}, \tau]$, where $\hat{C}$ is the *partner* and τ is a sequence of messages. The sequence of messages τ is meant to act like a transcript; however, since in DoS-resistant protocols a server may not store state early in the protocol, portions of τ could have been forged by an adversary. Accepted presessions must be unique within a party. Additionally, since the protocol may be used for another purpose – key agreement, electronic voting, etc. – we do not require that the protocol terminate after accepting, and indeed expect that it may continue to perform some additional application-level functionality.

To provide DoS resistance, a protocol will typically include some test so the server can decide, based on the proposed presession $[\hat{C}, \hat{S}, \tau]$ and its secret ρ, whether to accept or reject based on some DoS countermeasure in the protocol. It is the adversary's goal to cause a server to accept without the adversary having faithfully followed the protocol.

The Adversary

The adversary controls all communication links and can send, create, modify, delay, and erase messages to any participants. Additionally, the adversary can learn private information from parties or cause them to perform certain actions.

The following queries model how the adversary interacts with the parties:

- $\mathsf{Send}(\hat{U}, i, M)$: The adversary sends message M to instance i of $\hat{U}$ who performs the appropriate protocol action (either $\mathsf{CAction}j(\hat{U}, i, m, M)$ or $\mathsf{SAction}j(\hat{U}, i, m, M)$ based on the instance's last completed action $j - 1$), updates its state, and returns its outgoing message, if any.
- $\mathsf{Expose}(\hat{S})$: The adversary obtains $\hat{S}$'s secret value $\rho_{\hat{S}}$; mark $\hat{S}$ as *exposed*.

Security Definition

The basic idea of the security definition is as follows: the amount of credit the adversary gets in terms of accepted presessions should not be greater than the amount of work the adversary itself did. An important part of the definition below is solutions from legitimate clients.

An instance $\Pi_i^{\hat{S}}$ that has accepted a presession $[\hat{C}, \hat{S}, \tau]$ is said to be *fresh* provided that $\hat{S}$ was not exposed before $\hat{S}$ accepted this presession, and there does not exist an instance $\Pi_j^{\hat{C}}$ which has a matching conversation [11] for τ. (Intuitively, a "fresh" instance is an attackable instance, one that has not been trivially solved by exposing the server's private information.)

Let k be a security parameter, let $n \geq 1$, and let $\mathcal{A}$ be a probabilistic algorithm. The security experiment $\mathsf{Exec}^{\mathrm{DOS}}_{\mathcal{A},n,P}(k)$ for DoS resistance of a protocol P is defined as follows:

- $\mathsf{Exec}^{\mathrm{DOS}}_{\mathcal{A},n,P}(k)$: Run $\mathsf{GlobalSetup}(k)$. For each $\hat{S} \in \mathsf{Servers}$, run $\mathsf{ServerSetup}(\hat{S})$. Run $\mathcal{A}(\mathit{params})$ with oracle access to Send and Expose. If, summing over all servers, the number of fresh instances accepted is n, then return **true**, otherwise return **false**.

A protocol is DoS-resistant if the probability that an adversary with bounded runtime can cause a server to accept n fresh presessions is bounded:

Definition 6.10 (Denial-of-Service-Resistant Protocol). Let $\epsilon_{k,n}(t)$ be a family of functions that are monotonically increasing in t, where $\epsilon_{k,n}(t) \leq \epsilon_{k,1}(t/n)$ for all t, n such that $\epsilon_{k,n}(t) \leq 1$. Fix a security parameter k. Let $n \geq 1$. We say that a protocol P is $\epsilon_{k,n}(\cdot)$-*denial-of-service-resistant* if

1. For all probabilistic algorithms $\mathcal{A}$ running in time at most t,

$$\Pr\left(\mathsf{Exec}^{\mathrm{DOS}}_{\mathcal{A},n,P}(k) = \textbf{true}\right) \leq \epsilon_{k,n}(t) + \mathrm{negl}(k), \text{ and}$$

2. No call to $\mathsf{SAction}j_P(\hat{S}, i, m, M)$ results in an expensive operation unless $\Pi_i^{\hat{S}}$ has accepted.

Remark 6.11. This definition of DoS resistance contains two aspects. The first aspect addresses the ability of an adversary to cause the server to accept a presession: the inequality in part 1 provides a bound on the ability of an adversary to cause the server to accept n presessions when the adversary has only done t operations. The requirement that $\epsilon_{k,n}(t) \leq \epsilon_{k,1}(t/n)$ enforces the idea that the amount of work required to cause n presessions to be accepted should be n times the amount of work required to cause one presession to be accepted.

The second aspect addresses the idea that a server should not perform expensive operations unless the countermeasure has been passed. As the notion of 'expensive' can vary from setting to setting, we leave it vague, but it can easily be formalised, for example by using Meadows' cost-based framework [43].

Though a DoS countermeasure does not provide explicit authentication, we still wish to avoid impersonations. For example, suppose a client $\hat{C}$ sends messages meant to prove its legitimate intentions in communicating with server $\hat{S}$. It should not be possible for an adversary to easily use those messages to cause another server $\hat{S}'$ to perform expensive operations, nor should it be possible for an adversary to easily use those messages to convince $\hat{S}$ that a different client $\hat{C}'$ intended to communicate with $\hat{S}$.

This is prevented by the model since party names are included in the presession identifiers. If an adversary observed a presession $[\hat{C}, \hat{S}, \tau]$ and then tried to use that information to construct a presession $[\hat{C}', \hat{S}, \tau']$ of another user $\hat{C}'$ with the same server, then this new presession would be unexposed and the adversary would be prohibited from easily causing a server to accept it by Definition 6.10. This in effect requires a binding of values in the DoS countermeasure transcript τ to the parties – $\hat{C}$ and $\hat{S}$ – in question.

We follow the approach of Stebila and Ustaoglu [61] in dealing with replay attacks, where replay attacks are avoided by uniqueness of presession identifiers of accepted presessions. This does mean that the server has to store a table of presession identifiers, but this does not constitute a vector for a DoS attack because the server only stores a presession identifier after it accepts a presession, so it is doing an expensive operation only after the DoS countermeasure has been passed.

6.4.1.7 Building DoS-Resistant Protocols from Client Puzzles

Having defined what it means for a protocol to be denial-of-service, how can we build such a protocol and prove that it satisfies that definition? Generically, we can do so using client puzzles. In this section, we present a generic technique that transforms any protocol P into a DoS-resistant protocol $D(P)$. Our technique uses strongly difficult interactive client puzzles as a DoS countermeasure and message authentication codes for integrity of stateless connections [7]. We prove that the combined protocol $D(P)$ is a DoS-resistant protocol.

The client and server each provide nonces and construct the string *str* using their names, nonces, and any additional information, such as a timestamp or information from a higher-level protocol. The server generates a puzzle from *str*, authenticates the puzzle using the message authentication code (to avoid storing state), and sends it to the client. The client solves the puzzle using its own string *str* and sends the solution to the server. The server checks the message authentication code and the correctness of the solution. Finally, the server checks that the presession is unique and accepts. The messages for the DoS countermeasure are interleaved, where possible, with the messages of the main protocol, and after the countermeasure has accepted the main protocol continues as needed.

$D(P)_{\mathsf{Puz},d,\mathsf{MAC},k}$ – $\mathsf{Send}(\hat{U}, i, M)$ protocol specification		
Client $\hat{C}$		Server $\hat{S}$ long-term secret: $\rho_{\hat{S}} = mk_{\hat{S}}$
$\underline{\mathsf{CAction1}_{D(P)}}$:		
$N_C \leftarrow_R \mathsf{NonceSpace}$		
$(M_1, m_1) \leftarrow \mathsf{CAction1}_P()$	$\xrightarrow{\hat{C}, N_C, M_1}$	$\underline{\mathsf{SAction1}_{D(P)}}$:
		$N_S \leftarrow_R \mathsf{NonceSpace}$
		$(M_1', m_1') \leftarrow \mathsf{SAction1}_P(M_1)$
		$str \leftarrow (\hat{C}, \hat{S}, N_C, N_S, M_1, M_1')$
		$puz \leftarrow \mathsf{GenPuz}(\perp, d, str)$
$\underline{\mathsf{CAction2}_{D(P)}}$:	$\xleftarrow{N_S, M_1', puz, \sigma}$	$\sigma \leftarrow \mathsf{MAC}_{mk_{\hat{S}}}(str, puz)$
$str \leftarrow (\hat{C}, \hat{S}, N_C, N_S, M_1, M_1')$		
$soln \leftarrow \mathsf{FindSoln}(str, puz, t)$		
$(M_2, m_2) \leftarrow \mathsf{CAction2}_P(m_1, M_1')$	$\xrightarrow{str, puz, \sigma, soln}$	$\underline{\mathsf{SAction2}_{D(P)}}$:
		reject if $\sigma \neq \mathsf{MAC}_{mk_{\hat{S}}}(str, puz)$
		reject if $\neg\mathsf{VerSoln}(\perp, str, puz, soln)$
		$\tau \leftarrow (N_C, N_S, M_1, M_1', puz, soln)$
		verify no stored presession $[\hat{C}, \hat{S}, \tau]$
		accept and store presession $[\hat{C}, \hat{S}, \tau]$
continue with $\mathsf{CActionj}_P$		continue with $\mathsf{SActionj}_P$

Fig. 6.6 $D(P)_{\mathsf{Puz},d,\mathsf{MAC},k}$ DoS countermeasure protocol

Specification

Let P be a protocol such that $\mathsf{SAction1}_P$ does not involve any expensive operations. Let k be a security parameter. Let $\mathsf{MAC} : \{0,1\}^k \times \{0,1\}^* \to \{0,1\}^k$ be a family of secure message authentication codes [10]. Let $\mathsf{Puz} = (\mathsf{Setup}, \mathsf{GenPuz}, \mathsf{FindSoln}, \mathsf{VerSoln})$ be a strongly difficult interactive client puzzle with long-term secret key space $\mathsf{sSpace} = \{\perp\}$ (there is no long-term secret key for puzzles). Although this may seem restrictive, many puzzles satisfy this constraint, including the hash-based puzzle earlier in this section. Fix a DoS difficulty parameter $d \in \mathsf{diffSpace}$.

Let $D(P)_{\mathsf{Puz},d,\mathsf{MAC},k}$ be the protocol consisting of the following algorithms:

- $\mathsf{GlobalSetup}(1^k)$: Set $\rho\mathsf{Space} \leftarrow \{0,1\}^k$ and $\mathsf{NonceSpace} \leftarrow \{0,1\}^k$
- $\mathsf{ServerSetup}(\hat{S} \in \mathsf{Servers})$: Set $mk_{\hat{S}} \leftarrow_R \{0,1\}^k$ and $\rho_{\hat{S}} \leftarrow mk_{\hat{S}}$
- $\mathsf{CAction}j_{D(P)}(\ldots)$, $\mathsf{SAction}j_{D(P)}(\ldots)$: as specified by the protocol in Fig. 6.6

Remark 6.12. The construction $D(P)$ requires that $\mathsf{SAction1}_P$ not involve any expensive operations, as $\mathsf{SAction1}_P$ is called by $\mathsf{SAction1}_{D(P)}$ before the server instance has accepted. If $\mathsf{SAction1}_P$ does in fact involve expensive operations, then P would need to be rewritten so that the expensive operation is delayed until $\mathsf{SAction2}_P$. In other words, the $D(P)$ construction may result in an additional round being added before the P protocol is run; this should not be surprising.

Additionally, $\mathsf{SAction1}_P$ may result in a private output m'_1 which the server instance needs to store until the next message is received. If state storage is considered an expensive operation (as it could be a vector for a resource depletion DoS attack), then there are two options: use a stateless connection [7] to encrypt m'_1 and send it to the client who must return it in the following round, or, as above, rewrite P so as to delay the operation until $\mathsf{SAction2}_P$.

Theorem 6.13. *Let P be a protocol such that $\mathsf{SAction1}_P$ does not involve any expensive operations. Suppose that Puz is an $\epsilon_{d,k,n}(t)$-strongly difficult interactive puzzle with long-term secret key space $\mathsf{sSpace} = \{\bot\}$ and that MAC is a family of secure message authentication codes. Then $D(P)_{\mathsf{Puz},d,\mathsf{MAC},k}$ is an $\epsilon'_{d,k,n}(t)$-denial-of-service-resistant protocol, for $\epsilon'_{k,n}(t) = \epsilon_{d,k,n}(t + t_0 q_{\mathsf{Send}}) + \mathrm{negl}(k)$, where q_{Send} is the number of Send queries issued and t_0 is a constant depending on the protocol, assuming $t \in \mathrm{poly}(k)$.*

The proof of Theorem 6.13 follows by a sequence of games, first replacing the message authentication code with a MAC challenger, and then replacing the puzzles with a Puz challenger. Fresh accepted presessions correspond to valid solutions to the Puz challenger, yielding the bound relating the protocol and the puzzle.

6.4.2 Game Theory

A challenge in the client-puzzle approach is deciding on the difficulty of the puzzle to be sent. One approach suggested by Feng et al. [24] is to adjust the puzzle difficulty proportional to the current load on the server. Juels and Brainard [29] suggested that the difficulty of the puzzle be scaled uniformly for all clients according to the severity of the attack on the server. In both these approaches, the quality of service to legitimate users is not considered. Alternatively, the server can generate puzzles of varying difficulties based on a probability distribution. Such an approach based on game theory can be seen in [13, 22].

Though there have been several works that formally analyse denial-of-service attacks using game theory [2, 13, 22, 35, 39, 40, 55], only a few of them analyse the client-puzzle approach [13, 22, 39]. Bencsath et al. [13] modelled the client-puzzle approach as a single-shot strategic game and identified the server's optimal strategy. Fallah [22], on the other hand, used an infinitely repeated game to come up with puzzle-based defense mechanisms. He also proposed extensions to tackle distributed attacks. Recently, Jun-Jie [39] applied game theory to puzzle auctions.

Game theoretic defense mechanisms against DoS attacks focus on *fine-tuning the parameters* of the system in such a way that the server is not overloaded by the attacker. Our work builds on the game theoretic model and defense mechanisms proposed by Fallah.

In addition to the basic properties of a good puzzle listed in Sect. 6.3.2, we introduce the following requirement: *the difficulty of the puzzle should not be determined by the attacker without expending a minimal amount of computational*

Client		Defender
	$\xrightarrow{Request}$	$X = H(S, N_s, M)$
		$Y = H(X)$
	$\xleftarrow{(X', Y), N_s}$	$X' = X \oplus (I_1, I_2, ..., I_{k-1}, 1, 0_{k+1}, ..., 0_n)$
Find rp such that	$\xrightarrow{rp, N_s}$	$X = H(S, N_s, M)$
$H(rp) = Y$		$H(rp) \stackrel{?}{=} H(X)$

Fig. 6.7 Hidden difficulty puzzle 1. Here, H is a cryptographic hash function and I is a binary number chosen uniformly at random

effort. In the rest of this section, we propose three concrete puzzles that satisfy this requirement. Using game theory, we show that defense mechanisms are more effective when the puzzle difficulty is hidden from the attacker.

6.4.2.1 Hidden Difficulty Puzzle (HDP)

The difficulty of a client puzzle is said to be hidden if it cannot be determined by an attacker without expending a minimal amount of computational effort. We first introduce a modified version of the hash-reversal puzzle [29], which satisfies this requirement. The puzzle generation and verification are detailed in Fig. 6.7.

A preimage X is generated by hashing a server secret S, a server nonce N_s and a session parameter M together. The server nonce is used to check whether the puzzle is recent and the session parameter allows the server to be stateless [58]. The preimage is again hashed to obtain Y and some of the first k bits of X are randomly inverted. Let X' be the resultant binary string. The puzzle consisting of X' and Y is sent to the client along with the server nonce.

Note that k determines the difficulty of the puzzle and is unknown to the client. In order to solve the puzzle he would have to carry out an exhaustive search and arrive at the solution after testing up to 2^k possible preimages. The solution to the puzzle along with the received nonce is sent back to the defender. The defender recomputes the preimage X and verifies the solution.

Here, puzzle generation takes 2 hash computations, while the verification takes 3 hash computations. Further, the client needs to compute an average of $\frac{(2^k+1)}{2}$ hashes to solve the puzzle.

Assume the defender uses two instances of the described hidden difficulty puzzle, P_1 and P_2, with difficulty levels k_1 and k_2, respectively. On receiving a puzzle, the attacker does not know whether it is P_1 or P_2. Any solution to P_1 would have the k_1^{th} bit inverted, while any solution to P_2 will have the k_2^{th} bit inverted. Clearly, the solution spaces of the two puzzles do not overlap. To solve the puzzle, the attacker could first test possible preimages for one of the puzzles and if it is not solved, test preimages for the other. He could also try out preimages for both puzzles simultaneously. In any case, the attacker would know the puzzle difficulty only after putting in the effort required to solve one of the puzzles. Clearly, the attacker cannot determine the puzzle difficulty without minimal resource expenditure.

6.4.2.2 Game Theoretic Analysis of HDP

We shall now see how a hidden difficulty puzzle can make a defense mechanism more effective. We assume the network consists of a server, a set of legitimate clients/users, and an attacker. The attacker seeks to mount a denial-of-service attack on the client-server protocol by overloading the computational resources of the server. The client-puzzle approach is used as a defense mechanism against the attack. The interaction between the attacker and the defender during a denial-of-service attack is viewed as a two-player game. We use the same notations as in [22] to model the game.

Rational Attacker. Our primary assumption is that the attacker is rational. The objective of the attacker is to maximise the resource expenditure of the defender with minimum computational effort. This is reasonable from the point of view of the proof of work paradigm, where a rational attacker is the **strongest** attacker. On the other hand, if the attacker is not rational and takes nonoptimal decisions, it would be in the interest of the defender.

6.4.2.3 Model

Consider a game between an attacker and a defender. We categorise the puzzles used by the defender as either easy or difficult. A puzzle is easy if the time taken to solve it is lesser than the time taken by the defender to provide the requested service and is difficult if the time taken to solve the puzzle is greater than the service time. Assume that the defender uses an easy puzzle P_1 and a difficult puzzle P_2 to defend himself. (We later show in Sect. 6.4.2.6 that two puzzles are sufficient for an effective defense mechanism.)

Let T be a reference time period. Let α_m be the fraction of the time T that the defender spends in providing the service, α_{PP} be the fraction of T he takes to produce a puzzle, and α_{VP} be the fraction of T he takes to verify it. Let α_{SP_1} be the fraction of T that the attacker is expected to spend to solve P_1 and let α_{SP_2} be the fraction of T to solve P_2. As mentioned earlier, the defender chooses the puzzles P_1 and P_2 such that $\alpha_{SP_1} < \alpha_m < \alpha_{SP_2}$.

Attacker Actions. On receiving a puzzle, the attacker may choose from one amongst the following actions: (i) correctly answer the puzzle (CA), (ii) randomly answer the puzzle (RA), and (iii) try to answer the puzzle, but give up if it is too hard (TA). In the case of TA, the attacker gives a correct answer if the puzzle is solved and a random answer if he gives up. Note that TA is relevant only when the puzzle difficulty is hidden. If the attacker knows the difficulty of the puzzle on receiving it, he can immediately decide on whether to answer it correctly or randomly.

Attacker Payoff. Let u_2 denote the payoff of the attacker. Attacker's action is profitable if the defender expends computational resource, else it is a loss when he himself incurs an expenditure. Let P_i, $i = 1, 2$ be the puzzle received by the attacker. If he chooses CA, he incurs a cost α_{SP_i} in solving the puzzle, while the defender expends resources in generating and verifying the puzzle and providing the requested service. His payoff is therefore

$$u_2(P_i; CA) = \alpha_{PP} + \alpha_{VP} + \alpha_m - \alpha_{SP_i}.$$

If the attacker chooses RA, the attacker incurs no cost, while the defender incurs a cost in generating and verifying the puzzle.

$$u_2(P_i; RA) = \alpha_{PP} + \alpha_{VP}.$$

If the attacker's response is TA, his payoff depends on when he gives up.

Try and Answer. When the attacker receives puzzle P_1, he is better off answering it correctly, rather than answering it randomly. This is because $u_2(P_1; CA) > u_2(P_1; RA)$ (as $\alpha_{SP_1} < \alpha_m$). On the other hand, when he receives P_2, $u_2(P_2; CA) < u_2(P_2; RA)$ (as $\alpha_{SP_2} > \alpha_m$), and hence RA would be a better choice than CA. A decision on RA and CA can be made only if the puzzle difficulty is known. In the case of HDPs, the attacker is sure that the puzzle is not P_1 only when he fails to solve it after expending α_{SP_1} amount of resource. Hence, when the attacker chooses TA, he puts in the (minimal) effort required to solve P_1 and gives up when he realises the puzzle is P_2. If the puzzle sent is P_1, the attacker would solve it with the minimal effort and give the correct answer, while if it is P_2, he would give up and send a random answer. His payoff for the action TA is given by

$$u_2(P_1; TA) = \alpha_{PP} + \alpha_{VP} + \alpha_m - \alpha_{SP_1} \text{ and}$$
$$u_2(P_2; TA) = \alpha_{PP} + \alpha_{VP} - \alpha_{SP_1}.$$

6.4.2.4 Analysis of Attacker Payoff

Let $0 < p < 1$ be the probability with which the attacker receives puzzle P_1. ($1 - p$ is the probability with which he receives P_2.) We denote the corresponding mixed strategy of the defender as α_1. If the difficulty of the puzzle is hidden, the expected payoff of the attacker for his actions is given by

$$U_2(\alpha_1; CA) = \alpha_{PP} + \alpha_{VP} + \alpha_m - p\alpha_{SP_1} - (1 - p)\alpha_{SP_2}, \tag{6.1}$$
$$U_2(\alpha_1; RA) = \alpha_{PP} + \alpha_{VP} \text{ and} \tag{6.2}$$
$$U_2(\alpha_1; TA) = \alpha_{PP} + \alpha_{VP} + p\alpha_m - \alpha_{SP_1}. \tag{6.3}$$

The attacker's choice is influenced by the probability p and the values of α_{SP_1} and α_{SP_2}. The attacker would prefer RA over TA only if $p < \frac{\alpha_{SP_1}}{\alpha_m} = p_t$. He would prefer RA over CA when $p < \frac{\alpha_{SP_2} - \alpha_{SP_1}}{\alpha_{SP_2} - \alpha_m} = p_c$. From the payoffs, it is evident that the action CA is more beneficial than TA when $\alpha_{SP_2} - \alpha_{SP_1} < \alpha_m$.

On the other hand, if the difficulty of the puzzle is known to the attacker, he would choose CA if the puzzle is P_1 and RA if it is P_2 [22]. We represent the corresponding strategy as (CA, RA) and his expected payoff is

$$U_2(\alpha_1; (CA, RA)) = \alpha_{PP} + \alpha_{VP} + p(\alpha_m - \alpha_{SP_1}). \tag{6.4}$$

We now show that **the attacker receives lower payoff when HDPs are used**. Consider each of the attacker's actions:

(i) ***RA***. The attacker chooses RA when (a) $\alpha_{SP_2} - \alpha_{SP_1} < \alpha_m$ and $p < p_t$ or (b) $\alpha_{SP_2} - \alpha_{SP_1} > \alpha_m$ and $p < p_c$. From (6.2) and (6.4), for $0 < p < 1$,

$$U_2(\alpha_1; RA) < U_2(\alpha_1; (CA, RA)). \tag{6.5}$$

(ii) ***TA***. The attacker chooses TA when $\alpha_{SP_2} - \alpha_{SP_1} < \alpha_m$ and $p > p_t$. From (6.3) and (6.4), for $0 < p < 1$,

$$U_2(\alpha_1; TA) < U_2(\alpha_1; (CA, RA)). \tag{6.6}$$

(iii) ***CA***. The attacker chooses CA when $\alpha_{SP_2} - \alpha_{SP_1} > \alpha_m$ and $p > p_c$. From (6.1) and (6.4), as $\alpha_{SP_2} < \alpha_m$ and $0 < p < 1$,

$$U_2(\alpha_1; CA) < U_2(\alpha_1; (CA, RA)). \tag{6.7}$$

In all three cases, the attacker is benefitted less when the puzzle difficulty is hidden than when it is known to him.

Effectiveness. We define the effectiveness of a defense mechanism using proof of work as *the difference between the amount of work done by the attacker and the amount of work done by the defender*. Clearly, a defense mechanism would be more effective when it uses a HDP.

6.4.2.5 Defense Mechanisms

We propose two defense mechanisms against DoS attacks based on the concept of Nash equilibrium. Hidden difficulty puzzles are used in both the defense mechanisms. As in [22], the Nash equilibrium is used in a prescriptive way, where the defender selects and takes part in a specific equilibrium profile and the best thing for the attacker to do is to conform to his equilibrium strategy. Initially, we assume that the attack takes place from a single machine and later propose an extension to handle distributed attacks.

Table 6.1 Cost incurred by the players and the legitimate user when action profile a is chosen

a	$\psi_1(a)$	$\psi_2(a)$	$\psi_u(a)$
$(P_l; RA)$	$\alpha_{PP} + \alpha_{VP}$	0	α_{SP_l}
$(P_i; TA)$	$\alpha_{PP} + \alpha_{VP} + \alpha_m$	α_{SP_i}	α_{SP_i}
$(P_j; TA)$	$\alpha_{PP} + \alpha_{VP}$	α_{SP_k}	α_{SP_j}
$(P_l; CA)$	$\alpha_{PP} + \alpha_{VP} + \alpha_m$	α_{SP_l}	α_{SP_l}

Here $1 \leq l \leq n$, $1 \leq i \leq k$ and $k+1 \leq j \leq n$

6.4.2.6 Strategic Game

The attacker is unaware of the difficulty of a puzzle when he receives it and the defender is unaware of the attacker's response when he sends the puzzle. We therefore model the interaction between an attacker and defender during a denial-of-service attack as a strategic game.

Defender's Actions. We assume the defender uses n puzzles $P_1, P_2, \ldots, P_n$ such that $\alpha_{SP_1} < \ldots < \alpha_{SP_k} < \alpha_m < \alpha_{SP_{k+1}} < \ldots < \alpha_{SP_n}$, where α_{SP_i} is the cost incurred by an attacker in solving puzzle P_i. The generation and verification costs are same for all puzzles and equal to α_{PP} and α_{VP}, respectively. (This assumption is reasonable as generation and verification time for a good client-puzzle is negligible [58].)

Defender's Payoff. The defender seeks to maximise the effectiveness of the defense mechanism and minimise the cost to a legitimate user. We introduce a balance factor $0 < \eta < 1$ that allows him to strike a balance between the two. His payoff is therefore given by $u_1 = (1-\eta)(\text{effectiveness}) + \eta(-\text{legitimate user cost})$. Let $\psi_1(a)$ and $\psi_2(a)$ be the cost incurred by the defender and attacker, respectively, when the action profile a is chosen. Let $\psi_u(a)$ be the corresponding cost to a legitimate user. Hence,

$$u_1(a) = (1 - \eta)(-\psi_1(a) + \psi_2(a)) + \eta(-\psi_u(a)).$$

The costs incurred by the players and the legitimate user for the various action profiles are tabulated in Table 6.1.

A legitimate user always solves the given puzzle and incurs a cost α_{SP_i} for a puzzle P_i. Here, it is assumed that the attacker and a legitimate user take equal time to solve a puzzle. The model can be easily extended to distributed attacks, where the computational power of the attacker is considered much higher than that of a legitimate user [48].

For the puzzles P_1, $P_2, \ldots$ and P_k, the attacker is better off giving the correct answer, while for puzzles $P_{k+1}, \ldots$ and P_n, the attacker is better off giving a random answer. When the puzzle difficulty is unknown, the attacker may choose to try and answer (TA), where the maximum effort he puts in is the effort required to solve P_k. If the puzzle is solved with a maximum resource expenditure of α_{SP_k}, he sends a correct answer. Otherwise, he gives up and sends a random answer.

Proposition 6.14 ([48]). *In the strategic game of the client-puzzle approach, the best response of the defender to the attacker's action TA is the puzzle P_k or the puzzle P_{k+1} or a lottery over both.*

Let P_1 and P_2 be the two puzzles corresponding to Proposition 6.14.

Analysis of Defender Payoff

Let us consider the defender's mixed strategy α_1, where he chooses P_1 with probability $0 < p < 1$ and P_2 with probability $1 - p$. A legitimate user would always incur a cost $\psi_u = p\alpha_{SP_1} + (1-p)\alpha_{SP_2}$. If the puzzle difficulty is hidden, the attacker would choose an action $a_2 \in \{RA, TA, CA\}$. The defender's payoff would then be

$$u_1(\alpha_1; a_2) = (1 - \eta)(-u_2(\alpha_1; a_2)) + \eta(-\psi_u).$$

As discussed earlier, the attacker would choose $a_2 = (CA; RA)$ [22] if the puzzle difficulty is not hidden and the corresponding payoff to the defender would be

$$u_1(\alpha_1; (CA; RA)) = (1 - \eta)(-u_2(\alpha_1; (CA; RA))) + \eta(-\psi_u).$$

For the same value of η, it is seen from (6.5), (6.6), and (6.7) that $u_1(\alpha_1; a_2) > u_1(\alpha_1; (CA; RA))$ for all $a_2 \in \{RA, TA, CA\}$. Hence, **the defender receives higher payoff while using HDPs**.

Nash Equilibrium

We now analyse the existence of Nash equilibria in the game of the client-puzzle approach. One possible Nash equilibrium is where the attacker chooses the action TA. The conditions for such an equilibrium are given in the following proposition.

Proposition 6.15. *In the strategic game of the client-puzzle approach, for $0 < \eta < \frac{1}{2}$, a Nash equilibrium of the form $(p \circ P_1 \oplus (1 - p) \circ P_2; TA)$[5], $0 < p < 1$, exists if $\eta = \frac{\alpha_m}{\alpha_m + \alpha_{SP_2} - \alpha_{SP_1}}$, $\alpha_{SP_2} - \alpha_{SP_1} > \alpha_m$ and $p > \frac{\alpha_{SP_1}}{\alpha_m}$.*

Proof. Let us prove the existence of a Nash equilibrium, where the defender uses a mixed strategy $\alpha_1 = p \circ P_1 \oplus (1 - p) \circ P_2$, where $0 < p < 1$ and the attacker uses the pure strategy TA. The profile $(\alpha_1; TA)$ is a Nash equilibrium if

$$u_1(P_1; TA) = u_1(P_2; TA), \tag{6.8}$$

$$u_2(\alpha_1; TA) > u_2(\alpha_1; RA) \text{ and} \tag{6.9}$$

$$u_2(\alpha_1; TA) > u_2(\alpha_1; CA). \tag{6.10}$$

[5] The notation $p_1 \circ a_1 \oplus p_2 \circ a_2 \oplus ... \oplus p_n \circ a_n$ denotes a lottery over the set of actions $\{a_1, a_2, ..., a_n\}$, where $p_1 + p_2 + ... + p_n = 1$.

Equation 6.8 is satisfied when

$$\eta = \frac{\alpha_m}{\alpha_m + \alpha_{SP_2} - \alpha_{SP_1}}, \tag{6.11}$$

(6.9) is satisfied when $p > \frac{\alpha_{SP_1}}{\alpha_m}$ and (6.10) is satisfied when

$$\alpha_{SP_2} - \alpha_{SP_1} > \alpha_m. \tag{6.12}$$

From (6.11) and (6.12), it can be easily seen that the maximum value that η can take is less than $\frac{1}{2}$. Hence, $0 < \eta < \frac{1}{2}$. □

We now construct a defense mechanism against a DoS attack by prescribing the Nash equilibrium given in Proposition 6.15. A Nash equilibrium allows us to predict the behaviour of a rational attacker during a DoS attack, but does not prevent the flooding attack from being successful.

Mitigating DoS Attack

Let N be the maximum number of requests that an attacker can send in time T. It is assumed that the defender has a resource r_p for puzzle generation and verification and another resource r_m for providing the requested service [22]. As per the property of a good client puzzle, the generation and verification time must be negligible. In fact, the verification time can be minimised by using a table lookup [65]. Hence, it is reasonable to assume that r_p is not exhausted in an attack, that is $N(\alpha_{PP} + \alpha_{VP}) < 1$. On the other hand, the attack is successful when r_m is exhausted before all requests are serviced. If β is the probability with which the attacker solves a given puzzle, $N\beta$ is the expected number of attack requests for which the defender would provide service. When $N\beta\alpha_m > 1$, the defender is overwhelmed and the attack is successful. In order to mitigate an attack, we need to ensure that

$$N\beta\alpha_m \leq 1 \text{ or } \beta \leq \frac{1}{N\alpha_m}.$$

In the prescribed Nash equilibrium, $\beta = p$ and the following condition must hold for an attack to be unsuccessful: $\frac{\alpha_{SP_1}}{\alpha_m} < p < \frac{1}{N\alpha_m}$. Note that this is possible only if $\alpha_{SP_1} < \frac{1}{N}$.

The probability p is chosen such that the defender can provide service for all attack requests. It has to be remembered that even legitimate requests need to be serviced during an attack. Hence, out of the total number of requests that the defender can service, we take $\frac{1}{\alpha_m}$ as the number of requests allocated to the defense mechanism, while the rest are for the legitimate users.

We propose a defense mechanism based on the prescribed Nash equilibrium. (We call it the **strategic game defense mechanism**.) The idea is to fine-tune the various parameters such that the conditions for the equilibrium are satisfied.

1. For a desirable balance factor η, $0 < \eta < \frac{1}{2}$, choose two puzzles P_1 and P_2 such that

$$\alpha_{SP_1} < \frac{1}{N} < \alpha_m < \alpha_{SP_2},\ \alpha_{SP_2} - \alpha_{SP_1} > \alpha_m \text{ and}$$
$$\eta = \frac{\alpha_m}{\alpha_m + \alpha_{SP_2} - \alpha_{SP_1}}.$$

2. Choose a value for p such that

$$\frac{\alpha_{SP_1}}{\alpha_m} < p < \frac{1}{N\alpha_m}.$$

3. On receiving a request, generate a random variable $\mathbf{x}$ such that $Pr(\mathbf{x} = 0) = p$ and $Pr(\mathbf{x} = 1) = 1 - p$. If $\mathbf{x} = 0$, send puzzle P_1. Otherwise, send puzzle P_2.

The other possible Nash equilibria in the strategic game have been discussed in the full version of this work [48].

6.4.2.7 Infinitely Repeated Game

During a denial-of-service attack, the attacker repeatedly sends requests to the defender. Since the game is played repeatedly, this scenario can be modelled as a repeated game. Also, the probability of arrival of a request is non-zero at any point in time, and hence the game is infinitely repeated [22].

Threat of Punishment

In repeated games, the payoff of a player is not only influenced by his decision in the current game, but is also influenced by his decisions in all periods of the game. A player would therefore be willing to take suboptimal decisions for the current game if it would give him a higher payoff in the long run. Deviation of a player from a desired strategy profile can be prevented if he is threatened with sufficient punishment in the future. Therefore, Nash equilibria with high payoffs can be achieved in an infinitely repeated game if a player is patient enough to see long-term benefits over short-term gains.

Minmax Payoff

The minmax payoff of a player is the minimum payoff that he can guarantee himself in a game, even when the opponents play in the most undesirable manner. The minmax payoff of player i in a strategic game is given by

$$v_i^* = \min_{\alpha_{-i} \in \Delta(A_{-i})} \max_{a_i \in A_i} u_i(a_i, \alpha_{-i}),$$

where $\Delta(X)$ is the set of probability distributions over X, A_i is the set of permitted actions for player i, and u_i is his payoff function.

Nash Equilibrium

Consider a two-player infinitely repeated game. Let v_1^* be the minmax payoff of player 1 and v_2^* be the minmax payoff of player 2. Let the mixed strategy profile resulting in v_1^* and v_2^* be $M^1 = (M_1^1; M_2^1)$ and $M^2 = (M_1^2; M_2^2)$ respectively. Here, M_1^2 is player 1's minmax strategy against player 2 and M_2^1 is player 2's minmax strategy against player 1.

Let $(\alpha_1; \alpha_2)$ be a strategy profile such that $v_1 = u_1(\alpha_1; \alpha_2) > v_1^*$ and $v_2 = u_2(\alpha_1; \alpha_2) > v_2^*$. Fudenberg and Maskin [25] show that an equilibrium where each player i receives an average payoff of v_i is possible through threat of punishment. The following repeated game strategy for player i is a Nash equilibrium.

(A) Play α_i each period as long as $(\alpha_1; \alpha_2)$ was played last period. After any deviation from phase (A), switch to phase (B).
(B) Play M_i^j, $j \neq i$, τ times (say) and then start phase (A) again. If there are any deviations while in phase (B), restart phase (B).

A description of their theorem along with the calculation of τ has been detailed in the full version of this work [48].

Interpretation

A possible equilibrium in the game of the client-puzzle approach consists of two phases:

Normal Phase (A). The defender and attacker choose a strategy profile, where each of them receives a payoff greater than the minmax payoff. Note that the strategy played may not be the optimal choice of the players in the given period. However, if either of them deviate, the game switches to the punishment phase (B).

Punishment Phase (B). In this phase, each player chooses a minmax strategy against the other player. This phase remains for τ periods, after which the game switches to the normal phase. Again, the minmax strategy may not be the optimal strategy of a player in the current period. But, any deviation from this strategy would restart the phase.

Any deviation in the normal phase is deterred by the threat of switching to the punishment phase. Similarly, a deviation from the punishment phase is deterred by the threat of prolonged punishment. Note that the **punishment period** τ must be sufficiently long for the equilibrium to exist.

The following propositions identify some minmax strategies in the game of the client-puzzle approach.

Proposition 6.16. *In the game of the client-puzzle approach, when* $\alpha_{SP_2} - \alpha_{SP_1} < \alpha_m$*, one of the defender's minmax strategy against the attacker is*

$$p_1 \circ P_1 \oplus (1 - p_1) \circ P_2,$$

where $p_1 = \frac{\alpha_{SP_2} - \alpha_m}{\alpha_{SP_2} - \alpha_{SP_1}}$.

Proof. Let $\alpha_1 = p_1 \circ P_1 \oplus (1 - p_1) \circ P_2$, $0 < p_1 < 1$, be the defender's minmax strategy against the attacker. By our assumption, $\alpha_{SP_2} - \alpha_{SP_1} < \alpha_m$. Clearly, the attacker would prefer CA over TA. Therefore, the attacker's minmax payoff is $max(U_2(\alpha_1; CA), U_2(\alpha_1; RA))$, where U_i is the expected payoff of player i for $i = 1, 2$. Note that $U_2(\alpha_1; CA) > U_2(\alpha_1; RA)$ when $\alpha_{PP} + \alpha_{VP} + \alpha_m - p_1\alpha_{SP_1} - (1 - p_1)\alpha_{SP_2} > \alpha_{PP} + \alpha_{VP}$ or $p_1 > \frac{\alpha_{SP_2} - \alpha_m}{\alpha_{SP_2} - \alpha_{SP_1}}$. The higher the value of p_1 above $\frac{\alpha_{SP_2} - \alpha_m}{\alpha_{SP_2} - \alpha_{SP_1}}$, the higher is the attacker's payoff. If $p_1 \leq \frac{\alpha_{SP_2} - \alpha_m}{\alpha_{SP_2} - \alpha_{SP_1}}$, the attacker's payoff is minimum and equal to $\alpha_{PP} + \alpha_{VP}$. This is the attacker's minmax payoff, and hence the attacker is minmaxed when the defender chooses the mixed strategy $p_1 \circ P_1 \oplus (1 - p_1) \circ P_2$, where $p_1 = \frac{\alpha_{SP_2} - \alpha_m}{\alpha_{SP_2} - \alpha_{SP_1}}$. □

Proposition 6.17 ([49]). *In the game of the client-puzzle approach, when* $\alpha_{SP_2} - \alpha_{SP_1} < \alpha_m$ *and* $0 < \eta < \frac{1}{2}$*, the attacker's minmax strategy against the defender is*

$$p_2 \circ CA \oplus (1 - p_2) \circ RA,$$

where $p_2 = \frac{\eta}{1-\eta}$.

Proof. Let the attacker's minmax strategy against the defender be $\alpha_2 = q_1 \circ RA \oplus q_2 \circ CA \oplus q_3 \circ TA$, where $q_1 + q_2 + q_3 = 1$. By our assumption, $\alpha_{SP_2} - \alpha_{SP_1} < \alpha_m$. When the attacker chooses CA, the defender would receive equal or lower payoff than when the attacker chooses TA. Hence, the attacker's minmax strategy against the defender should assign non-zero probabilities to CA and RA and zero probability to TA, that is $q_1 = p_2$, $q_2 = 1 - p_2$, and $q_3 = 0$, where $0 < p_2 < 1$. When $0 < \eta < \frac{1}{2}$, the defender's best response for the attacker's pure strategy RA is P_1 and that for CA is P_2. For the attacker's mixed strategy α_2, the defender's best response is P_1 only if $U_1(P_1; \alpha_2) > U_1(P_2; \alpha_2)$. This is possible when $p_2((1 - \eta)\alpha_{SP_1}) - \eta\alpha_{SP_1} > p_2((1 - \eta)\alpha_{SP_2}) - \eta\alpha_{SP_2}$ or $p_2 < \frac{\eta}{1-\eta}$. The lower the value of p_2 below $\frac{\eta}{1-\eta}$, the higher is the defender's payoff. Similarly, if $p_2 > \frac{\eta}{1-\eta}$, the defender would prefer P_2 over P_1 and his payoff increases as p_2 increases. Clearly, the defender is minmaxed when $U_1(P_1; \alpha_2) = U_1(P_2; \alpha_2)$ or $p_2 = \frac{\eta}{1-\eta}$. □

Punishment Phase Strategies. During the punishment phase, the defender chooses the mixed strategy $p_1 \circ P_1 \oplus (1 - p_1) \circ P_2$, while the attacker chooses the mixed strategy $p_2 \circ CA \oplus (1 - p_2) \circ RA$. It can be shown that the corresponding strategy profile is a Nash equilibrium, where both the players receive their minmax payoff. This means that each player's strategy in the punishment phase is the best response to the opponent's strategy. Clearly, deviations in the punishment phase are not profitable.

Normal Phase Strategies. The following strategy profile is chosen during the normal phase:

$$(p \circ P_1 \oplus (1-p) \circ P_2; TA),$$

where $0 < p < 1$. For a Nash equilibrium, this profile must give each player a payoff greater than his minmax payoff. This is possible when

$$\frac{\alpha_{SP_1}}{\alpha_m} < p < \frac{\alpha_{SP_1} - \eta(\alpha_{SP_2} - \alpha_m + \alpha_{SP_1})}{\alpha_m - \eta(\alpha_{SP_2} + \alpha_m - \alpha_{SP_1})}.$$

It can be shown that the defender receives higher payoff in the Nash equilibrium of the repeated game than in the Nash equilibrium of the single-shot strategic game described in subsection 6.4.2.6.

Mitigating DoS Attack

As seen in the previous defense mechanism, the existence of a Nash equilibrium does not necessarily prevent flooding. In the **normal phase**, flooding is prevented if

$$N\alpha_m p < 1 \text{ or } p < \frac{1}{N\alpha_m}.$$

In the **punishment phase**, even if the attacker chooses to answer all the puzzles correctly, his average resource expenditure would be $p_1\alpha_{SP_1} + (1-p_1)\alpha_{SP_2} = \alpha_m$. Since he cannot overload the defender, flooding is not possible in this phase.

The defense mechanism based on the described Nash equilibrium is given below. (We call it the **repeated game defense mechanism**.)

1. For a desirable balance factor η, $0 < \eta < \frac{1}{2}$, choose two puzzles P_1 and P_2 such that $\alpha_{SP_1} < \frac{1}{N} < \alpha_m < \alpha_{SP_2}\ and\ \alpha_{SP_2} - \alpha_{SP_1} < \alpha_m$.
2. Choose a value for p such that

$$\frac{\alpha_{SP_1}}{\alpha_m} < p < min\left(\frac{1}{N\alpha_m}, \frac{\alpha_{SP_1} - \eta(\alpha_{SP_2} - \alpha_m + \alpha_{SP_1})}{\alpha_m - \eta(\alpha_{SP_2} + \alpha_m - \alpha_{SP_1})}\right)$$

   and determine the value of p_1 according to

$$p_1 = \frac{\alpha_{SP_2} - \alpha_m}{\alpha_{SP_2} - \alpha_{SP_1}}.$$

3. Determine an appropriate value for τ.
4. Phase (A) and phase (B) of the defense mechanism have been described in Fig. 6.8, where **x** and **y** are random variables and $\phi(msg)$ is the phase corresponding to the puzzle whose solution has been received.

Here, we have considered only one possible minmax strategy profile in the defense mechanism. The other strategy profiles have been analysed in [48].

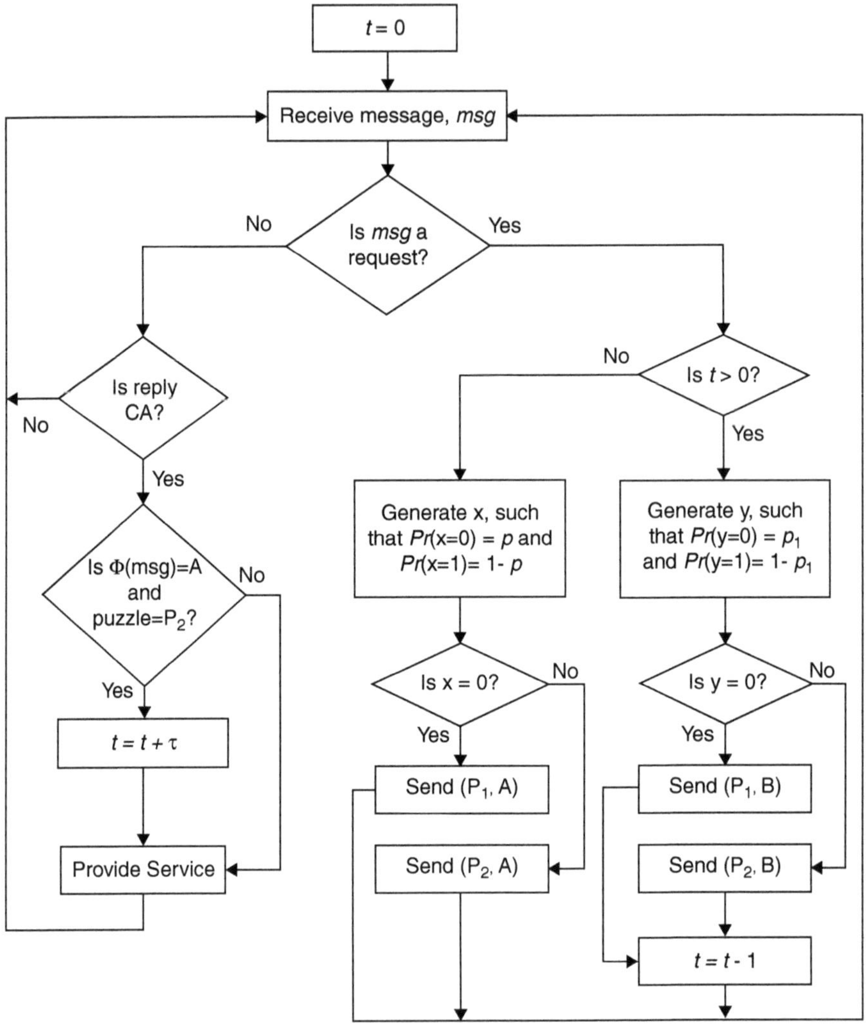

Fig. 6.8 Closed-loop defense mechanism

6.4.2.8 Comparison with Previous Work

We now compare the proposed HDP-based defense mechanisms with the corresponding defense mechanisms proposed by Fallah [22], where puzzle difficulties are known to the attacker. Previously, while analysing the payoffs of the attacker and defender, we did not consider the existence of a Nash equilibrium in the game. We now present an analysis of the attacker and defender payoffs under the equilibrium conditions given by the various defense mechanisms.

For the sake of convenience, the HDP-based strategic game defense mechanism shall be referred to as $HDM1$, while the HDP-based repeated game defense mechanism shall be referred to as $HDM2$. The corresponding puzzle-based defense mechanisms in [22] would be referred to as $PDM1$ and $PDM2$, respectively.

Proposition 6.18. *The expected equilibrium payoff of an attacker in HDM1 is lower than that in PDM1, while the expected equilibrium payoff of the defender is the same in both defense mechanisms.*

Proof. Let α^1 be the equilibrium strategy profile used in HDM1. Under equilibrium conditions, the expected payoff of the defender is $U_1(\alpha^1) = (1-\eta)(-\alpha_{PP} - \alpha_{VP} - \alpha_m + \alpha_{SP_1}) - \eta\alpha_{SP_1}$. The attacker, on the other hand, receives an average payoff of

$$U_2(\alpha^1) = \alpha_{pp} + \alpha_{VP} + p\alpha_m - \alpha_{SP_1}. \tag{6.13}$$

In the case of PDM1 [22], the strategy profile $\alpha^2 = (p \circ P_1 \oplus (1 - p) \circ P_2; (CA,RA))$ corresponds to an equilibrium when $\alpha_{SP_1} < \alpha_m < \alpha_{SP_2}$, $\alpha_{VP} < \alpha_m - \alpha_{SP_1}$, $\alpha_{VP} < \alpha_{SP_2} - \alpha_m$, and $\eta = \frac{\alpha_m - \alpha_{SP_1}}{\alpha_m - 2\alpha_{SP_1} + \alpha_{SP_2}}$, where $0 < \eta < \frac{1}{2}$ and $0 < p < 1$. Under the equilibrium conditions, the expected payoff for the defender is $U_1(\alpha^2) = (1-\eta)(-\alpha_{PP} - \alpha_{VP} - \alpha_m + \alpha_{SP_1}) - \eta\alpha_{SP_1}$ and that for the attacker is

$$U_2(\alpha^2) = \alpha_{PP} + \alpha_{VP} + p(\alpha_m - \alpha_{SP_1}). \tag{6.14}$$

Note that we are looking at two games with different equilibrium conditions. For a given value of $0 < \eta < \frac{1}{2}$, we choose the same values for α_{PP}, α_{VP}, α_{SP_1}, and p and different values for α_{SP_2} in the two games such that equilibrium conditions are satisfied. Clearly, $U_1(\alpha^1) = U_1(\alpha^2)$, while from (6.13) and (6.14), for $0 < p < 1$,

$$U_2(\alpha^1) < U_2(\alpha^2).$$

Thus, the expected payoff for the defender is same in both the defense mechanisms, while the attacker's expected payoff is lower in HDM1. □

Note that the equilibrium conditions are different in the two defense mechanisms. As a consequence, the defender's expected payoff does not change when HDPs are used. However, the attacker is benefitted less in $HDM1$ making it more effective than $PDM1$.

Proposition 6.19. *The minmax payoff of the defender in HDM2 is higher than that in PDM2, while the minmax payoff of the attacker is the same in both defense mechanisms.*

Proof. In PDM2, the minmax strategy profile in the infinitely repeated game is $(P_2; RA)$ [22]. The corresponding minmax payoff for the defender is

$$U_1(P_2; RA) = (1-\eta)(-\alpha_{PP} - \alpha_{VP}) - \eta\alpha_{SP_2} \tag{6.15}$$

and that for the attacker is $U_2(P_2; RA) = \alpha_{PP} + \alpha_{VP}$. In the case of HDM2, the minmax strategy profile is $(\alpha_1; \alpha_2)$, where $\alpha_1 = p_1 \circ P_1 \oplus (1-p_1) \circ P_2$ and

$\alpha_2 = p_2 \circ CA \oplus (1-p_2) \circ RA$. The defender's minmax payoff in HDM2 is given by $U_1(\alpha_1;\alpha_2) = p_1(p_2u_1(P_1;CA)+(1-p_2)u_1(P_1;RA))+(1-p_1)(p_2u_1(P_2;CA)+(1-p_2)u_1(P_2;RA))$, which reduces to

$$U_1(\alpha_1;\alpha_2) = (1-\eta)(-\alpha_{PP} - \alpha_{VP}) - \eta\alpha_m. \qquad (6.16)$$

Similarly, the attacker's minmax payoff in HDM2 is given by $U_2(\alpha_1;\alpha_2) = p_1(p_2u_2(P_1;CA) + (1-p_2)u_2(P_1;RA)) + (1-p_1)(p_2u_2(P_2;CA) + (1-p_2)u_2(P_2;RA)) = \alpha_{PP} + \alpha_{VP}$. It is clear that the attacker's minmax payoff is same in both defense mechanisms. For a given value of $0 < \eta < \frac{1}{2}$, choosing same values for α_{PP}, α_{VP} and α_{SP_1} in Eqs. 6.15 and 6.16 and different values for α_{SP_2} satisfying the corresponding equilibrium criteria, we have

$$U_1(\alpha_1;\alpha_2) > U_1(P_2;RA)$$

and $U_2(\alpha_1;\alpha_2) = U_2(P_2;RA)$. Thus, the defender's minmax payoff is higher in HDM2 when compared to PDM2, while the attacker's minmax payoff is same in both the defense mechanisms. □

Since the minmax payoff is a lower bound on the defender's payoff, the defender is better off in HDM2. Moreover, in $PDM2$, only P_2 puzzles are used by the defender in the punishment phase. Whereas, in $HDM2$, the defender's minmax strategy against the attacker is a lottery over P_1 and P_2. Clearly, **a legitimate user is hurt less in the punishment phase of HDM2 as he has a chance of receiving P_1.**

6.4.2.9 Client-Puzzles

The requirements of a puzzle in our defense mechanisms are given below.

1. **Hidden difficulty:** The difficulty of the puzzle should not be determined without a minimal number of computations.
2. **High puzzle resolution:** The granularity of puzzle difficulty must be high. This allows us to fine-tune the parameters of the defense mechanisms.
3. **Partial solution:** Submission of partial solutions should be possible without increasing the verification time. This requirement allows the defender to determine whether the attacker chose RA or TA.

We now introduce two new puzzles that conform to these requirements.

Puzzle 2:

The puzzle is described in Fig. 6.9. Here, the client can submit a partial solution by giving the correct answer for the first part ($rp1$) of the puzzle and a random

Client		Defender
	$\xrightarrow{Request}$	$X = H(S_1, N_s, M)$
		$Y = H(X)$
		$a = H(S_2, N_s, M) \bmod D + l$
		$X' = X - a$
		$Z = H(X')$
	$\xleftarrow{(X'', Y, Z), N_s}$	$X'' = X' \oplus (I_1, ..., I_{k-1}, 1, 0_{k+1}, ..., 0_n)$
Find $rp1$ such that $H(rp1) = Z$. Find a' such that $H(rp2) = Y$, where $rp2 = rp1 + a'$.	$\xrightarrow{rp1, rp2, N_s}$	$X = H(S_1, N_s, M)$
		$a = H(S_2, N_s, M) \bmod D + l$
		$H(rp1) \stackrel{?}{=} H(X - a)$
		$H(rp2) \stackrel{?}{=} H(X)$

Fig. 6.9 Hidden difficulty Puzzle 2. Here, S_1 and S_2 are server secrets, N_s is a server nonce, M is a session parameter, D and k are difficulty parameters, l is a constant, and I is a binary number chosen uniformly at random

answer for the second part ($rp2$). It takes 4 hash computations for generating a puzzle and a maximum of 6 hash computations for verifying the puzzle solution. When $l = 1$, the average number of hash computations required to solve the puzzle is $\frac{(2^k+1) + (D+1)}{2}$. Thus, the difficulty of the puzzle can be varied exponentially by adjusting k and linearly by adjusting D. Also, the client is not aware of the difficulty of the puzzle when he receives it.

Puzzle 3:

Figure 6.10 contains the puzzle description. When $l = 1$, the average number of hash computations required to solve this puzzle is $\frac{(D_a+1) + (D_b+1)}{2}$, and hence the difficulty varies linearly with D_a and D_b. The production of the puzzle requires 4 hash computations and the verification requires a maximum of 6 hash computations. Even here, the puzzle difficulty is hidden from the client.

6.5 Denial-of-Service Resistance in Key Agreement Protocols

Authenticated key agreement protocols provide the important cryptographic properties of authentication and confidentiality, but, because of the computationally intensive nature of public key operations, can be vectors of attack for

Client		Defender
	$\xrightarrow{Request}$	$X = H(S_1, N_s, M)$
		$Y = H(X)$
		$a = H(S_2, N_s, M) \bmod D_a + l$
		$X' = X - a$
		$Z = H(X')$
	$\xleftarrow{(X'', Y, Z), N_s}$	$X'' = X' - b$
Find b' such that $H(rp1) = Z$, where $rp1 = X'' + b'$.		
Find a' such that $H(rp2) = Y$, where $rp2 = rp1 + a'$.	$\xrightarrow{rp1, rp2, N_s}$	$X = H(S_1, N_s, M)$
		$a = H(S_2, N_s, M) \bmod D_a + l$
		$H(rp1) \stackrel{?}{=} H(X - a)$
		$H(rp2) \stackrel{?}{=} H(X)$

Fig. 6.10 Hidden difficulty Puzzle 3. Here, b is a value chosen uniformly at random from $\{1, \ldots, D_b\}$

denial-of-service (DoS) attacks: An attacker could initiate many key agreement sessions and cause a server to exhaust its computational resources.

As a result, a number of key agreement protocols [3, 31, 47, 59, 61] have been designed with some resistance to such denial-of-service attacks. Generally, servers employing these protocols use some or all of the techniques described in Sect. 6.3:

- Deferring expensive operations until the client's identity is confirmed (a form of *gradual authentication*);
- Not storing state on the server, possibly by offloading storage to the client (*stateless connections*);
- Requiring the client to echo back a given string to confirm its existence (*cookies*);
- *Balancing* the computational requirements of the client and the server
- Requiring the client to perform some computationally expensive task to demonstrate its legitimate intentions (*client puzzles*).

However, few of these protocols have had their denial-of-service resistance analysed rigourously; only the DoS-CMQV protocol of Stebila and Ustaoglu [61] has any formal analysis. Indeed, only recently have models [18, 61] been developed to analyse denial-of-service resistance as a cryptographic property in a provable security framework.

Our goal in this section is to examine the denial-of-service resistance of several key exchange protocols using the provable security model of Sect. 6.4.1. We observe that the majority of the protocols that were designed with ad hoc denial-of-service resistance have flaws that allow the DoS countermeasure to be circumvented. Only

Protocol	DoS counter-measure	DoS resistance Section 6.4.1 model	Key agreement security
JFKi [3]	cookies	weak DoS-resistance	ephemeral key reuse
JFKi w/puzzles [59]	hash preimage	not DoS-resistant	ephemeral key reuse
HIP [47]	hash preimage	strong DoS-resistance	ephemeral key reuse
IKEv2 [31]	cookies	authentication flaw ⇒ not DoS-resistant	CK01 model [15]
DoS-CMQV [61]	hash preimage	strong DoS-resistance	eCK [36]

Fig. 6.11 Denial-of-service resistance and security of key exchange protocols considered in this report

the Host Identity Protocol (HIP) [47] and DoS-CMQV [61] provably offer a high level of denial-of-service resistance. Our results are summarised in Fig. 6.11.

6.5.1 Analysing Common Countermeasures

No protection. If a protocol does not use any denial-of-service countermeasure – no cookies, no client puzzles – and therefore immediately accepts every presession, then, in the language of Sect. 6.4.1, it is an $\epsilon_{k,n}(\cdot)$-denial-of-service-resistant protocol with $\epsilon_{k,n}(t) \approx 1$.

Cookies. Let μ be the length of the cookie in bits. If a protocol uses cookies, then the adversary should respond to the challenge issued by the server by echoing it back, and otherwise should not be able to succeed with non-negligible probability. In this case, a protocol should be able to achieve $\epsilon_{k,n}(\cdot)$-denial-of-service resistance with

$$\epsilon_{k,n}(t) \approx \begin{cases} \text{negl}(\mu), & \text{if } t < n, \\ 1, & \text{if } t \geq n. \end{cases} \tag{6.17}$$

Hash-based puzzles. Let μ be the difficulty, in bits, of a hash function puzzle: for example find a preimage of a hash value given all but μ bits of the preimage, or find a value the hash of which begins with μ zeros. Here, a protocol could achieve $\epsilon_{k,n}(\cdot)$-denial-of-service resistance with

$$\epsilon_{k,n}(t) \approx t/n2^{\mu} + \text{negl}(k). \tag{6.18}$$

6.5.2 Just Fast Keying (JFKi)

The Just Fast Keying (JFKi) protocol [3] is an authenticated key exchange protocol that was proposed as a replacement for the Internet Key Exchange (IKE) protocol [26]. It offers authenticated key exchange between two parties, with identity

Modified JFKi

	Client $\hat{C}$		Server $\hat{S}$
			ρ, r, g^r, $\sigma \leftarrow \mathsf{Sign}_{\hat{S}}(g^r, g)$, $\boxed{\ell}$
1.	$N_C \leftarrow_R \{0,1\}^\lambda$		
2.	$N'_C \leftarrow H(N_C)$		
3.	$i \leftarrow_R \mathbb{Z}_q$		
4.	compute g^i		
5.		$\xrightarrow{N'_C, g^i, \hat{S}'}$	
6.			$N_S \leftarrow_R \{0,1\}^\lambda$
7.			$t \leftarrow H_\rho(g^r, N_S, N'_C, IP_C)$
8.		$\xleftarrow{N'_C, N_S, g^r, g, \hat{S}, \sigma, t, \boxed{\ell}}$	
9.	verify σ		
10.	$K_e \leftarrow H(g^{ir}, N'_C, N_S, 1)$		
11.	$K_a \leftarrow H(g^{ir}, N'_C, N_S, 2)$		
12.	$sk \leftarrow H(g^{ir}, N'_C, N_S, 0)$		
13.	$\sigma' \leftarrow \mathsf{Sign}_{\hat{C}}(N'_C, N_S, g^i, g^r, \hat{R}, sa)$		
14.	$m \leftarrow \{\hat{C}, sa, \sigma'\}^{K_e}_{K_a}$		
15.	$\boxed{\text{find } sol \text{ s.t. } [H(t\|sol)]_\ell = 0\dots0}$		
16.		$\xrightarrow{N_C, N_S, g^i, g^r, t, m, \boxed{sol}}$	
17.			verify $t = H_\rho(g^r, N_S, H(N_C), IP_C)$
18.			$\boxed{\text{verify } sol}$
19.			$K_e \leftarrow H(g^{ir}, N'_C, N_S, 1)$
20.			$K_a \leftarrow H(g^{ir}, N'_C, N_S, 2)$
21.			verify MAC of m under K_a
22.			decrypt m using K_e
23.			verify σ'
24.			$sk \leftarrow H(g^{ir}, N'_C, N_S, 0)$
25.			$\sigma'' \leftarrow \mathsf{Sign}_{\hat{S}}(N'_C, N_S, g^i, g^r, \hat{C}, sa, sa')$
26.			$m' \leftarrow \{\sigma''\}^{K_e}_{K_a}$
27.			accept
28.		$\xleftarrow{m'}$	
29.	verify MAC of m' under K_a		
30.	decrypt m' using K_e		
31.	verify σ''		
32.	accept		

Fig. 6.12 The JFKi protocol [3] (omitting commands in boxes) and the Modified JFKi protocol with client puzzles [59] (including commands in boxes)

protection for the initiator; there is a similar protocol, JFKr, that offers identity protection for the responder.

The protocol diagram for JFKi appears in Fig. 6.12.

JFKi also aims to offer limited denial-of-service resistance by having the server

- Use stateless connections to avoid storing state early in the protocol
- Reuse ephemeral keys to avoid public key operations in each run of the protocol
- Use an authenticator to verify the existence of the client.

This authenticator acts like a cookie in that it requires the client to respond with a value related to a previously sent value.

Based on the observation above about the effectiveness of cookies as a DoS countermeasure, we expect that JFK is an $\epsilon_{k,n}(\cdot)$-denial-of-service-resistant protocol, where $\epsilon_{k,n}(t)$ is as in Eq. 6.17. Given that this cookie is the only countermeasure before the server does an expensive group operation g^{ir}, the server's computational costs are not well-balanced by the computational costs of the client.

We note as well a peculiarity in the JFKi protocol. In the first flow from the initiator to the responder, the initiator sends its ephemeral public key g^i. The responder's action does not make use of the initiator's ephemeral public key at all, so in fact this value need not be sent by the initiator in its first flow.

6.5.3 Modified JFKi with Client Puzzles

Smith et al. [59] analysed JFKi using Meadows' cost-based framework [43], observing a disparity in the cost of the work required by the client versus the cost of the work done by the server to describe a denial-of-service attack. They proposed the use of client puzzles to prevent the attack.

The modified version of JFKi is presented in Fig. 6.12 with the differences from JFKi indicated using boxes.

This modification of JFKi is aimed to protect the server having to do many expensive public key operations by requiring the client to solve a hash-based client puzzle first.

We observe that this modified JFKi protocol with client puzzles remains susceptible to a denial-of-service attack identified via the model from Sect. 6.4.1. In particular, the client's ephemeral key g^i is not bound to the denial-of-service countermeasure, nor is the client's logical identity bound to client's network address via the denial-of-service countermeasure. As a result, an adversary who can spoof network addresses can steal the client's work in solving the puzzle, use its own ephemeral key $g^{i'}$, and establish a secure connection without the adversary having solved the puzzle itself.

Thus, rather than achieving denial-of-service resistance with a $\epsilon_{k,n}(t)$ of the form given in Eq. 6.18, as a protocol employing puzzles ought to, this protocol only offers DoS resistance with $\epsilon_{k,n}(t)$ of the form given in Eq. 6.17, equivalent to a protocol employing cookies.

6.5.4 Host Identity Protocol (HIP)

The Host Identity Protocol (HIP) [47] is a networking protocol aimed to decouple the transport layer from the internetworking layer in TCP/IP by introducing a new addressing space for hosts, separate from IP addresses, using cryptographic keys.

HIP with denial of service resistance parameter λ			
	Client $\hat{C}$		Server $\hat{S}$
			$\rho,\ x \leftarrow_R \mathbb{Z}_q,\ X \leftarrow g^x,\ s \leftarrow \mathsf{Sign}_{\hat{S}}(X, \ldots)$
1.		$\xrightarrow{\text{“hello”}, \hat{C}}$	
2.			$i \leftarrow f_\rho(\hat{C}, \hat{S})$
3.		$\xleftarrow{i,k,X,s}$	
4.	verify signature s		
5.	find j s.t. $[H(i, \hat{C}, \hat{S}, j)]_k = 0 \ldots 0$		
6.	$y \leftarrow_R \mathbb{Z}_q$		
7.	$Y \leftarrow g^y$		
8.	$\sigma \leftarrow X^y$		
9.	$s' \leftarrow \mathsf{Sign}_{\hat{C}}(\text{packet})$		
10.	$K_1 \leftarrow H(\sigma, \text{sort}(\hat{C}, \hat{S}), i, j, 1)$		
11.	$K_2 \leftarrow H(\sigma, \text{sort}(\hat{C}, \hat{S}), i, j, 2)$		
12.	$K_3 \leftarrow H(\sigma, \text{sort}(\hat{C}, \hat{S}), i, j, 3)$		
13.	$M \leftarrow \mathsf{MAC}_{K_2}(\text{packet})$		
14.		$\xrightarrow{j, Y, \hat{C}, \hat{S}, M, s'}$	
15.			$i \leftarrow f_\rho(\hat{C}, \hat{S})$
16.			verify $[H(i, \hat{C}, \hat{S}, j)]_k = 0 \ldots 0$
17.			$\sigma \leftarrow Y^x$
18.			$K_1 \leftarrow H(\sigma, \text{sort}(\hat{C}, \hat{S}), i, j, 1)$
19.			$K_2 \leftarrow H(\sigma, \text{sort}(\hat{C}, \hat{S}), i, j, 2)$
20.			$K_3 \leftarrow H(\sigma, \text{sort}(\hat{C}, \hat{S}), i, j, 3)$
21.			verify M
22.			verify s'
23.			$M' \leftarrow \mathsf{MAC}_{K_3}(\text{packet}')$
24.			$s'' \leftarrow \mathsf{Sign}_{\hat{R}}(\text{packet}')$
25.		$\xleftarrow{M', s''}$	
26.	verify M'		
27.	verify s''		
28.	session key $\leftarrow K_1$		session key $\leftarrow K_1$
29.	accept		accept

Fig. 6.13 Host Identity Protocol (HIP) [47]

The base protocol was designed with a denial-of-service countermeasure in the form of a hash function zero-preimage puzzle, which allows the responder to avoid committing state until a valid puzzle solution is received in the second flow from an initiator. However, this comes at the cost of the responder needing to reuse the ephemeral key across multiple sessions to avoid performing expensive computations.

Strictly speaking, the puzzle generation mechanism is unspecified by the HIP protocol. However, Appendix A of RFC 5201 [47] does specify a puzzle generation mechanism, and this is the one we consider in this section as in Fig. 6.13.

During the setup phase, a server $\hat{S}$ picks a secret key ρ. Upon receiving a request from client $\hat{C}$, it constructs a challenge $i = f_\rho(\hat{C}, \hat{S})$ using a pseudorandom function family f. This means that only one puzzle can be issued per client-server pair. If we were to allow multiple acceptances of the puzzle solution, then we would allow a denial-of-service attack against the server since the solution to one puzzle could be reused many times.

As it is, this technique opens the client up to a possible denial-of-service attack: The adversary could prevent legitimate clients from being able to form connections by spoiling that single client-server pair. An adversary could, for example, alter the client's signature s' or the client's authentication tag M; the server accepts the puzzle solution (and then disallows all other puzzle solutions from $\hat{C}$), but then rejects because s' or M fails to verify.

A simple solution for this problem is for puzzle generation to include a random nonce (provided either by the client or by the server). This allows a client to have multiple sessions with a server provided it solves the corresponding number of puzzles.

Since the puzzle solution j involves the party identities $\hat{C}$ and $\hat{S}$, puzzle solutions cannot be stolen by an adversary.

In our phrasing of the HIP protocol, we have omitted the transcript from the definition of the presession identifier $[\hat{C}, \hat{S}]$, because the HIP puzzle generation mechanism $i \leftarrow f_\rho(\hat{C}, \hat{S})$ can only generate a single puzzle per client-server pair, as noted above. Thus, only one presession can be accepted per client-server pair, at least until the server changes its private key ρ, which is outside the scope of this model.

HIP does not include the ephemeral public keys in the puzzle solving process. If the ephemeral public key were included in the transcript, and the transcript were included in the puzzle solution, then this would allow multiple sessions per client-server pair.

It is not hard to see that the client puzzle used in HIP is an interactive strongly difficult client puzzle. In particular, suppose H is a random oracle and f_ρ is a random function; let q be the number of queries to H. Then the client puzzle used in HIP is an $\epsilon_{\lambda,k,n}$(q)-interactive-strongly difficult client puzzle, where $\epsilon_{\lambda,k,n}(q) = (q+n)/(n2^k)$.

We can then show that HIP is a denial-of-service-resistant protocol:

Theorem 6.20. *Let f be a family of random functions and let H be a random oracle. Fix a puzzle difficulty k. Then HIP is an $\epsilon_{\lambda,n}(q)$-denial-of-service-resistant protocol, for $\epsilon_{\lambda,n}(q) = (q+n)/(n2^k) + \mathrm{negl}(\lambda)$.*

6.5.5 Internet Key Exchange v2 Protocol (IKEv2)

The Internet Key Exchange version 2 (IKEv2) protocol [31] is a key exchange protocol for the IPsec suite [34] which in turn provides security services for the IPv4

IKEv2			
	Client $\hat{C}$		Server $\hat{S}$
			ρ
1.	$N_C \leftarrow_R \{0,1\}^\lambda$		
2.	$i \leftarrow_R \mathbb{Z}_q$		
3.	compute g^i		
4.		$\xrightarrow{sa, g^i, N_C}$	
5.			$c \leftarrow H(N_C, IP_C, sa, \rho)$
6.		$\xleftarrow{c}$	
7.		$\xrightarrow{c, sa, g^i, N_C}$	
8.			verify c
9.			$N_S \leftarrow_R \{0,1\}^\lambda$
10.			$r \leftarrow_R \mathbb{Z}_q$
11.			compute g^r
12.		$\xleftarrow{sa', g^r, N_S}$	
13.	$s \leftarrow H(N_C, N_S, g^{ir})$		
14.	$K_e, K_a, sk \leftarrow \mathsf{PRF}(s, N_C, N_S, sa, sa')$		
15.	$\sigma \leftarrow \mathsf{Sign}_{\hat{C}}(\text{packets})$		
16.	$m \leftarrow \{\hat{C}, \hat{S}, \sigma, sa\}_{K_a}^{K_e}$		
17.		$\xrightarrow{m}$	
18.			$s \leftarrow H(N_C, N_S, g^{ir})$
19.			$K_e, K_a, sk \leftarrow \mathsf{PRF}(s, N_C, N_S, sa, sa')$
20.			verify MAC of m under K_a
21.			decrypt m using K_e
22.			verify σ
23.			$\sigma' \leftarrow \mathsf{Sign}_{\hat{S}}(\text{packets})$
24.			$m' \leftarrow \{ID_S, \sigma'', sa'\}_{K_a}^{K_e}$
25.			accept
26.		$\xleftarrow{m'}$	
27.	verify MAC of m' under K_a		
28.	decrypt m' using K_e		
29.	verify σ'		
30.	accept		

Fig. 6.14 Internet Key Exchange version 2 (IKEv2) protocol with cookies

and IPv6 protocols. IPsec is a very complex protocol and we omit a long discussion of its details, but refer the reader to a useful overview of IPsec for cryptographers given by Paterson [50].

The protocol diagram for IKEv2 appears in Fig. 6.14.

DoS resistance design. IKEv2 optionally uses cookies to test for liveness of the initiator: The responder only continues with a connection if the initiator correctly echoes back the cookie. Although generation of the cookies is not standardised, the RFC recommends hashing the initiator's nonce, IP address, and session identifier under the responder's secret hash key; this avoids the responder needing to store state.

6.5.5.1 Authentication Flaw Implies No DoS Resistance

IKEv2 was also designed to provide client-to-server and server-to-client authentication based on certificates. Mao and Paterson [41] discovered an authentication flaw in IKEv2 that can result in a type of denial of service.

The attack is as follows: An adversary $\mathcal{M}$ who controls a popular IKEv2 responder can relay traffic from a client $\hat{C}$ intended for $\mathcal{M}$ to another server $\hat{S}$. After relaying this information between $\hat{C}$ and $\hat{S}$, the client $\hat{C}$ (who thinks it is interacting with $\mathcal{M}$) will abort; however, the server $\hat{S}$, thinking it is interacting with $\hat{C}$, will accept and establish a session. In this attack, $\mathcal{M}$ only expends bandwidth but can cause $\hat{S}$ to initiate a large number of connections. The attack is possible due to mismatched identities: the server $\hat{S}$ accepts without knowing that the client actually intended to communicate with it. In other words, there is a lack of binding between the authentication and the identities of the parties involved in the interaction.

We can model this attack in our model of DoS-resistant protocols by considering the pre-session to be the entire session establishment protocol, and the expensive operation to be the establishment of a session. The attacker described by Mao and Paterson can cause a server to accept a large number of sessions. The sessions accepted by the server will be fresh because the last message from the server to the client is dropped by the adversary (so no matching conversation exists for the session).

6.5.6 DoS-CMQV

The DoS-CMQV protocol [61] was designed as a denial-of-service resistance key exchange protocol using client puzzles as the denial-of-service countermeasure and CMQV [63] as the underlying key exchange protocol. It was accompanied by a model for DoS-resistant key exchange that extended the eCK model for secure key exchange [36] and the protocol was proven to be DoS-resistant.

The model we have given in Sect. 6.4.1 for denial-of-service resistance is stronger than the Stebila-Ustaoglu model in which DoS-CMQV was originally proven DoS-resistant: the later model considers the ability of an adversary to cause multiple sessions to accept, not just a single session. Not all key exchange protocols that are DoS-resistant in the Stebila-Ustaoglu model are DoS-resistant in the Stebila et al. model. However, the DoS-CMQV protocol happens to be DoS-resistant in the stronger model since it uses a straightforward hash function client puzzle involving finding a preimage of a string with sufficiently many zeros. Thus, for DoS-CMQV, one can show that DoS-CMQV is a DoS-resistant protocol in the model of Sect. 6.4.1.

6.6 Conclusion

Authentication protocols are a key countermeasure to DoS attacks, for they can be used to discriminate illegitimate requests for service. However, when used to mitigate DoS attacks, it is imperative that these protocols be not vulnerable to denial of service themselves. This chapter has been concerned with improving the resistance of authentication protocols to DoS attacks. The work has focused on two aspects. The first is the development of mechanisms, primarily client puzzles, which can be used in authentication protocols to mitigate denial of service. The second is the development of appropriate formal models to define and measure the effectiveness of the proposed mechanisms once they are integrated with authentication protocols.

The goal of the computational security model has been to improve security definitions for client puzzles and DoS-resistant protocols by considering a multiuser network setting in order to account for the effects of an adversary who has enough resources to solve more than one puzzle. This definition is sufficiently general to be useful for analysing and proving the difficulty of a wide range of computation- and memory-bound puzzle constructions.

The new security model has been used to analyse recent key establishment protocols. For the majority of the protocols considered, the denial-of-service countermeasures were originally only addressed in an ad hoc fashion, without any rigourous security analysis. Given that only a few of the considered protocols do actually achieve denial-of-service resistance in a formal sense, this chapter has demonstrated the value of the provable security approach to DoS resistance.

Game theory has been successfully applied to study the effect of hiding the difficulty of client-puzzles from denial-of-service attackers. Three concrete puzzles that satisfy this requirement have been analysed and shown to be more effective than existing ones.

There is considerable scope for extensions to the work described in this chapter. In the area of DoS mitigation mechanisms, puzzles suited for deployment on low-power devices, such as sensors and smart cards, can be designed by taking into account client-side computational constraints. Other issues include basing strong puzzles on different sorts of computational problems, extensions of current designs to password-based protocols, and consideration of DoS protection for broadcast protocols.

On the theoretical side it would be interesting to study the applicability of the analytical models of this chapter to distributed defenders, such as a coalition of network routers cooperating against distributed denial-of-service attacks. More generally, it would be useful to integrate the computational models with game-theoretic models in a meaningful way in order to maximise the benefits of both approaches.

Finally it must be noted that, while the emphasis has been on authentication protocols, the same core techniques and approaches are applicable to other types of protocols, as exemplified in Chapter 7 where client puzzles are implemented within web services protocols.

References

1. Abadi, M., M. Burrows, M. Manasse, and T. Wobber. 2003. Moderately hard, memory-bound functions. In *the 10th Annual Network and Distributed System Security Symposium*, San Diego, 6–7 Feb 2003.
2. Agah, A., and S.K. Das. 2007. Preventing dos attacks in wireless sensor networks: A repeated game theory approach. *International Journal of Network Security* 5(2): 145–153.
3. Aiello, W., S.M. Bellovin, M. Blaze, R. Canetti, J. Ioannidis, A.D. Keromytis, and O. Reingold. 2004. Just fast keying: Key agreement in a hostile Internet. *ACM Transactions on Information and System Security* 7(2): 1–30.
4. Aura, T., and P. Nikander. 1997. Stateless connections. In *Proceeding of the International Conference on Information and Communications Security (ICICS'97)*, eds. Y. Han, T. Okamoto, and S. Qing, *LNCS*, vol. 1334, 87–97, Beijing, China, Nov 1997. Springer.
5. Aura, T., P. Nikander, and J. Leiwo. 2000. DoS-resistant authentication with client puzzles. In *Security Protocols Workshop 2000*, 170–181. Cambridge, Apr 2000.
6. Aura, T., P. Nikander, and J. Leiwo. 2001. DOS-resistant authentication with client puzzles. In *Revised Papers from the 8th International Workshop on Security Protocols, Lecture notes in computer science*, vol. 2133, 170–177. Springer-Verlag.
7. Aura, T., and P. Nikander. 1997. Stateless connections. Technical report A46, Helsinki University of Technology, Digital Systems laboratory, Espoo, Finland.
8. Back, A. 1997. A partial hash collision based postage scheme. http://www.hashcash.org/papers/announce.txt. Accessed 31 Aug 2011.
9. Back, A. 2004. Hashcash. http://www.hashcash.org/docs/hashcash.html#stamp_format__version_1__. Accessed 31 Aug 2011.
10. Bellare, M., J. Kilian, and P. Rogaway. 2000. The security of the cipher block chaining message authentication code. *Journal of Computer and System Sciences* 61(3): 362–399.
11. Bellare, M., and P. Rogaway. 1994. Entity authentication and key distribution. In *Proceedings of the 13th Annual International Cryptology Conference on Advances in Cryptology, CRYPTO '93*, 232–249, London. Springer-Verlag.
12. Bellare, M., and P. Rogaway. 1993. Random oracles are practical: a paradigm for designing efficient protocols. In *Proceedings of the 1st ACM conference on Computer and communications security, CCS '93*, 62–73, New York, 1993. ACM.
13. Bencsath, B., I. Vajda, and L. Buttyan. 2003. A game based analysis of the client puzzle approach to defend against DoS attacks. In *Proceedings of the 2003 International Conference on Software, Telecommunications and Computer Networks*, 763–767, 2003.
14. Canetti, R., S. Halevi, and M. Steiner. 2005. Hardness amplification of weakly verifiable puzzles, In J. Kilian (ed.), Theory of Cryptography Conference (TCC), LNCS 3378, pp. 17–33. Springer, 2005.
15. Canetti, R., and H. Krawczyk. 2002. Security analysis of IKE's signature based key-exchange protocol. In M. Yung (ed.), Advances in Cryptology – Proc. CRYPTO, LNCS 2442, pp. 27–52. Springer, 2002.
16. Castelluccia, C., E. Mykletun, and G. Tsudik (2006). Improving secure server performance by re-balancing SSL/TLS handshakes. In *ASIACCS '06: Proceedings of the 2006 ACM Symposium on Information, Computer and Communications Security*, 26–34, New York, 2006. ACM Press.
17. Chan, E., C.A. Gunter, S. Jahid, E. Peryshkin, and D. Rebolledo. 2008. Using rhythmic nonces for puzzle-based DoS resistance. In *Proceedings of the 2nd ACM Workshop on Computer Security Architectures*, 51–58, New York, 2008. ACM Press.
18. Chen, L., P. Morrissey, N.P. Smart, and B. Warinschi. 2009. Security notions and generic constructions for client puzzles. In M. Matsui (ed.), Advances in Cryptology – Proc. ASIACRYPT 2009, LNCS 5912, pp. 505–523. Springer, 2009.
19. Dwork, C., A. Goldberg, and M. Naor. 2003. On memory-bound functions for fighting spam. In *the 23rd Annual International Cryptology Conference (CRYPTO 2003)*, 426–444, Aug 2003.

20. Dwork, C., and M. Naor. 1992. Pricing via processing or combatting junk mail. In *the 12th Annual International Cryptology Conference on Advances in Cryptology, Lecture notes In Computer Science*, vol. 740, 139–147, 1992. Springer-Verlag.
21. Dwork, C., M. Naor, and H. Wee. 2005. Pebbling and proofs of work. In *CRYPTO*, 37–54, 2005.
22. Fallah, M. 5555. A puzzle-based defense strategy against flooding attacks using game theory. *IEEE Transactions on Dependable and Secure Computing* 99(2): 5555.
23. Feng, W., E. Kaiser, W. Feng, and A. Luu. 2004. The design and implementation of network layer puzzles. Technical report 04-003, OGI CSE, Aug 2004.
24. Feng, W., E. Kaiser, and A. Luu. 2005. Design and implementation of network puzzles. In *INFOCOM 2005. 24th Annual Joint Conference of the IEEE Computer and Communications Societies. Proceedings IEEE*, vol. 4, 2372–2382, March 2005.
25. Fudenberg, D. and E. Maskin. 1986. The folk theorem in repeated games with discounting or with incomplete information. *Econometrica* 54(3): 533–54.
26. Harkins, D. and D. Carrel. 1998. The internet key exchange (IKE), November 1998. Obsoleted by RFC 4306, updated by RFC 4109.
27. Jakobsson, M., and A. Juels. Proofs of work and bread pudding protocols (extended abstract). In B. Preneel (ed.), Proceedings of the IFIP TC6/TC11 Joint Working Conference on Secure Information Networks: Communications and Multimedia Security, volume 152 of IFIP Conference Proceedings, pp. 258–272. Kluwer, 1999.
28. Jakobsson, M., and A. Juels. 1999. Proofs of work and bread pudding protocols. In *The IFIP TC6 and TC11 Joint Working Conference on Communications and Multimedia Security (CMS '99)*. Also available as http://citeseer.nj.nec.com/238810.html
29. Juels, A., and J. Brainard. 1999. Client Puzzles: A cryptographic defense against connection depletion attacks. In *Proceedings of the Network and Distributed System Security Symposium (NDSS '99)*, 151–165, San Diego, Feb 1999. Internet Society Press, Reston.
30. Karn, P.R., and W.A. Simpson. 1999. Photuris: Session-key management protocol. RFC 2522, IETF.
31. Kaufman, C. 2005. Internet key exchange (IKEv2) protocol. RFC 4306.
32. Kennell, R., and L.H. Jamieson. 2003. Establishing the genuinity of remote computer systems. In *12th USENIX Security Symposium*, 295–308, 2003.
33. Kent, S., and R. Atkinson. 1998. Security architecture for the internet protocol. Standards track RFC 2401, IETF. http://www.ietf.org/rfc/rfc2401.txt. Accessed 31 Aug 2011.
34. Kent, S., and K. Seo. 2005. Security architecture for the internet protocol, December 2005.
35. Komathy, K., and P. Narayanasamy. 2008. Secure data forwarding against denial of service attack using trust based evolutionary game. In *Vehicular Technology Conference, 2008. VTC Spring 2008. IEEE*, 31–35, May 2008.
36. LaMacchia, B., K. Lauter, and A. Mityagin. 2007. Stronger security of authenticated key exchange. In W. Susilo, J.K. Liu, and Y. Mu (eds), First International Conference on Provable Security (ProvSec), LNCS 4784, pp. 1–16. Springer, 2007.
37. Leiwo, J., P. Nikander, and T. Aura. 2000. Towards network denial of service resistant protocols. In *the 15th Annual Working Conference on Information Security (SEC2000)*, vol. 175, Beijing, China, Aug 2000.
38. Lemon, J. 2002. Resisting SYN flood DoS attacks with a SYN cache. In *the BSDCon 2002*, 89–97, Berkley, 11–14 Feb 2002.
39. Lv, J.-J. 2008. A game theoretic defending model with puzzle controller for distributed dos attack prevention. In *2008 International Conference on Machine Learning and Cybernetics*, vol. 2, 1064–1069, July 2008.
40. Mahimkar, A., and V. Shmatikov. 2005. Game-based analysis of denial-of-service prevention protocols. In *CSFW '05: Proceedings of the 18th IEEE Workshop on Computer Security Foundations*, 287–301, Washington, DC, 2005. IEEE Computer Society.
41. Mao, W., and K.G. Paterson. 2002. On the plausible deniability feature of Internet protocols. Manuscript. http://citeseer.ist.psu.edu/678290.html. Accessed 31 Aug 2011.

42. Matsuura, K., and H. Imai. 2000. Modification of internet key exchange resistant against denial-of-service. In *Pre-Proceeding of Internet Workshop 2000 (IWS2000)*, 167–174, Feb 2000.
43. Meadows, C. 1999. A formal framework and evaluation method for network denial of service. In *Proc. 12th IEEE Computer Security Foundations Workshop (CSFW) 1999*, 4, 1999.
44. Meadows, C. 2001. A cost-based framework for analysis of denial of service in networks. *Journal of Computer Security* 9(1): 143–164.
45. Menezes, A.J., P.C. van Oorschot, and S.A. Vanstone. 1997. *Handbook of applied cryptography*. CRC Press series on discrete mathematics and its applications. CRC Press. ISBN 0-8493-8523-7.
46. Mirkovic, J., and P. Reiher. 2004. A taxonomy of DDoS attack and DDoS defense mechanisms. *SIGCOMM Computer Communication Review* 34(2): 39–53.
47. Moskowitz, R., P. Nikander, P. Jokela, and T.R. Henderson. 2008. Host identity protocol, Apr 2008. RFC 5201.
48. Narasimhan, H., V. Varadarajan, and C.P. Rangan. 2009. Game theoretic resistance to denial of service attacks using hidden difficulty puzzles. Cryptology ePrint Archive, Report 2009/350. http://eprint.iacr.org/. Accessed 31 Aug 2011.
49. Narasimhan, H., V. Varadarajan, and C.P. Rangan. 2010. Game theoretic resistance to denial of service attacks using hidden difficulty puzzles. In *ISPEC*, 359–376, 2010.
50. Paterson, K.G. 2006. A cryptographic tour of the IPsec standards. Cryptology ePrint Archive, Report 2006/097. http://eprint.iacr.org/2006/097.pdf. Accessed 31 Aug 2011.
51. Price, G. 2003. A general attack model on hash-based client puzzles. In *Cryptography and Coding, 9th IMA International Conference, Cirencester, UK, December 16–18, 2003, Proceedings*, ed. K. Paterson, *Lecture notes in computer science*, 319–331, vol. 2898. Springer-Verlag.
52. Rabin, M.O. 1979. Digitalized signatures and public-key functions as intractable as factorization. Technical report MIT/LCS/TR-212, Massachusetts Institute of Technology.
53. Rivest, R.L, A. Shamir, and L. Adleman. 1978. A method for obtaining digital signatures and public-key cryptosystems. *Communications of the ACM* 21(2): 120–126.
54. Rivest, R.L., A. Shamir, and D.A. Wagner. 1996. Time-lock puzzles and timed-release crypto. Technical report TR-684, Massachusetts Institute of Technology, Cambridge, 10 Mar 1996.
55. Sagduyu, Y.E., and A. Ephremides. 2009. A game-theoretic analysis of denial of service attacks in wireless random access. *Wireless Networks* 15(5): 651–666.
56. Shankar, U., M. Chew, and J.D. Tygar. 2004. Side effects are not sufficient to authenticate software. In *Proceedings of the Thirteenth USENIX Security Symposium*, 89–102, Aug 2004. USENIX.
57. Simpson, W.A. 1999. IKE/ISAKMP considered harmful. *USENIX ;login* 24(6).
58. Smith, J. 2007. *Denial of service: Prevention, modelling and detection*. PhD thesis, Queensland University of Technology, Brisbane.
59. Smith, J., J.M. González Nieto, and C. Boyd. 2006. Modelling denial of service attacks on JFK with Meadows's cost-based framework. In ACSW Frontiers '06: Proceedings of the 2006 Australasian workshops on Grid computing and e-research, pp. 125–134, Darlinghurst, Australia, Australian Computer Society, Inc.
60. Stebila, D., L. Kuppusamy, J. Rangasamy, C. Boyd, and J. Gonzalez-Nieto. 2011. Stronger difficulty notions for client puzzles and denial-of-service-resistant protocols. In *Topics in Cryptology – CT-RSA 2011*, ed. A. Kiayias *Lecture notes in computer science*, 284–301, vol. 6558, 2011. Springer, Berlin.
61. Stebila, D., and B. Ustaoglu. 2009. Towards denial-of-service-resilient key agreement protocols. In C. Boyd and J. González Nieto (eds), Proc. 14th Australasian Conference on Information Security and Privacy (ACISP), LNCS 5594, pp. 389–406. Springer, 2009.
62. Tritilanunt, S., C. Boyd, J. González Nieto, and E. Foo. 2007. Toward non-parallelizable cryptographic puzzles. In *of 6th International Conference on Cryptology and Network Security (CANS 2007)*, Singapore, 8–10 December 2007.

63. Ustaoglu, B. 2008. Obtaining a secure and efficient key agreement protocol from (H)MQV and NAXOS. *Designs, Codes and Cryptography* 46(3): 329–342.
64. Wang, X. and M.K. Reiter. 2003. Defending against denial-of-service attacks with puzzle auctions. In *Proceedings of the 2003 IEEE Symposium on Security and Privacy*, 2003. IEEE Computer Society.
65. Waters, B., A. Juels, J.A. Halderman, and E.W. Felten. 2004. New client puzzle outsourcing techniques for dos resistance. In *CCS '04: Proceedings of the 11th ACM Conference on Computer and Communications Security*, 246–256, New York, 2004. ACM.

Chapter 7
Denial of Service Defence Appliance for Web Services

S. Suriadi, A. Clark, H. Liu, D. Schmidt, J. Smith, and D. Stebila

Service-oriented architectures (SOAs), implemented using web services, seek to use open and interoperable standards to facilitate easier enterprise application integration, provide application flexibility and facilitate the dynamic composition of applications from component services. As with traditional distributed computing environments such as common object request broker architecture (CORBA), remote procedure call (RPC) and remote method invocation (RMI), the exposure of information resources via computer networks to remote users and applications requires that those resources be adequately protected.

Like any network-enabled application, resource protection requirements will include confidentiality, integrity and availability. In this chapter, we concentrate on the protection of the *availability* of resources from Denial of Service (DoS) and distributed denial of service (DDoS) attacks in web services applications.

Although the loosely coupled and dynamic nature of web services applications provides many beneficial features, it has also introduced complexities in their design and implementation:

- Web services use eXtensible Markup Language (XML)-based messages which in itself require a substantial amount of resources to process.
- To support dynamic binding of web services applications, a web services provider needs to expose its services' metadata to unauthenticated users.
- As applications are exposed through a web services interface, application-layer filtering is required.

These factors could easily be exploited by attackers to cause a DoS.

Nevertheless, despite the importance of protecting web services from DoS attacks, there are few tools available to assist those who choose to deploy them.

S. Suriadi (✉) • A. Clark • H. Liu • D. Schmidt • J. Smith • D. Stebila
Information Security Institute, Queensland University of Technology, Brisbane, Australia
e-mail: s.suriadi@qut.edu.au; a.clark@qut.edu.au; hua.liu@connect.qut.edu.au; schmidda@qut.edu.au; j4smith@qut.edu.au; stebila@qut.edu.au

S.V. Raghavan and E. Dawson (eds.), *An Investigation into the Detection and Mitigation of Denial of Service (DoS) Attacks: Critical Information Infrastructure Protection*,
DOI 10.1007/978-81-322-0277-6_7, © Springer India Pvt. Ltd. 2011

This fact is acknowledged in the 'Guide to Secure Web Services' report by the National Institute of Standards and Technology (NIST) who state that although there are many available technologies designed to prevent many of the threats in a web services environment (such as integrity and confidentiality protections), there is an evident lack in the technologies that one can use to prevent DoS attacks [33, Table 2-2, Section 2.6, pp. 2–17].

While the existing technologies to mitigate DoS attacks in web applications can be used to some extent, they are not sufficient as the service-oriented architecture (SOA) paradigm has shifted the nature of web interactions from the one meant to be used by humans through some presentation layer to the one meant for applications integration and sharing with a much greater degree of direct program-to-program interactions. In addition, the distinct properties of web services (XML-based, dynamic binding, application-layer operations) open up new challenges to traditional DoS mitigation methods.

The main goals of this chapter are:

- To explain and practically validate the DoS vulnerability in main web services technologies
- To detail our proposal of an architecture employing cryptographic client puzzles and other related techniques to mitigate DoS attacks in web services
- To demonstrate the effectiveness of the cryptographic client puzzles technique in mitigating DoS attacks in a web services environment

DoS attacks come in various forms and shapes, despite them having the common goal of denying the availability of services to legitimate consumers. Given this fact, we limit our experiments with cryptographic client puzzles to a few well-known web services DoS attacks. We acknowledge that this strategy is not likely to mitigate/prevent all web services DoS vulnerabilities.

This chapter is organized as follows: an introduction to web services technologies is provided in Sect. 7.1. Section 7.2 details the various points of DoS vulnerabilities in web services applications. Section 7.3 describes our experiments and their results in validating the identified points of DoS vulnerabilities. Section 7.4 explains a comprehensive DoS mitigation architecture for web services. Section 7.5 describes our implementation of cryptographic client puzzles (from Chapter 6) and the results of our experiments in assessing their effectiveness in mitigating DoS attacks. Finally, conclusions and future work are provided in Sect. 7.6.

7.1 Web Services Primer

In this section, we explain the basics of web services and the related concept known as the service-oriented architecture (SOA). We then give a brief overview of the various web services security technologies that are designed to secure web services applications (but which often, at the same time, introduce Denial of Service (DoS) vulnerabilities).

7.1.1 *Web Services: Basic Concepts*

SOAs and web services use Internet standardised technologies such as XML, HTTP and TCP to implement platform-independent and interoperable distributed computing services. A consumer of a web service is able to invoke a service by the construction of a request (formatted in eXtensible Markup Language (XML) indicating the remote method to be called and the parameters needed for that method invocation). The producer of a web service will process the request and format the return value (or output of the method invocation) as an XML response.

7.1.1.1 Service Description and Discovery

In SOAs a provider of a service *publishes* a description of the service they can provide in a standard web service description language (WSDL) document. A client who needs to consume the service only needs to retrieve the relevant WSDL document which contains enough information for the client to be able to form a compliant service request message.

A client may have the option to find which service they want to invoke at runtime (referred to as late binding) by searching a web services registry, known as the universal description, discovery and integration (UDDI). However, the adoption of UDDI has been less than ideal; as a result, UDDI is currently not as widely used as it was intended to be. Instead, clients often discover and obtain the necessary WSDL documents through other channels.

Web Services Description Language

The WSDL describes the functions of a web service, how it communicates and where it can be found. It is similar to an interface definition language (IDL) used in common object request broker architecture (CORBA).

A WSDL specification consists of four major components, each of which are described in the remainder of this section. An overview of the components contained in a WSDL specification is depicted in Fig. 7.1[1].

WSDL definitions provide the name of the WSDL and indicate the namespaces of the WSDL document and other namespaces as needed, depending on the content of the WSDL.

The WSDL Operations component of the specification includes two sections delimited by `message` and `portType` tags, respectively, and details what the web

[1]The figures depicting a WSDL specification are adapted from [31, Chapter 2].

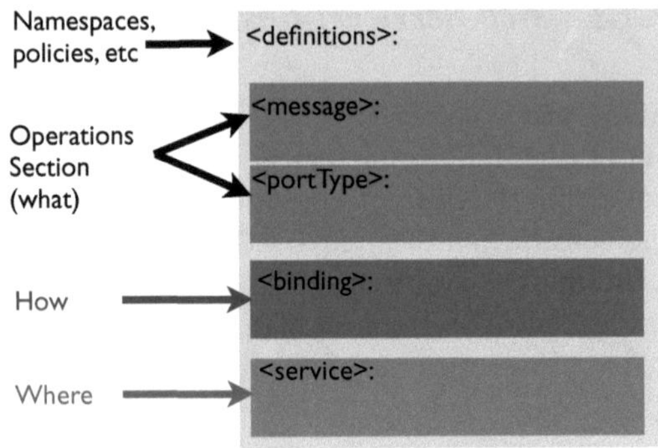

Fig. 7.1 WSDL overview

service does. In particular, the `message` element defines several input and output messages, while the `portType` element defines the operation name that the service provides and the type of messages (defined in `message` tag) to be used. Often, a WSDL has another tag, called `types` (not shown in Fig. 7.1) which defines the schema for the elements that will be used in the `messages` tag.

The WSDL Bindings component of the specification describes how the service binds together a port, a transport and the operations of the service. For example, in the `bindings` tag, it may state that an operation *X* must be called using HTTP transport, and that the input and output messages (as defined in `messages` and `portType`) must be encoded using document/literal style.

The WSDL services component of the specification indicates where the service resides (shown as a URL).

7.1.1.2 Service Invocation

Web services applications execute as a series of method calls and responses. These calls and responses are encoded in XML messages adhering to the SOAP specification[2] of the W3C [14, 15].

Message Construction

A web service consumer interacts with a web service producer by exchanging XML messages. SOAP specifies how XML-based messages can be used to exchange structured information in decentralised and distributed environments. The structure

[2]Earlier versions of the SOAP specification indicate that SOAP was an acronym for *simple object access protocol*; recent versions of the standard no longer use the term SOAP as an acronym.

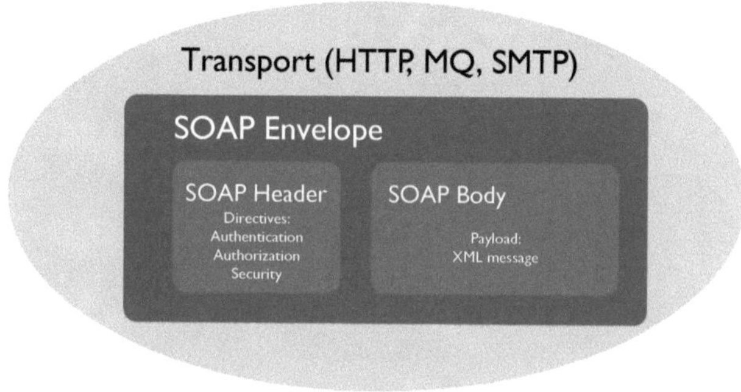

Fig. 7.2 SOAP structure overview

of a SOAP message is depicted in Fig. 7.2, which shows that a SOAP message consists of:

- An envelope, which is the outermost element of a SOAP message and contains definitions of the namespaces used by the message
- A header, which will contain directives related to authentication and authorisation
- A body, which is the payload of the message (method request or response, for example)

Message Transport

SOAP messages may be transported over a number of different communications protocols including the hypertext transfer protocol (HTTP), the simple mail transfer protocol (SMTP) or message queueing (MQ) systems. In other words, web services messages are operated purely on the application layer stack. A SOAP message may be handled by an arbitrary number of intermediaries before reaching its ultimate destination.

Service Invocation Summary

A summary of web service invocation is depicted in Fig. 7.3. A service provider will optionally publish a description of the services it provides in a UDDI service registry. A web service consumer can discover services using a UDDI registry, or retrieve a WSDL specification for a service if the WSDL location is known a priori (via some out-of-band mechanism).

Once the consumer is in possession of the WSDL, it can encode method invocation requests (including required parameters) and receive service response in XML-encoded SOAP messages.

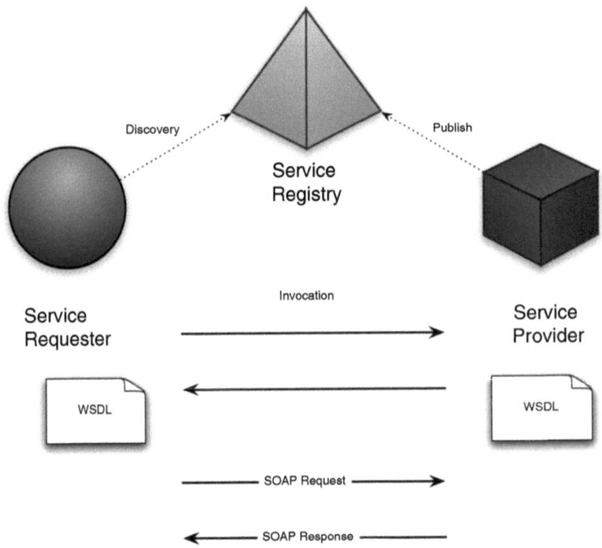

Fig. 7.3 Web service discovery and invocation

7.1.1.3 Development Frameworks

Service-oriented architectures and web services may be viewed as an extension of existing *n*-tier enterprise applications, so it should be unsurprising that the dominant development frameworks for enterprise application development, namely Java Metro, Apache Axis and Microsoft's .NET, are popular with the web services development community. All of our web services experiments are conducted against these popular platforms.

7.1.2 Web Services Security Technologies

Message layer security provides an end-to-end security, therefore enabling protection to SOAP messages from the original sender to the final recipient, regardless of how many hops the messages have to pass in transit. In addition, message layer security protects the messages both while they are in motion (being transported in the communication channel) and at rest. This complements the lack of at-rest message protection of transport layer security (such as SSL).

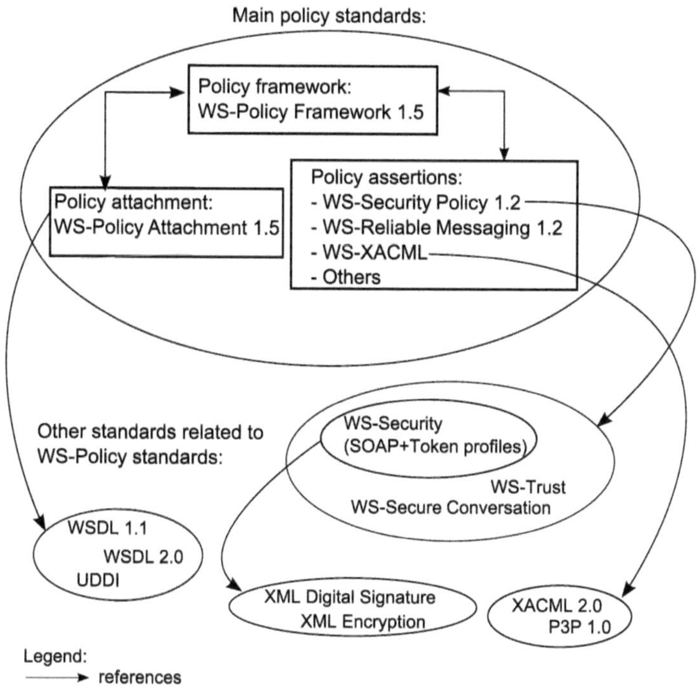

Fig. 7.4 Overview of WS-Policy and its related standards

7.1.2.1 WS-* Standards

There are many aspects that one needs to consider in securing web services messages (token generation, message encryption, message signing, etc.). As a result, to ensure compatibility between various security technologies and platforms, there are an abundance of standards that have been developed and implemented, all with the goal of securing web services. An overview of the standards, and their relationships, is provided in Fig. 7.4.

The main technology that provides security on the web services message layer is the OASIS *Web Services Security: SOAP Message Security 1.1 (WS-Security 2004)* standard [20] (henceforth referred to as the WS-Security standard) which specifies the standard mechanism to secure web services messages. This specification extends the basic SOAP message structure [6, 14, 15, 26] that was standardised by W3C. WS-Security relies on the W3C XML Digital Signature and Encryption standards [12, 13].

In addition to the WS-Security standard, there are also other standards to govern other aspects of security. WS-Trust standard [18] has been provided by OASIS for the purpose of enabling the issuance and dissemination of credentials within different trust domains. Specifically, WS-Trust, extending the WS-Security

```
<soap:Envelope>
  <soap:Header>...
    <ws:Security ws:actor="..." ws:mustUnderstand="...">...
    </ws:Security>
  </soap:Header>...
</soap:Envelope>
```

Fig. 7.5 WS-Security element and placement

standard, provides methods for issuing, renewing and validating security tokens, as well as to broker trust relationships. WS-Trust describes how entities who may not have any prior trust relationship can obtain security tokens from a common trusted entity and exchange them securely.

Furthermore, we also need a standard to describe the security requirements of both clients and web services providers. For example, a service provider may require its clients to be able to process some cryptographically signed and/or encrypted message to secure its service. Conversely, a client may also require the service provider to provide its X.509 certificate signed by certain issuer as its identity information.

In a web services environment, there should be an interoperable mechanism to express these policies so that a service provider and a client who have never interacted before can still communicate successfully in a policy-compliant manner. The WS-Policy framework [40] and its related standards [19, 39] have been recommended by the World Wide Web Consortium (W3C) as the mechanism to facilitate the expression and the exchange of security policy requirements (see Fig. 7.4).

It is beyond the scope of this book to give a comprehensive explanation of the security standards mentioned above. However, in this book, a considerable attention is given to the DoS vulnerability of the WS-Security standard. Therefore, further description of this standard is provided below.

7.1.2.2 WS-Security

WS-Security fields are contained within the `<ws:Security>` elements and added to the SOAP Header section of the SOAP Message (see Fig. 7.5). Multiple WS-Security elements can be included in a SOAP Header to accommodate multiple recipients. The recipients are indicated by either the ‘actor’ or ‘role’ attribute in the security header. If the attribute ‘mustUnderstand’ has the value of ‘1’ or ‘true’, then the recipient of the SOAP header must be able to process the security header, else, a SOAP fault must be returned.

WS-Security secures SOAP messages at the message level itself, thus ensuring that the SOAP messages are secured independently of the lower layer security mechanisms (such as SSL and IPSec). WS-Security provides several message layer security services, such as identification and authentication, message integrity and message confidentiality. How WS-Security enables such services will be described briefly in the next few sections.

```
<soap:Envelope>
 <soap:Header>...
  <ws:Security ws:actor="..." ws:mustUnderstand="...">
   <ws:BinarySecurityToken ...
    ValueType="...#X509v3 EncodingType="...#Base64Binary">YLDCD98XAw...
   </ws:BinarySecurityToken>...
  </ws:Security>
 </soap:Header>....
</soap:Envelope>
```

Fig. 7.6 X.509 token inside WS-Security header

Identification and Authentication

WS-Security header can include identity information of the SOAP message sender. Additionally, this information can be used to determine key information necessary for the WS-Security digital signature and encryption processing. This identity information is normally carried in the form of a *security token* specified by the WS-Security standard and includes X.509 certificate, Kerberos ticket, SAML Assertion, Username token and other token types. For each of these token types, a corresponding token profile is provided to standardise the methods to use them [22–24].

For illustration, how we can include an X.509 certificate into a SOAP message is explained in this section. The X.509 certificate token can be included into WS-Security header for the purpose of *authentication* and *key establishment*. The standard methods to do this are laid out in the OASIS *Web Services Security X.509 Certificate Token Profile 1.1* [21]. An example of the placement of an X.509 certificate token into WS-Security header is given in Fig. 7.6.

The X.509 certificate token is embedded into the WS-Security header as a binary security token element (defined in the main WS-Security standard document [20]). The X.509 Token profile document defines several values to be used with the 'ValueType' attribute, depending on the X.509 certificate used; they include X.509v3, X.509PKIPathv1 and PKCS7. X.509 certificate token can be used to identify the signing and encryption keys. Nevertheless, using X.509 certificate means having to deal with its existing problems, such as path processing and validation, certificate revocation and several others.

Confidentiality and Integrity

WS-Security uses XML-Encryption and XML-Signatures to provide message layer confidentiality and integrity protection.

One of the main uses of WS-Security is to protect the message integrity of SOAP messages. The WS-Security standards uses the XML-Signature standard [13] in conjunction with security tokens (as discussed in Sect. 7.1.2.2) to achieve the message integrity protection capability. While there are several variations on how a SOAP message is to be signed, the following example as shown in Fig. 7.7 provides the basic concept of how a digital signature is achieved in WS-Security.

```
<soap:Envelope>
 <soap:Header>...
   <ws:Security ws:actor="..." ws:mustUnderstand="...">
    <ws:BinarySecurityToken ... ValueType="...#X509v3
      EncodingType="...#Base64Binary">YLDCD98XAw...
    </ws:BinarySecurityToken>
    <ds:Signature>
      <ds:SignedInfo>
        <ds:CanonicalizationMethod Algorithm="..."/>
        <ds:SignatureMethod Algorithm="..."/>
        <ds:Reference URI="#body">
         <ds:Transforms>...</ds:Transforms>
         <ds:DigestMethod Algorithm="...".>
          <ds:DigestValue>...</ds:DigestValue>
        </ds:Reference>
        <ds:Reference URI="#X509Token">...</ds:Reference>
       </ds:SignedInfo>
       <ds:SignatureValue>ABBCD1234...</ds:SignatureValue>
       <ds:KeyInfo>
        <ws:SecurityTokenReference><ws:Reference URI="#x509Token"/>
        </ws:SecurityTokenReference>
      </ds:KeyInfo>
    </ds:Signature> ...
   </ws:Security>
  </soap:Header>
 <soap:Body wsu:Id="body"> ... </soap:Body>
</soap:Envelope>
```

Fig. 7.7 WS-Security digital signature

All elements with namespace 'ds' are defined in the XML Digital Signature document. The example uses the binary security token, which is of type X.509 as the key to sign the message. The token ID is 'x509Token'. The <ds:SignedInfo> element provides the necessary information to ensure that the signature can be verified. This includes the canonicalisation method, the signing algorithm and the reference information to indicate the parts of the SOAP message that are included in the signature – the signature scope. The <ds:Reference> element points to URI 'body' and 'x509Token' to indicate that the signature scope includes the elements with ID 'body' and 'x509Token', which in this case are the <soap:Body> and <ws:BinarySecurityToken> elements. Further information is given to indicate the transformation and digest algorithm used in order to make sure that a correct digest value can be reproduced. The <ds:KeyInfo> provides the information about the key needed to verify the signature. In the example provided in Fig. 7.7, it uses the binary security token included in the SOAP header with the ID 'x509Token'.

WS-Security can also be used to protect message confidentiality. It uses the XML Encryption standard. The WS-Security standard document specifies the way to enable a secure *key transport* and methods to format the encrypted message to allow successful decryption. A simple example of how WS-Security encryption is implemented is depicted in Fig. 7.8.

The <xenc:EncryptedKey> element is the main element that is used to provide a secure *key transport* mechanism. The information provided includes

```
<soap:Envelope>
 <soap:Header>...
  <ws:Security ws:actor="..." ws:mustUnderstand="...">
   <xenc:EncryptedKey>
    <ds:KeyInfo>
     <ws:SecurityTokenReference>
      <ds:X509IssuerSerial>
       <ds:X509IssuerName>...</ds:X509IssuerName>
       <ds:X509SerialNumber>12345698</ds:X509SerialNumber>
      </ds:X509IssuerSerial>
     </ws:SecurityTokenReference>
    </ds:KeyInfo>
    <xenc:CipherData>
     <xenc:CipherValue>...</xenc:CipherValue>
    </xenc:CipherData>
    <xenc:ReferenceList><xenc:DataReference URI="#enc1"/></xenc:ReferenceList>
   </xenc:EncryptedKey>
  </ws:Security>
 </soap:Header>
 <soap:Body wsu:Id="body">
  <xenc:EncryptedData ...wsu:Id="enc1">...
   <xenc:EncryptionMethod .../>
   <xenc:CipherData>
    <xenc:CipherValue>...</xenc:CipherValue>
   </xench:CipherData>
  </xench:EncryptedData>
 </soap:Body>
</soap:Envelope>
```

Fig. 7.8 WS-Security encryption

the key information, the *encrypted key* cipher text as well as the reference list. The `<ds:KeyInfo>` provides the key information needed in order to successfully decrypt the *encrypted key*. The reference list points to an element with the ID 'enc1', which means that the *encrypted key*, once decrypted, is to be used to decrypt the encrypted element with the ID 'enc1'. In Fig. 7.8, the element with the ID 'enc1' is in the SOAP Body.

7.2 Denial of Service Vulnerabilities in Web Services

The focus of this book is to assess the Denial of Service (DoS) vulnerabilities in main web services technologies. Therefore, in this section, we identify the points of DoS vulnerability in the main web services technologies, including:

- eXtensible Markup Language (XML) technology
- Web service description language (WSDL)
- Simple object access protocol (SOAP)
- WS-Security standards.

We acknowledge that there exist many more DoS vulnerabilities in other web services (including other WS-* standards) technologies which we have not considered in this book.

7.2.1 General Web Services DoS Vulnerabilities

In this section, several general vulnerabilities in web services that could result in DoS are identified.

7.2.1.1 Unauthenticated Requests

There is a close relationship between authentication and DoS. Authentication can be used as a technique to mitigate DoS attacks [4, 30] whereby if a server only responses to requests that are authenticated, then arbitrary attackers cannot simply flood a server with numerous request for services with the intention to exhaust the server's resources.

However, at the same time, the requirement to authenticate every request itself can be exploited by attackers as the *source* of DoS attack in itself due to the heavy processing requirement in some authentication systems: an attacker could launch a brute-force attack by sending massive amount of requests, each of which needs to be authenticated. Consequently, a server's resources could be exhausted in an attempt to authenticate each of the requests. This is especially true for those authentication systems that are based on public-key cryptography. Efficient authentication mechanisms are needed to prevent them from being exploited for DoS purpose. Section 7.5 looks into the use of an efficient authentication mechanism (in the form of cryptographic client puzzles) in web services.

7.2.1.2 Infrastructure/Platform Vulnerabilities

Infrastructure vulnerabilities have their root cause in the underlying platform or operating system of the host of the web service. These vulnerabilities are well-documented and are not web service specific.

Nevertheless, certain infrastructure vulnerabilities can be exploited by attackers to launch a DoS attack by providing invalid or malicious input to the web service platform/server. The root cause of these vulnerabilities is a lack of proper validation, whether it be of user input, database input, the size of input from any source, method parameters, schemas or even lack of validation of strings in the web service's sourcecode. This lack of validation can be exploited by an attacker as part of a denial of service attack, to force the web service to accept malicious data which may compromise its function to legitimate users, to execute web service methods with malicious parameters, to execute buffer overflow attacks, to inject malicious code or SQL or any number of other objectives. Input manipulation vulnerabilities are the most common root cause of unique attacks reported to the OSVDB. They are best addressed at design time through ensuring that all input from all sources is properly validated.

For example, vulnerabilities have been discovered in multiple vendors servers (such as IBM Websphere Application Server version 5[3], Oracle Application Server[4]) capable of processing simple object access protocol (SOAP) messages. These vulnerabilities could be exploited by attackers to cause DoS.

Certain platform/server vulnerabilities could be exploited to cause memory leak. Memory leak is a situation whereby a system's memory is not properly freed up (intentionally or otherwise) such that over time, the system runs out of available memory to process service requests – hence a DoS. For example, a known vulnerability in Microsoft RPC could be exploited by attackers to cause DoS attack.[5]

The types of attack described can be classified as semantic attacks which could lead to a DoS.

Application Layer Firewall

Still related to the infrastructure vulnerabilities, the use of an application layer firewall could be exploited by attackers to cause DoS attacks. Because web services applications run at the message layer, and because web services messages commonly contain parameters and arguments that would be supplied directly to the web services applications, it is important that web services messages are sanitised.

However, application layer message sanitisation requires substantial processing overhead as it needs to understand the semantics, ordering and the flow of service messages. Such processing requirements could be exploited by attackers as a source of DoS attacks whereby an application layer firewall could be flooded with numerous web services messages, each requiring proper sanitisation by the firewall. As a result, firewall's resources could be exhausted, hence a DoS is achieved.

Therefore, it is crucial that application firewall is designed such that it can sanitise large amount of web services messages in an efficient manner, hence reducing its likelihood to become a DoS target.

Domain Name System (DNS) Server Vulnerabilities

A web services DoS could also be caused by exploiting or compromising DNS servers. This is mainly due to the fact that, as detailed in Section 7.2.2, XML allows schemas to be referenced externally. Therefore, a compromised DNS server would result in an entity connecting to an incorrect, and potentially malicious, server for the purpose of obtaining the relevant XML schema. This would result in either the failure of schema retrieval due to wrong endpoint connection or malicious payload

[3] http://www.securityfocus.com/bid/9185/info.

[4] http://xforce.iss.net/xforce/xfdb/15270.

[5] http://securitytracker.com/alerts/2004/Apr/1009758.html.

is sent (instead of legitimate schema), which could potentially be used to exploit some known vulnerabilities of the victim's system. Similar attack could be launched against web service description language (WSDL) endpoints.

7.2.2 *XML Vulnerabilities*

Web services messages are expressed using the XML. Due to its inherent vulnerability, using XML leaves web services applications vulnerable to DoS attacks.

We identify the following points of DoS vulnerability that arise from the use of XML.

7.2.2.1 Oversized Payload

Attackers could send a very large valid SOAP payload to a web services provider. As a result, the provider would potentially spend considerable amount of resources to parse the given SOAP payload. This is a type of semantic attack to cause a DoS.

Even when a web services provider has a powerful processing capability, it is difficult to prevent the flooding of oversized SOAP messages which could easily exhaust the provider's resources, hence, causing a DoS.

7.2.2.2 Deeply Nested Payload

This type of attack is a variant of the oversized-payload attack. An attacker could craft a SOAP message such that it contains excessive level of nested XML elements – also called XML-bombing. The message may need to be parsed by an XML parser which will continuously parse the deeply nested elements.

When a server is flooded with this type of XML message, computing resources could be exhausted resulting in a DoS.

7.2.2.3 Buffer Overflow Attack

Improper parsing of XML messages could open up an opportunity for attackers to launch a buffer overflow attack which may result in DoS [29]. This is especially true when the document object model (DOM)-style parsing of XML document is used. In this style of XML parsing, the entire XML document is loaded into the memory before being processed. An attacker could exploit this by sending a large XML payload (with malicious code) such that a memory overflow is triggered and, subsequently, the execution of the malicious code that may result in a DoS. This is a type of semantic attack.

7.2.2.4 External Entity Reference Vulnerability

The nature of XML schema specification is such that a document can reference XML namespace defined externally. As a result, an XML parser may attempt to contact the referenced location to obtain the schema. This style of processing could result in various types of DoS attacks, of which we have identified four.

Recursive External Referencing

A malicious web services provider could provide a SOAP document which recursively references external XML documents. Over time, this will exhaust the recipient computing resources who needs to hold the state of the previously parsed XML documents and establish endless connection with external entities. This could therefore result in a DoS. This type of attack exploits the design of the XML technology.

Malicious and/or Oversized Content Referencing

A malicious provider could point the schema location to bogus location such that instead of a valid XML schema being retrieved, it actually retrieves some payload which could either be very large or contain malicious payload. In either case, these opportunities could be exploited by attackers to cause DoS. This type of attack exploits the structure of the XML technology.

Flooding Victim with (Nonexistent) Schema Request

An attacker could cause a distributed denial of service (DDoS) attack at the victim by having a large number of requests being sent to the target victim requesting this particular schema (existing scheme or some bogus nonexistent scheme) with the intention of overwhelming the victim's system such that it crashes or reboots, hence a DoS. This attack is an example of brute-force attack.

Referencing Nonexistent Schema

Another type of attack is for attacker to put a reference to schema (in an XML document) located at a location under its control. When the reference schema is being requested, the request could be rejected, with the end result of the user being denied a service due to his/her inability to retrieve all the necessary schema to process a SOAP message. This is another variant of semantic attack against XML technology.

Many of these vulnerabilities could be overcome by using and processing only those schemas that are stored locally and have been validated.

7.2.2.5 Unicode Attack

XML allows representation of a character in various encoding styles. As a result, this can be exploited by attacker by sending multiple (same) requests encoded in different Unicode styles such that the firewall perceives those requests as different requests, therefore failing to recognise them as repetitive requests which it would have blocked otherwise. These repetitive requests can be exploited to cause DoS [29].

7.2.3 *SOAP Vulnerabilities*

SOAP technology is one of the most commonly used methods to encapsulate web services messages. As SOAP is commonly used, its design is such that, predictably, has to be flexible enough to accommodate the dynamic nature of web services message exchange pattern whereby a message could be sent to a recipient, which could then be further forwarded to other entities in a web services message orchestration or choreography style to accomplish some business tasks.

Nevertheless, as has been explained in the beginning of this chapter, the dynamic nature of web services, in this case of SOAP technology, comes with its own DoS vulnerabilities.

In this section of the report, we have identified several DoS vulnerabilities that could result from the use of the SOAP technology. Because SOAP is XML-based, these vulnerabilities are *in addition* to those that result from the use of XML, as detailed in Sect. 7.2.2.

A SOAP header can be seen as consisting of one or more SOAP header blocks. Each SOAP header block is used to encode self-contained information related to some processing. For example, a SOAP header could be made up of two main blocks (or XML elements): the first block contains information related to message encryption, while the second block contains information related to quality of service.

Because a SOAP message can be received and potentially processed by multiple intermediaries before arriving at the ultimate receiver, the SOAP specification [26] defines several attributes that can be used within SOAP headers element to indicate which recipient should process the SOAP headers:

- The `role` attribute can be used to indicate which recipient of the message should process the SOAP header block in which the `role` attribute exists. The SOAP standard defines three types of roles: ‘next’, ‘none’ and ‘ultimateReceiver’:
  - The ‘next’ attribute refers to all intermediary and the ultimate destination nodes.
  - When the ‘none’ attribute is used, it means that no SOAP nodes (including the sender, intermediary nodes and end receiver nodes) are referred to.
  - The ‘ultimateReceiver’ refers to the end receiver of the SOAP message.

Other types of application-specific roles can be used. Absence of the `role` attribute is equivalent to `ultimateReceiver`.

- The `mustUnderstand` attribute indicates whether the SOAP header intended for the recipients as indicated in the `role` attribute must be able to process the associated SOAP header accordingly. When the value is 'true', then all SOAP intermediary and end receiver nodes *must* process the header. Otherwise, if it is 'false' (or absent - which is equivalent to 'false'), then both intermediary or end receiver nodes may process it, though not mandatory in this case.
- The `relay` attributes is used to indicate whether a SOAP header block should be relayed to the next SOAP node or not. The default SOAP processing rules are as follows:
  - If a SOAP header is processed by a node as targeted by the `role` attribute, then it should be removed from the outbound message. However, it may be reinserted unchanged or modified if, from the processing of other SOAP header blocks, it is deemed required to.
  - If a SOAP header block targeted at a SOAP node is not processed, then it must be removed from the SOAP header before being relayed further.

  There are situations when the SOAP sender uses newly implemented SOAP header logic in which it just wants any capable SOAP intermediary to process. Using the default processing rule, if an intermediary node does not process the header block (because it does not understand it), then it will remove the header from the outgoing message. In order to ensure that the header gets relayed, the `relay` attribute can be set to 'true' thereby forcing the SOAP node to relay the header. When such an attribute is absent, it is equivalent to 'false', which means that the default behaviour of not relaying unprocessed header is used.

From the behaviour of the SOAP processing rules as just explained, we identify several DoS vulnerabilities:

7.2.3.1 SOAP Header Explosion

A semantic attack can be launched on the standard SOAP header processing rules such that a DoS could happen. An attacker could manipulate the use of `role` attribute to exhaust a SOAP node's resources. In its simplest case, a malicious SOAP sender could send (or modify) a SOAP message such that it has numerous SOAP header blocks with the `role` attribute set to 'ultimateReceiver' or absent. In this case, the end receiver node would try to process all of the SOAP header blocks, which may exhaust all of its available resources. Of course, in this scenario, the receiver has the option of ignoring the SOAP header blocks.

Nevertheless, if the malicious sender also includes the `mustUnderstand` attribute, and sets it to 'true', then the end receiver must attempt to process all of the headers, which will exhaust all of its resources and ultimately prevent it from functioning properly.

This type of attack does not require a man-in-the-middle to be able to intercept a session.

7.2.3.2 Modification of `role, mustUnderstand and relay` Attributes

A semantic attack variant on the SOAP processing model to cause a DoS is through a simple modification of the `role`, `mustUnderstand` and `relay` attributes. This type of attack is a variant of semantic-based attack, although in this case, it requires an attacker to be either a malicious legitimate SOAP intermediary or is able to intercept a session.

An attacker could set the `role` attribute to 'next'. In this case, *every intermediary node* will try to process the numerous SOAP header blocks. This will result in either the crashing or failure of intermediary nodes. Or, even if the intermediary nodes have extremely high processing capacity or apply some hard limit on number of SOAP header blocks to process, the fact that each intermediary node has to process these header blocks will very likely lead to a delay in service delivery. This can be considered a DoS for time-critical applications.

Such an attack could be made worse if the `relay` attribute is set to 'true' in all of the headers. In this case, no header blocks will be discarded in the process from one node to another.

Another variant is for an attacker to modify the value of the `role` attribute from 'next' to 'ultimateReceiver'. By doing so, when the end receiver receives the message, the message may not be processed properly because some prior processing to the message that should have been applied to the message by one or more of the SOAP intermediary nodes have not been realised.

Similar attack can also be used by changing the `mustUnderstand` value. When it is altered from 'true' to 'false', this may cause some missing processing to the message, which will ultimately cause a SOAP processing failure at the subsequent nodes. When it is altered from 'false' to 'true', it may cause an intermediary or end receiver node to generate a fault in the event of it not understanding the header block(s). In either case, message processing fails where it should not have, and this could lead to a DoS.

Malicious Intermediary SOAP Nodes

When SOAP messages are relayed through several intermediary nodes, a degree of trust in those nodes is almost crucial. From the explanation of the DoS attack in Sect. 7.2.3.2, we can see that when a SOAP intermediary is malicious, many security and privacy violations could occur [26, Section 7.2], including DoS attacks.

One of the distinguishing features of a service-oriented architecture (SOA) is the ability to dynamically compose a business service through consumption of several web services applications. In addition, the message exchange pattern has shifted from the traditional remote procedure call (RPC) request-response flow to a more dynamic one which can be as simple as the traditional request-response to a complex flow-pattern as commonly used in web services orchestration and choreography technologies. Such message processing pattern normally involves multi-hop message processing concurrent processing, and many other flow patterns.

As a consequence, it is very likely that a web services message will have to go through multiple SOAP intermediary nodes in the course of consuming a service, and these SOAP intermediary nodes may be dynamically chosen. The threat from such an approach is that messages are exposed to the risk of being manipulated by malicious intermediary SOAP nodes. Depending on the algorithm applied to choose services in composing a service, a malicious service provider could manipulate the algorithm used (such as advertising high quality of service (QoS) with low cost) so that the choreographed service will, with a high probability, include the malicious node as one of the intermediaries.

The need to trust SOAP intermediaries is specific to web services applications. Therefore, proper methods to address this problem is needed.

7.2.3.3 Non-relay of Messages by a Malicious SOAP Intermediary

A malicious SOAP intermediary can always, upon receiving a message, withhold the message and not further relay it to the subsequent SOAP nodes. This could negatively impact preceding intermediary SOAP nodes as they may have to retain state information related to the processing of the SOAP message.

When the malicious intermediary SOAP node also controls the originator of the SOAP message, a brute-force attack against these other intermediary nodes could easily be launched.

7.2.3.4 Large SOAP Attachment

The SOAP specification allows attachments to be included in a SOAP message [28]. This feature opens up two opportunities for attackers to cause a DoS.

First, an attacker could insert some malicious binary data as SOAP attachment which is used to exploit a known vulnerability in the recipient system's platform which could trigger a DoS (such as server failure). This is a semantic-based attack. Or, an attacker could use a brute-force strategy by sending massive numbers of SOAP messages with huge attachments with the intention of exhausting the recipient's resources [29].

In either cases, the use of attachment should be properly secured to avoid DoS attacks.

7.2.3.5 SOAP Fault Messages

The SOAP specification allows nodes to generate SOAP fault messages when it encounters errors in processing the SOAP message. This specification can be easily manipulated by attackers to generate traffic for a DoS attack: an attacker could craft some invalid SOAP messages such that a recipient node will, with a high probability, generate a SOAP fault[6]. The attacker would then spoof the SOAP message source address to the target victim's address and either send this SOAP message with a spoofed source address to multiple entities or multicast/broadcast it. Each of the recipient nodes would generate SOAP fault messages, which would all be returned to the victim's address. This is a brute-force-based attack, and it could result in a DoS.

7.2.4 WSDL Vulnerabilities

In this section, we detail the various points of DoS vulnerabilities in WSDL. As explained in Section 7.1, web services applications are described using the WSDL. However, despite the crucial role that WSDL plays in web services, the design of WSDL is unnecessarily complicated, and there has been an evident lack of emphasis on the need to secure the exchange of WSDL documents in the standards [8, 9]. These are several ingredients in a WSDL that could be exploited by attackers to cause DoS.

7.2.4.1 Unauthenticated WSDL Requests Flooding

As we have established, because WSDL is a metadata, in most cases, it is publicly accessible which means that no authentication of the requester is performed. As a result, DoS could be caused by launching a brute-force attack against a web services server by sending copious amount of WSDL requests which could result in exhaustion of the server's resources, and hence a DoS is accomplished.

7.2.4.2 Inconsistent SOAP Binding in WSDL

In a WSDL document, one could specify the binding requirements for web services messages such that they conform to certain SOAP encoding style, namely: rpc/encoded, rpc/literal, document/encoded and document/literal [7]. Correct and consistent encoding of SOAP messages is important to ensure interoperability. Nevertheless, various encoding styles exist, and, combined with various ways that

---

[6]It is not clear from the SOAP standard whether SOAP fault messages must always be generated or not. However, from the Web Services Interoperability Organization (WSI) Basic Profile document, one of the requirements (R1027, [5, Section 3.2.2]) is that a recipient must generate a SOAP fault if it cannot process a mandatory SOAP header block (those with the `mustUnderstand` attribute set).

```
1 <wsdl:binding name="..." type="tns:TestPort">
2   ...
3   <wsdl:operation name="method1">
4     <soap:operation soapAction=""> OR
5     <soap:operation soapAction="http://some.uri.org/etc/etc"> OR
6     <soap:operation>
7     ....
8   </wsdl:operation>
9 </wsdl:binding>
```

Fig. 7.9 SOAPAction specification in WSDL

clients and servers process web services messages (depending on the platforms used), these could easily result in a DoS, whether intentional or not.

The details of these various encoding styles are beyond the scope of this chapter; however, inconsistent encoding style of web services messages into the wire could result in messages not being understood by recipients, thus resulting in a DoS. Such inconsistencies could be the result of poorly designed WSDL documents on the server side, improper interpretation of WSDL document which results in incorrect stub classes (on the client's side) being used, or deliberate modification of WSDL document by attackers.

The last threat (modification by attacker) could be easily launched due to the fact that many WSDL documents are not integrity-protected. As WSDL describes a service metadata, many web services' WSDL documents can be accessed freely without any integrity or confidentiality protection to the documents. As a result, it is trivial to modify such documents in transit. This is a type of semantic-based attack, and it assumes that attackers have the ability to intercept a live session.

7.2.4.3 `SOAPAction` Header

When a SOAP request message is transported using the hypertext transfer protocol (HTTP) protocol, the SOAP 1.1 specification [6] defines an hyper-text transfer protocol (HTTP) header, called the `SOAPAction` header. The purpose of this header is to specify the intent of the SOAP HTTP request. One of the goals is to allow filtering devices, such as firewalls, to be able to appropriately filter the SOAP message in the HTTP request. In addition, it can also be used by servers to direct the SOAP message to specific process based on the value of the `SOAPAction` header. However, the SOAP 1.2 specification removes the use of this header.

The problem with the use (or nonuse) of this header is that different client platforms treat this header differently based on the WSDL specification as shown in Fig. 7.9. In this figure, lines 4–6 show the several possible ways to specify the use of `soapAction` header in WSDL:

- Specify the use of the header but with empty value
- Specify the use of the header and the value that can be included in the header
- Do not specify the use of the header at all

When a client retrieves a WSDL document, depending on the tools and framework they use, there could be a variation of the request message generated:

- The client request message does not use the SOAP action header at all
- The client request message includes the SOAP action header with an empty string
- The client request message includes the SOAP action header with the given value *without enclosing them in a pair of quotes*
- The client request message includes the SOAP action header with the given value *enclosed in a pair of quotes*

An experiment was performed by Siddiqui [32] to evaluate various platforms ability to handle a request message with varying use of the `SOAPAction` header. The result of the experiment shows that there are cases where certain HTTP requests failed due to unexpected format of the header that the client sent to the server.

Of course, the discussion so far suggests that these are implementation issues that can be rectified programmatically. However, from an attacker point of view, such an interoperability issue could be exploited: an attacker can simply modify a WSDL document in transit (assuming that it is not integrity-protected) such that the `SOAPAction` header specification is corrupted, or an attacker could simply corrupt the value of the `SOAPAction` header in a request such that the request is rejected. Again, this is a semantic-based attack and it assumes that attackers can intercept live sessions.

An obvious way to mitigate this threat is to provide integrity protection to the HTTP request messages. Note that the Web services SOAP security (WSS) technology only secures messages at the SOAP message level; therefore, it does not cover the HTTP header. Transport layer security, such as SSL/TLS which provides integrity protection to HTTP header, is needed.

7.2.4.4 Other Corruptions to WSDL Specification

Given that web services applications rely heavily on WSDL to facilitate their normal operations, essentially, any unauthorised modifications to WSDL documents will very likely result in interoperability problems, and hence DoS. We have mentioned the modification to the SOAP binding style and the `SOAPAction` header. However, modifications to other parts of the WSDL documents, including the XML schema definition (schema poisoning, deeply nested schema, etc.), the service endpoint, the input and output messages and so on would result in similar DoS effects.

The core problem to such a vulnerability is due to the fact that access to WSDL documents are often not integrity-protected.

7.2.5 WS-Security Vulnerabilities

While web services security technologies provide many security benefits, they are also susceptible to DoS attacks. In this section, various DoS vulnerabilities in WS-Security standards (explained in Section 7.1.2.2) are detailed.

```
<soap:Envelope xmlns:soap="http://schemas.xmlsoap.org/soap/envelope/"
 xmlns:wsse="..." xmlns:ds="...">
  <soap:Header>
    <wsse:Security mustUnderstand=1>
      <!-- large number of ds:Signature element --->
      <ds:Signature>....</ds:Signature>
      ....
      <ds:Signature>....</ds:Signature>
    </wsse:Security>
  </soap:Header>
  <soap:Body> ... </soap:Body>
</soap:Envelope>
```

Fig. 7.10 Large number of signatures within one <wsse:Security> block

7.2.5.1 Heavy Cryptographic Processing

As explained in Sect. 7.1.2.2, the WSS standard [20] defines methods to provide confidentiality, integrity and authenticity protections to web services messages. Nevertheless, the flexible and *expandable* nature of the design introduces DoS vulnerability whereby semantic-based attack could be launched. In addition, attackers do not have to intercept a live session to launch this type of attack.

When a SOAP message uses the the WSS techniques, they are normally encapsulated within a SOAP header block represented by a <wsse:Security> element (see Sect. 7.2.3 for an explanation of a SOAP header block). The WSS specification states that, although multiple <wsse:Security> header blocks are permitted within a single SOAP header, each <wsse:Security> block must be targeted at unique actor or role only. In other words, no two or more <wsse:Security> header blocks with the same actor or role attribute value are permitted. Such an approach prevents a recipient from having to process multiple <wsse:Security> SOAP headers[7]. However, the content of one <wsse:Security> header block is sufficient to cause DoS.

When the WSS message signature technique is used, the specification allows multiple signature blocks to be included within a SOAP header. Therefore, an attacker could craft a SOAP message such that it has one <wsse:Security> header block, but within that header block, a massive number of <ds:Signature> elements are included (see Fig. 7.10). Theoretically, such a message is still valid. An entity which does not anticipate such a message, 'upon reading the mustUnderstand' attribute would process every <ds:Signature> element, and this could result in resource exhaustion as signature verification process involves heavy public-key cryptographic processing.

Similarly, for encryption, an attacker could craft a SOAP message such that a massive number of <xenc:DataReference> elements are included within a single <xenc:ReferenceList>. Since each <xenc:DataReference>

[7]Nevertheless, it is still possible for multiple non-<wsse:Security> SOAP header blocks to be targeted to the same recipient.

```
1  <soap:Envelope xmlns:soap="http://schemas.xmlsoap.org/soap/envelope/"
2   xmlns:wsse="..." xmlns:xenc="...">
3    <soap:Header>
4      <wsse:Security mustUnderstand=1>
5        <!-- large number of xenc:DataReference element --->
6        <xenc:ReferenceList>
7          <xenc:DataReference ... />
8          .....
9          <xenc:DataReference ... />
10       </xenc:ReferenceList>
11     </wsse:Security>
12   </soap:Header>
13   <soap:Body>...</soap:Body>
14 </soap:Envelope>
```

Fig. 7.11 Large number of encrypted data blocks

element represents a block of encrypted data, having a massive number of such an element means that there are massive number of encrypted message blocks to decrypt (see Fig. 7.11, lines 6–10). This situation is aggravated if the attacker uses public key encryption method to encrypt each of the message block (i.e., no session symmetric key is established). Of course, an attacker could put arbitrary data within each encryption block, but recipient has to follow proper decryption algorithm. If it is a public key encryption, then substantial resources are required, and eventually, this may also lead to resource exhaustion, hence a DoS.

Therefore, an entity, while processing SOAP WSS security header block, should anticipate such an attack and act accordingly (such as limiting number of signature or encryption blocks, or using a strict conformance of message format to some predefined policies).

7.2.5.2 Invalid/Modification of Parameters

The WSS technology requires the parameters (such as key information, signing or encryption algorithms used, canonicalisation algorithm used, etc.) that are needed to properly process the WSS-protected messages to be included in the SOAP messages themselves in clear text. As a result, correct processing of those messages depends on the correct parameters being received. This exposes a semantic-based attack to cause a DoS.

While such processing model is not unique to web services, the fact that WSS protects messages at message level means that malicious SOAP intermediaries can see those parameters in cleartext, and thus able to modify them such that recipients of those modified parameters are unable to correctly process the message.

In Sect. 7.2.3.2, we have explained the DoS problem related with SOAP intermediaries. It is therefore crucial that, in order to protect message from this type of attack (security parameter modifications), SOAP intermediaries must either be trusted entities, or other methods are required to mitigate such a threat.

7.3 Validating DoS Vulnerabilities in Web Services

In Sect. 7.2, we have described the potential DoS vulnerabilities in web services technologies. In this section, we validate some of the theoretical DoS vulnerabilities identified in the previous section.

To do so, we have deployed several web services applications developed over various platforms, including Java Metro, Microsoft .NET, Apache Axis as well as Ruby on our testbed. The details of our testbed are available in Chap. 4.

Several DoS vulnerabilities that we have validated are detailed, followed by a description of our approach in conducting the experiments as well as how the attacks are implemented. We then explain the result of the attacks against the web services victims (which can be web services applications or web service clients). Please note that the material in this section is the extended version of the paper that has been published by Suriadi et al. [37].

7.3.1 Vulnerabilities Validated: Attack Scenarios

We consider four types of attack scenarios.

7.3.1.1 Scenario 1: Deeply Nested SOAP Message

This scenario tests whether DoS can happen when an entity receives a SOAP message with deeply nested elements. Such an attack could be launched from a malicious web service provider P to unsuspecting web service client C, or vice versa. Assume that C is malicious; this scenario requires that C sends a SOAP message whose body contains a deeply nested XML document.

Therefore, to test this scenario, we need to do the following:

- Set up P with a simple web services application.
- Let C send a SOAP request to the service with deeply nested XML document.
- Let P receive and process the SOAP request from C which contains deeply nested elements

We then observe the CPU and memory utilisation at P.

7.3.1.2 Scenario 2: WSDL Request Flooding

Right from the very beginning of a web services interaction, entities are already exposed to DoS attacks. When a web service provider P exposes some web services applications, the consumers of such applications require the WSDL descriptions of those exposed services. As we have explained previously in Sect. 7.2.4.1, unauthenticated request to WSDL documents could be a source of DoS attack. Therefore, the first scenario that we should test is how easily could an attacker launch a DoS by flooding P with WSDL requests.

In order to test this scenario, we need to do the following:

1. Set up P with a simple web services application.
2. Expose a uniform resource locator (URL) where the application WSDL can be obtained.
3. Set up several attackers to simultaneously send massive number of WSDL requests.

We then observe the CPU and memory utilisation at P.

7.3.1.3 Scenario 3: Heavy Cryptographic Processing Scenario

This scenario attempts to test DoS vulnerabilities due to maliciously crafted WSS SOAP header as detailed in Sect. 7.2.5.1. The goal is to show that while WSS does offer some security protection, it can also at the same time be a source of DoS. To test this scenario, we need to show the following:

- Set up P with a simple web services application with WSS-enabled.
- Set up a malicious SOAP request by C such that numerous encryption and/or signature headers are included in a SOAP header with the `mustUnderstand` attribute set to 'true'.
- Let C send the crafted SOAP message to P.

We then observe the CPU and memory utilisation at P.

7.3.1.4 Scenario 4: Malformed Metadata

Alternatively, we could also test a scenario whereby a web service provider P is malicious, or one or more of the entities involved are malicious. In this scenario, when a client C requests for a WSDL document, P returns the WSDL document with possibly recursive links to external XML schema which may also reference a very large but legitimate-looking XML schema. This attack can also happen even when P is honest, but one or more of the schema being referenced in the WSDL document have been compromised by external attackers. This is a quite realistic scenario since it is not uncommon for a web service definition to reference several XML schema, some of which may be managed by an entity other than P who may have been compromised.

For simplicity, let us assume that P is malicious. To test this attack, we need to:

- Set up P with a simple web services application.
- Let P craft a malformed WSDL with recursive XML schema (pointing to a malformed schema).

- Expose a Uniform Resource Locator (URL) at P where the crafted WSDL can be obtained.
- Let a client C make a WSDL request to P.

C is expected to read the WSDL document and attempt to parse the possibly recursive and large XML schema being referenced in the WSDL. We then observe the memory and CPU utilisation at the client side.

7.3.2 Experimental Approach

One of the key features of a distributed denial of service (DDoS) attack is that it normally consists of a large amount of attackers, each requesting a service from the victim. As a server is servicing a flood of requests, it will inevitably use up a higher amount of resources. However, this is a normal observation and, therefore, does not mean that a particular web service is vulnerable to DoS attack.

To verify if each of the above scenarios (see Sect. 7.3.1) really are valid points of DoS vulnerability, we need to first study the *baseline* server behaviour when it is not under attack. By *baseline* behaviour, we mean the amount of resources being consumed by the web server when it receives large but legitimate requests (not, for example, malformed requests with malicious payloads). Once we have this baseline information, we can then get the attackers to send the same amount of requests, but this time with malicious payloads.

By comparing the baseline resource utilisation with the resource utilisation under attack, we can then conclude if a particular web service DoS vulnerability really does significantly amplify the resource utilisation on the server's side such that the server's resources can be easily exhausted.

Therefore, for every framework of each scenario being experimented, we need to run at least two experiments: the first one to obtain the baseline resource utilisation data, and the second one to obtain the resource utilisation data when the server is under attack.

7.3.3 Attack Implementation and Results

In this section, the details of the attack implementation and the results are explained.

7.3.3.1 Scenario 1: Deeply Nested SOAP Message

The details of this experiment are as follows.

```
#!/bin/bash
counter=0
k=0
while [ $counter -lt $1 ]; do
let i=0
while [ $i -lt 50 ]; do
wget --no-proxy -q --header='Accept: text/xml, multipart/related, text/html'
     --header='Content-Type: text/xml; charset=utf-8'
     --header='SOAPAction: "add"' --post-file=$2
     http://192.168.200.61:8080/CalculatorApplication/CalculatorApplication4op &
let i=i+1
let k=k+1
done
sleep 1s
let counter=counter+1
done
echo $k
```

Fig. 7.12 An example attack script

Web Service Provider – the Victim

In this scenario, the victim of the DoS attack is the server. We launch a deeply nested XML attack against several similar web service applications but based on different framework. These applications have already been deployed in the testbed.

In total, there are four separate framework being tested in Scenario 1, including the Java Metro, Apache Axis, Microsoft .NET and Ruby framework.

Attacker

Before we launch the attack, we need to obtain the baseline resource utilisation data. To do so, we need to send a legitimate request with a valid non-malicious payload to the relevant web service application (i.e., without the deeply nested XML elements). The generation of the attack payload is very simple – a simple script to generate a deeply nested XML payload is used.

Once we have both the valid non-malicious payload and the attack payload, we then need to create a script to mimic the launch of a request flood. An example of the script is provided in Fig. 7.12.

This script simply runs the simple Linux `wget` command in a loop. The script sends a burst of 50 `wget` requests followed by a sleep of 1 s, before another burst of 50 requests is made.

The attack script accepts two arguments: the first argument determines the total number of 50-request burst to be made; the second argument determines whether the script should send a valid non-malicious payload, or a malicious payload.

Experiments Details

For each platform, we use three attacker machines, each executing the same attack script (see Fig. 7.12 for an example) simultaneously. The experiments performed

Table 7.1 Scenario 1: experiment details

Type	Framework	No. of. reqs	Attack length	Observ.
Base	Metro	3 Attackers@5,500 reqs each	≈2 m 20 s each	10 min
	Axis	3 Attackers@5,500 reqs each	≈2 m 5 s each	
	WCF	3 Attackers@5,500 reqs each	≈2 m 10 s each	3 m 20 s
	Ruby	1 Attacker@500 reqs	≈4 min	10 min
Attack:	Metro	3 Attackers@5,500 reqs each	≈2 m 30 s each	10 min
100K-deep	Axis	3 Attackers@5,500 reqs each	≈2 m 30 s each	
nesting	WCF	3 Attackers@5,500 reqs each	≈2 m 1 s each	3 m 20 s
	Ruby	1 Attacker@500 reqs	>10 min	10 min

are summarised in Table 7.1. This table shows the type of experiment being performed (baseline experiment and attack experiment), the framework being tested, the duration of the attack (i.e., the length of time it takes for each of the attacker machines to launch the necessary number of web service requests) and the length of observation (i.e., the duration of the system performance data collected from the time the attack starts).

As we can see from Table 7.1, the attack duration for attacks on the Metro, Axis and WCF platform barely goes beyond 2 min and 30 s however, we observe the system's performance for a 10-min period due to the fact that the server might still be working overtime to serve the flood of requests. Furthermore, note that the attack length does not take into account the amount of time required to receive the last response from the server. When server is under heavy load, a web service request may time out; as a result, `wget` normally attempts several reconnections before it gives up. Therefore, an observation of 10-min is generally needed to observe the full range of the effects of the attacks, although in some cases, the effect goes well beyond the 10-min range.

For the Ruby platform, we notice that we can achieve a DoS effect with a much lower threshold. As we will discuss later, the details for Ruby attack shown in Table 7.1 are sufficient to cause a successful DoS on the Ruby server.

Result

Our experiments show that in most frameworks (namely, Metro, Axis and WCF), contrary to popular belief, a deeply nested XML attack does not severely exhaust memory. However, it does consume a lot of CPU resources. Figs. 7.13 and 7.14 show the CPU and memory utilisation of Metro, Axis and Ruby web services applications under a normal non-malicious load and under deeply nested XML attack. Note that the CPU utilisation value shown in those figures are the total utilization over two CPUs.

As we can see, there is an obvious increase in the amount of CPU utilisation for those web services running on Java Metro and Apache Axis frameworks when it is under attack (close to 100% mark). We have also observed that similar trend applies to .NET application (graphs not shown).

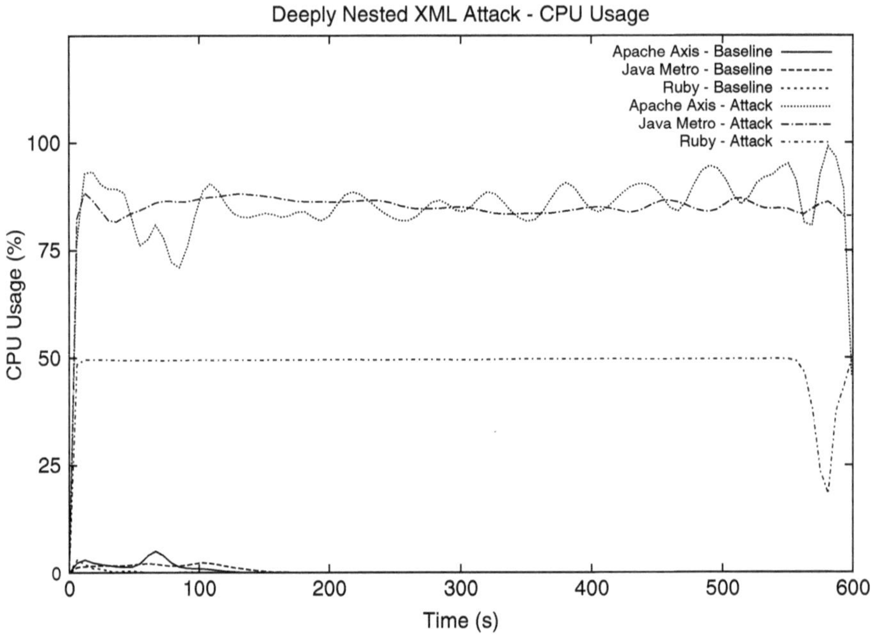

Fig. 7.13 Deep XML attack – CPU usage – base and under Attack – Metro, Axis and Ruby

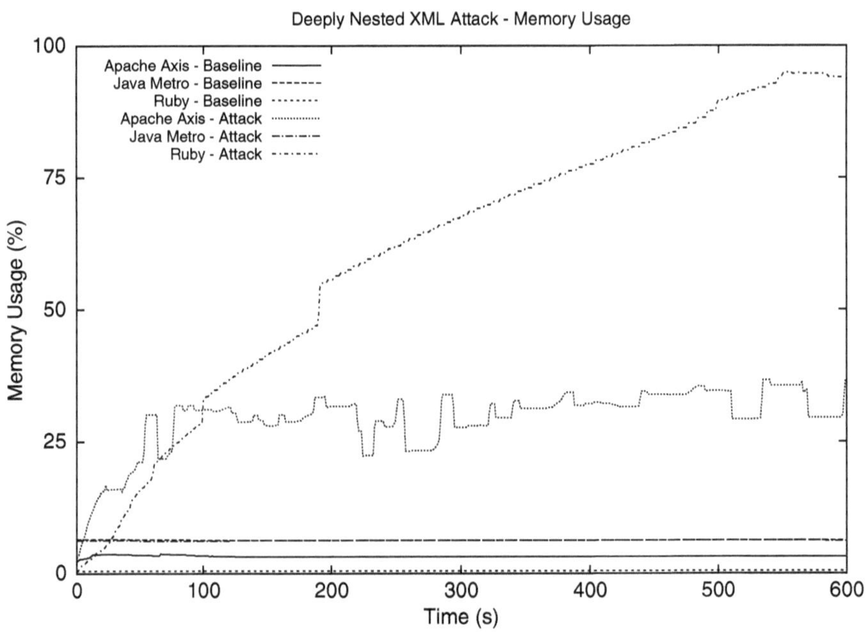

Fig. 7.14 Deep XML attack – memory usage – base and under attack – Metro, Axis and Ruby

For the Ruby web service, the CPU utilisation is also very high, around 50% of total CPU utilisation (which consists of two CPUs). In other words, due to the simple single-threaded application of Ruby, it cannot take advantage of the power of two CPUs. Thus, the maximum utilisation in this case will be around 100% of one CPU, or, 50% of two CPUs.

However, the memory utilisation at the Metro and Axis (and also in the .NET framework) does not seem to be affected much by such an attack (see Fig. 7.14). On the other hand, for the Ruby framework, such an attack does cause severe memory exhaustion. In our experiments, this attack resulted in the Ruby server being killed eventually.

7.3.3.2 Scenario 2: WSDL flooding

We have implemented this scenario, the setup in general is as follows.

Web Service Provider – The Victim

The victim of the denial of service attack in this scenario is the web service provider. The setup of the provider is just a simple echo web service application, similar to the ones used for Scenario 1 (see Sect. 7.3.3.1).

To provide the WSDL of a service, we use the `?wsdl` syntax. Such a syntax requires the server to perform some processing before the correct WSDL file can be provided. In other words, the WSDL file is generated dynamically, instead of using a static WSDL file.

We have tested this attack against the Metro, Axis and WCF framework.

Attackers

Similar to Scenario 1, to launch an attack, first we need to prepare the base non-malicious payload and the malicious payload. To measure the base performance of a server, the base payload in this case is a simple HTTP GET request, requesting a static HTML page, e.g. http://localhost:8080/index.html. For the attack payload, the attack payload is also a simple HTTP get request, but this time requesting the dynamically generated WSDL, e.g. http://localhost:8080/axis2/services/EchoService?wsdl.

The script for running the attack is very similar to the one shown in Fig. 7.12; i.e., a simple `wget` utility is used to repeatedly request for the static/dynamic page.

The details of the experiments being launched to test this DoS vulnerability is detailed in Table 7.2.

Table 7.2 Scenario 2: experiment details

Type	Framework	No. of. reqs	Attack length	Observ.
Base	Metro	3 Attackers@37,500 reqs each	≈1 m 43 s each	10 min
	Axis	3 Attackers@37,500 reqs each	≈1 m 58 s each	
	WCF	3 Attackers@37,500 reqs each	≈1 m 55 s each	3 m 20 s
Attack	Metro	3 Attackers@37,500 reqs each	≈4 m 27 s each	10 min
	Axis	3 Attackers@37,500 reqs each	≈4 m 20 s each	
	WCF	3 Attackers@37,500 reqs each	≈1 m 56 s each	3 m 20 s

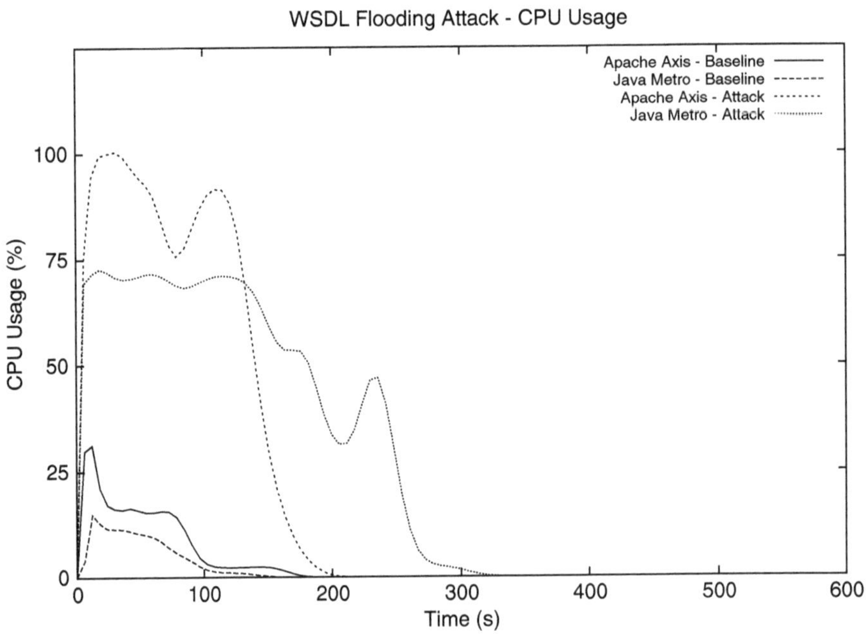

Fig. 7.15 WSDL flood attack – CPU usage – base and under attack – Metro and Axis

Result

As we can see from Fig. 7.15, this attack (which simply requests dynamically-generated WSDL) does cause the servers to use up a significantly higher CPU time. This is true for all of the framework tested, including the Metro, Axis and WCF frameworks (graphs not shown for the .NET framework). This is a problem in a web service application because WSDL is generally a public document, thus, restricting access may not be that simple. While it is true that in most frameworks there is a method to configure the server to return a static WSDL instead, the use of dynamically generated WSDL (i.e., the '?wsdl' syntax) seems to be the 'de facto' way in many development frameworks.

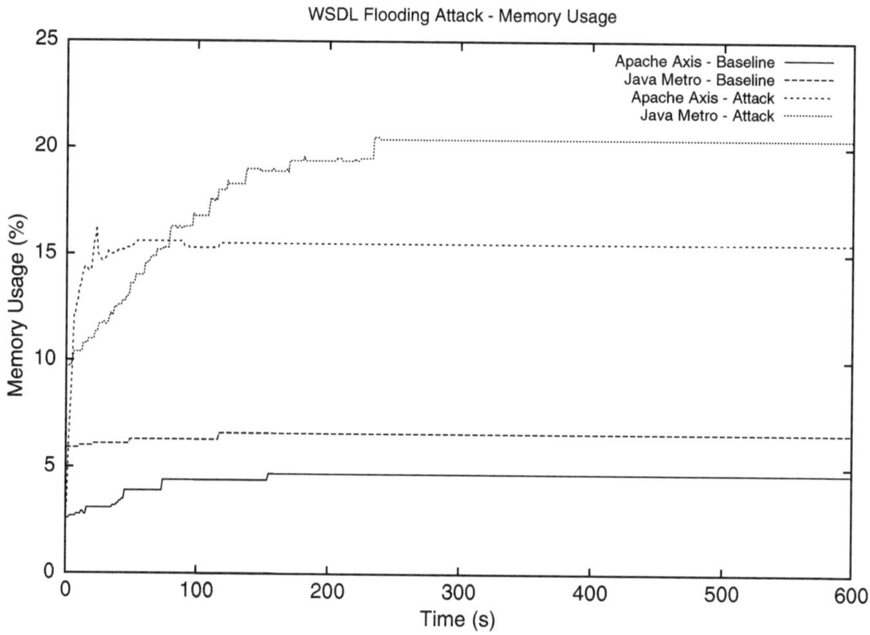

Fig. 7.16 WSDL flood attack – memory usage – base and under attack – Metro and Axis

However, as can be seen from Fig. 7.16, this attack does not seem to cause any significant increase in the memory usage, regardless of the framework being attacked.

7.3.3.3 Scenario 3: Heavy Cryptographic Processing

We have implemented this scenario, the setup in general is as follows.

Web Service Provider – the Victim

The victim of the denial of service attack in this scenario is the web service provider. The setup of the provider is a simple echo web service application, similar to the ones used for Scenario 1 (see Sect. 7.3.3.1). However, for this scenario, we configure the server to require WS-Security to be used. In most cases, we require at least a web service request to be signed using X.509 mutual authentication mechanism. Encryption is also used in some framework implementation.

We have tested this attack against the Metro, Axis and WCF framework.

Table 7.3 Scenario 3: experiment details

Type	Framework	No. of. reqs	Attack length	Observ.
Base	Metro	3 Attackers@5,500 reqs each	≈2 m 0 s each	10 min
	Axis	3 Attackers@5,500 reqs each	≈2 m 7 s each	
	WCF	3 Attackers@5,500 reqs each	≈2 m 10 s each	3 m 20 s
Attack:	Metro	3 Attackers@5,500 reqs each	≈2 m 3 s each	10 min
50 security	Axis	3 Attackers@5,500 reqs each	≈2 m 0 s each	
headers	WCF	3 Attackers@5,500 reqs each	≈2 m 22 s each	3 m 20 s

Attackers

Similar to Scenario 1, to launch an attack, first we need to prepare the base non-malicious payload and the malicious payload. To measure the base performance of a server, the base payload in this case is a valid SOAP request containing the necessary WS-Security header (also valid header).For the attack payload, we simply copy the WS-Security header multiple times within a SOAP header (for SOAP request) and set the `mustUnderstand` flag to true. The logic here is to force the server to process the WS-Security header multiple times to exhaust its resources (as message signature verification and decryption normally requires expensive public key operation).

We do not use timestamp security mechanism in our experiment simply for convenience. The default setting in most platform is to have a SOAP message freshness to be valid for 5 min after its generation. We have managed to successfully reply the SOAP message with timestamp; however, as expected, it became invalid after 5 min. Having to change the payload every time we need to run an experiment is counterproductive. Therefore, we design the server so that no timestamp is required.

Therefore, the existence or non-existence of timestamp information is not important in this experiment because the same type of attack that we launched here can be performed even with the timestamp information included.

The script for running the attack is similar to Scenario 1.

The details of the experiments being launched to test this DoS vulnerability is detailed in Table 7.3.

Result

As we can see from Fig. 7.17, different frameworks respond quite differently to this attack.

In the Axis framework, this attack actually causes the server to consume *less* CPU load in comparison with the baseline value. This is understandable because the server will not be able to understand the `mustUnderstand` directive, and thus send a SOAP fault message which is *not* signed. As a result, less resources are being consumed. When sending a valid payload, on the other hand, the server

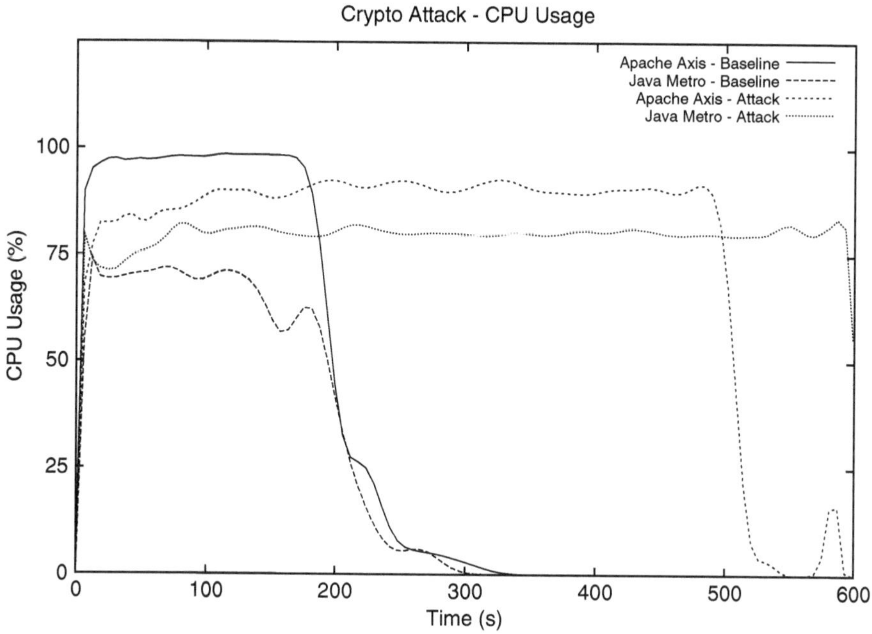

Fig. 7.17 Heavy cryptographic attack – CPU usage – base and attack – Metro and Axis

needs to sign and possibly encrypt, the return SOAP message, thus taking up a much higher resources. However, although the attack causes the server to use up less CPU time, the CPU utilisation does remain quite high for a much longer period than the baseline measurement.

The Java and .NET framework behaves quite similarly (the results of the .NET framework are not shown here). On these frameworks, the attack payload seems to cause an increase in the CPU utilisation time. We have not investigated the differences between the .NET/Metro framework and the Axis framework to determine what makes this type of attack work on the .NET/Metro framework but not on the Axis framework. This attack does not seem to affect CPU utilisation significantly, regardless of the platform used (see Fig. 7.18).

7.3.3.4 Scenario 4: Malformed Metadata

This scenario has also been implemented. The details are as follows.

Web Service Provider

In this scenario, we assume that the web service provider is the malicious entity. We assume that the provider has managed to get the client to retrieve a schema from

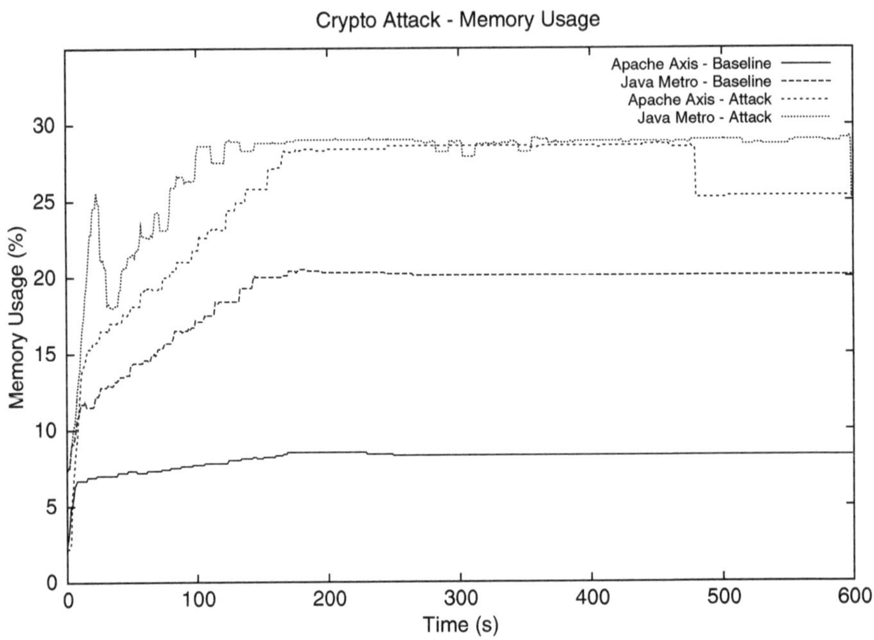

Fig. 7.18 Heavy cryptographic attack – memory usage – base and under attack – Metro and Axis

```
Schema1
=======
<schema>...<import namespace=".../assertion" schemaLocation="schema2.xml"/></schema>

Schema2
=======
<schema>...<import namespace=".../xmlenc#" schemaLocation="schema3.xml"/></schema>

Schema3
=======
<schema">...<import namespace=".../xmldsig3#"schemaLocation="largeSchema.xml"/></schema>
```

Fig. 7.19 Multi-stage schema referencing

an address suggested from the provider. This schema, in turn, imports several other schema, one of which is a *very large* schema which the provider has crafted.

In particular, the original schema references a schema, which in turn reference few other schema before the large schema is retrieved. Figure 7.19 shows that the first schema (Schema1) references another schema (Schema2), which in turn references another schema (Schema3). Schema3 finally references the very large schema as shown in Fig. 7.20.

```
<schema version="2.0" ...">
<complexType name="SignatureType1" abstract="true">
<sequence>
  <element name="SignatureMethod1" type="string" minOccurs="0" />
  <!-- many elements being defined -->
  </sequence>
  <attribute name="SId1" type="ID" use="optional" />
  <!-- many attributes being defined -->
  </complexType>
<complexType name="SignatureType2" abstract="true">
<sequence>
  <element name="SignatureMethod2" type="string" minOccurs="0" />
  <!-- many elements being defined -->
  </sequence>
  <attribute name="SId2" type="ID" use="optional" />
  <!-- many attributes being defined -->
  </complexType>
....
<!-- similar schema definition repeated for 100k times-->
</schema>
```

Fig. 7.20 A provider-crafted large XML schema

Client – the Victim

To simplify the process, we develop a client that directly goes to the pointed schema location and retrieves it. The client is developed using the Microsoft .NET WCF framework. In particular, we use the built-in `DiscoveryClientProtocol` library that .NET WCF framework provides. This library can be used to retrieve a web service metadata. The goal is to allow clients to dynamically discover a service and access it during runtime. When we launch the client, it will then try to retrieve the schema one by one, and it eventually retrieves the very large schema.

We have only implemented this attack on the .NET WCF framework. This is because this type of attack only makes sense if it is directed against a client during runtime. This type of attack can also occur when a client is being developed (such as when the client is generating stub classes based on a WSDL). However, the effect of such an attack on an off-line client may not be significant.

We have found a good support for a dynamic client that can read a WSDL document live during runtime and generate the necessary stubs and call the web services operations. We have not investigated how an equivalent behaviour can be achieved in the Java Metro framework and the Apache Axis framework.

Result

Our experiment shows that the .NET `DiscoveryClientProtocol` library will attempt to retrieve *all* of the schema being referenced from the original schema and it eventually downloads the very large schema. Furthermore, not only does it download the very large schema, it also attempts to parse the schema, resulting in a very high consumption of the memory at the client's side (see Fig. 7.21). The end result is the Windows throwing a memory exhaustion error (which is what happens

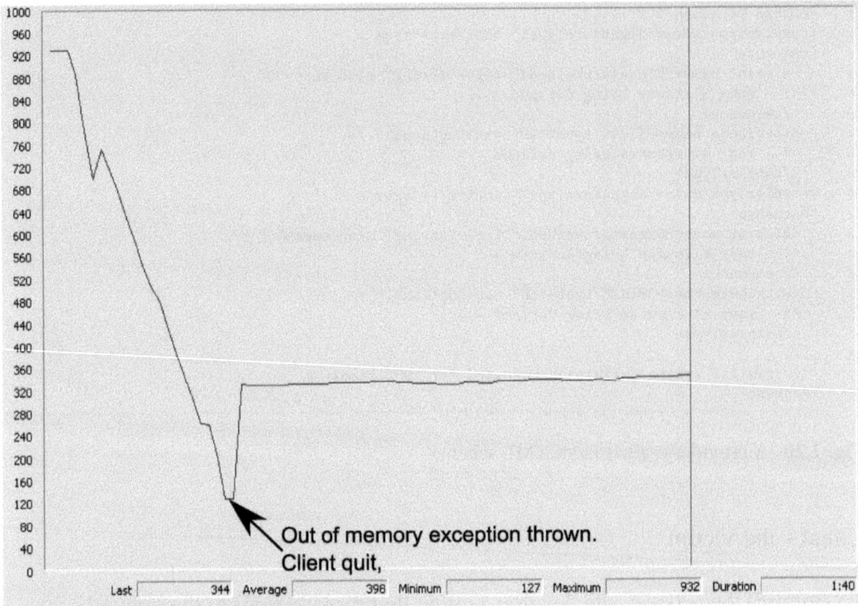

Fig. 7.21 Scenario 4: total available memory in MB

most of the time), and the client program quits. Therefore, the client has been denied the service it originally requested.

This vulnerability seems to stem from the insecure logic of the `DiscoveryClientProtocol` class of the .NET framework which does not check if the referenced XML schema is too large.

Practically, such a vulnerability can be exploited to create DoS in certain circumstances:

- Man in the middle attack: changing/adding reference to malicious XML schema doc in the WSDL being retrieved (assuming access to WSDL endpoint is unsecured).
- The external XML file being referenced is beyond the control of the referencing entity. For example, company A references a schema from company B which is not properly secured. As a result, company B's schema can be illegally modified, causing DoS on the company A's clients.

7.4 DoS Mitigation Strategy for Web Services

Having validated some of the known web services DoS vulnerabilities, in this section, we explain several mitigation strategies that could be employed in practice.

7.4.1 *Requirements of Web Services DoS Mitigation Strategies*

Based on the observed behaviours of web services servers and clients under the various attack scenarios which we have experimented in Sect. 7.3, we propose several features that effective DoS mitigation strategies for web services applications should deliver. In particular, our mitigation strategies should address three main areas:

Defending semantic-based web services attacks: the deeply nested XML attack, malformed metadata and the heavy-cryptographic processing attack are just three of many other forms of semantic-based attacks against web services applications. The techniques used to provide semantic-based attacks are elaborated later in this section.

Defending flood-based web services attacks on server-side: the WSDL flooding attack is just one of many flooding attacks that can affect web services applications. We consider basic network filtering approach and cryptographic client puzzle solution as the main mechanisms in defending our web services applications against flooding-based DoS attacks. These techniques are further elaborated in the remainder of this section.

Intelligent DoS attacks detection and coordination: having strategies to defend both semantic-based and flooding-based attacks is a reasonable starting point. However, such a defence-in-depth strategy needs to be properly coordinated to achieve the most efficient mitigation strategy.

The techniques that can be used to achieve the mitigation strategies mentioned above are detailed in the next section.

7.4.2 *Mitigation Techniques*

Several mitigation techniques are considered for web services DoS mitigation strategy.

7.4.2.1 Semantic-Based Attack: XML Payload Analysis and Metadata Investigator

Several techniques can be used to mitigate semantic-based web services DoS attacks.

XML Payload Analysis

Web services applications mostly rely on SOAP messages to deliver service request and response messages. Consequently, many DoS attacks on web services

```
<?xml version="1.0" ?>
<S:Envelope xmlns:S="http://schemas.xmlsoap.org/soap/envelope/">
 <S:Body>
  <ns2:echo xmlns:ns2="http://ws.tomcat/">
    <i>Echo Generated</i>
    <somethingwithverylargename>
     <somethingwithverylargename>
      <somethingwithverylargename>
       <somethingwithverylargename>
        <somethingwithverylargename>
         <somethingwithverylargename>
          <somethingwithverylargename>
           <somethingwithverylargename>
            <somethingwithverylargename>
             ..................
            </somethingwtihverylargename>
           </somethingwtihverylargename>
          </somethingwtihverylargename>
         </somethingwtihverylargename>
        </somethingwtihverylargename>
       </somethingwtihverylargename>
      </somethingwtihverylargename>
     </somethingwtihverylargename>
  </ns2:echo>
 </S:Body>
</S:Envelope>
```

Heavily nested document, up to thousands or more

Fig. 7.22 An example of deeply nested XML signature

applications target the vulnerabilities in the structure of web services SOAP messages themselves. The deeply nested SOAP message attack (detailed in Sect. 7.3.3.1) and heavy cryptographic processing attack (detailed in Sect. 7.3.3.3) are two examples of attacks exploiting SOAP message structure.

In this type of attack, the main 'differentiating' factor between a good non-malicious SOAP payload with a malicious SOAP payload rests solely on the structure of the SOAP payloads themselves. The rest of the information being received at the receiver's end (such as the IP header, TCP/UDP header, even the HTTP header) can be exactly identical between a good payload and a malicious payload.

Since SOAP messages are formatted in XML, a key requirement for web services DoS mitigation strategies should include a component that is able to perform XML packet analysis to detect if an incoming SOAP message contains malicious payload based on some known malicious payload structure (signature). Furthermore, this component should reside before the web service applications themselves so that any malicious SOAP payload can be detected and dropped. An example of a malicious payload signature is provided in Fig. 7.22. In this example, the key distinguishing feature for such a malicious payload is the unusual number of children elements in the XML payload.

To achieve the 'XML Payload Analysis' capability, it is inevitable that we need to do deep-packet analysis. One possible deep-packet analysis technique is to employ the XML clustering technique which has been a subject of research for quite some time [27, 38]. While XML schema in general is used to validate if a document conforms to its specification, it is not sufficient: a valid XML document conforming to its schema may still be malicious. Therefore, we look into the use of different XML analysis techniques. The XML clustering technique is one of them. The key concept with this approach is to analyse XML documents based on either their contents, structures or both to determine the classification of these documents. Our preliminary investigation into this technique [2, 17, 27, 38] shows that this is a rather promising technique that we can use to detect if an XML payload 'looks similar' to a benign payload or to a malicious payload.

Performing XML payload analysis can be a challenging task given that an analyser needs to first determine if a payload is indeed an XML document before it starts the analysis. To do so, a form of efficient XML parsing is required.

One strategy is to extend a well-known software-based network intrusion detection and prevention system called Snort [35] to perform deep-packet analysis by adding new rules that are specific for malicious web services payloads. While this is possible, we do realise that the performance of Snort may not be optimum due to the significant processing overhead required to perform XML payload analysis.

Alternatively, we can use a hardware-based solution. Fast network card, such as the Endace DAG7.5G2[8] ,is capable of both capturing, analysing and performing hardware-based packet filtering at a reasonable speed. Furthermore, preliminary investigation suggests that such a card has field-programmable gate array (FPGA) capability which can be programmed to filter out packets with specific characteristics. Encoding the filtering rules for malicious web services, XML payload on the Endace hardware card may improve the efficiency of the process.

While further investigation is required, the use of hardware-based XML payload analysis (using XML clustering technique) to filter out malicious web services messages seems to provide a promising outcome.

Metadata Investigator

The malformed metadata vulnerability (experimented in Sect. 7.3.3.4) requires a different mitigation strategy too. This is not a flooding style vulnerability. Rather, this DoS attack exploits the insecure logic of some XML importer which simply imports all documents regardless of their size. An ideal mitigation strategy to counter this attack would be to use an XML importer which can 'preview' the content of the XML document to be imported and check for any signs of malformed content (such as recursively called various XML documents, or excessively referencing other documents, and other similar features).

---

[8]http://www.endace.com/assets/files/resources/Endace_Datasheet_DAG7.5G2-G4_20100330.pdf.

This logic should be embedded in the dynamic metadata finder API of the respective programming framework. For example, in the Apache Axis framework, this logic would have to be included in its service client API (most likely by creating a new class which extends the main `org.apache.axis.client.Service` class).

7.4.2.2 Flooding Attack: Efficient Requests Filtering and Proof of Work

The WSDL flooding attack (see Sect. 7.3.3.2) is a different type of attack as compared to semantic-based attack. The WSDL-flooding attack is a type of a 'flooding attack', and normally there is no distinguishing factor between legitimate requests and malicious request payloads. They both contain legitimate payload requesting the publicly available WSDL documents. Therefore, preventing such an attack by analysing the request payload is useless.

In fact, the distinguishing characteristics for such an attack normally resides within the lower-layer protocol header information, such as:

- Repeated WSDL requests from the same clients (identified through the use of the same IP address, for example) over a period of time
- A sudden increase in the intensity of WSDL requests from multiple IP addresses

Change Point Analysis Detection and IP-Whitelisting

To counter the first instance, we need a mechanism that can weed out those clients that have repeatedly sent WSDL requests because this is an abnormal behaviour: a client normally only requires a WSDL document once before it starts consuming the service from the server. Of course, this assumes that the client is a dynamic client who can generate correct request messages by just interpreting the WSDL document *during runtime*. Another type of client, known as static client, normally only reads a server's WSDL once and is then statically configured to generate request based on some hard-coded programs. For static clients, they normally do not request for a server's WSDL before consuming the service. Therefore, because the probability of a normal legitimate client to repeatedly request a WSDL document is very low, any repeated WSDL requests from the same client should be flagged as an attack and should be *filtered out*.

The use of IP address white-listing in combination with the use of *change point analysis* technique [1] (to detect anomaly in the rate of change in the arrival of new IP addresses) is one of the key research topics already detailed in Chap. 5. The key feature of this technique is that when a server is under attack (detected through the use of the *change point analysis* technique), it will drop all packets that originate from any IP addresses which are not already in the white-list. The white-list is developed based on historical data. This same technique can, theoretically, be used to prevent WSDL flooding attack for a WSDL document request is, often, nothing more than a simple HTTP GET request. However, further experimentation is required to verify its effectiveness in mitigating WSDL flooding attack.

Updating the white-list when a system is under attack is an important feature, otherwise an attacker could trick the filtering system by behaving honestly before it starts an attack. That is, an attacker could behave 'normally' for a period of time before starting the attack, thus ensuring its IP address being included in the white-list. Then, the attacker could start launching attacks to the server and its packets will still be passed on by the filter because its IP address is already in the white-list.

Therefore, the use of IP address white-listing technique needs to be complemented with some 'intelligent controller' who will theoretically detect if, for example, a client has been performing excessive WSDL document requests. This DoS Mitigation Module (DMM) will then update the firewall to remove this client from its white-list, even under attack. Currently, such an 'intelligent controller' has not yet been implemented and is part of the future work.

Cryptographic Client Puzzle

To counter the second instance, whereby a set of clients simultaneously request for a server's WSDL document, we need a different mitigation mechanism. In this situation, it is likely that each client only makes very few WSDL requests, but there are *many* clients who make the same requests simultaneously. In this situation, it is difficult to determine which WSDL requests are legitimate and which ones are not. We do not want legitimate client to be denied the WSDL document, but we also do not want the server's resources to be depleted.

One strategy to address this situation would be to 'authenticate' each request so that only genuine clients who have the necessary credentials can access the resources. However, this poses a problem as a client and a server may not have known each other beforehand. Furthermore, traditional connection authentication techniques, such as Secure Socket Layer (SSL), require a large amount of computing resources to perform (this in itself becomes a DoS vulnerability).

Alternatively, one could use the client puzzle solution whereby a client needs to perform some amount of work to obtain solutions for a challenge that the server gives *before* the server allocates any resources for this client. Using the client puzzle mechanism, while we cannot authenticate the identity of the client, we can at least have some assurance that those clients who provide correct solution to the puzzle have legitimate intention. Besides, a server can also adjust the difficulty level of the client puzzle based on the load of the server at any given time, thus 'pacing' the client requests to protect its resources from being exhausted.

We have investigated the use of client puzzle in web services, and we have come up with two avenues in which this technique can be applied. First, we can use the client puzzle solution in an integrated manner with lower-layer network operation, such as integrating it with an OpenSSL solution (which is detailed in Chapter 6). By doing so, all web services requests (as well as all other application layer services) delivered through HTTP transport can benefit from the protection provided by the client puzzle technique. In particular, this integration approach protects web services applications from those DoS attacks which aim at exploiting transport (or lower) layer protocols.

Alternatively, we can use the client puzzle technique at the application layer, *independent of* lower-layer transport protocol. In particular, we seek to add a new security mechanism that can be advertised by a web service provider to its clients such that requests from clients with incorrect solutions can be dropped *even before* any resources are allocated by the server to perform nontrivial tasks, such as cryptographic operations processing (e.g. encryption and digital signature verification). By doing so, we can theoretically mitigate web services applications from DoS attacks at the application layer itself. This has the advantage of not having to rely on the lower layer network/transport protocols for DoS protection.

We have investigated and evaluated the effectiveness of client puzzle mechanism in mitigating web services flooding attack and the details of our integration mechanisms and experiments are elaborated in Sect. 7.5.

7.4.3 Mitigation Architecture Design

We have described several mitigation techniques that can be employed to counter semantic-based and flooding-based attacks. However, as described in Sect. 7.4.1, we not only need these techniques, but we also need to be able to coordinate the use of these techniques together in an intelligent manner such that we can have an effective defence-in-depth strategy. Such a defence-in-depth strategy is closely tied with the architecture in which these mitigation strategies are employed.

In this section, a DoS mitigation architecture, based on the mitigation techniques detailed in Sect. 7.4.2, is detailed. In particular, we propose an architecture which shows where each of the discussed DoS mitigation techniques should be implemented and how they interact with other network components and/or other mitigation techniques to achieve optimum results.

An overview of the architecture is provided in Sect. 7.4.3.1. This architecture is extended from the DMM architecture already discussed in Sect. 5.4.2. Using this extended architecture, an explanation of the DoS mitigation architecture for web services applications' perspective is provided in Sect. 7.4.3.2. Later in Sect. 7.4.4, we provide several use-case scenario to illustrate how the various components of the DoS defence architecture interact together to cohesively defend the web services applications against various types of DoS attacks.

7.4.3.1 Overview of DoS Mitigation Solution Architecture

As shown in Fig. 7.23, the DoS mitigation architecture is comprised of several key components: a network firewall, an application layer firewall, a reverse proxy

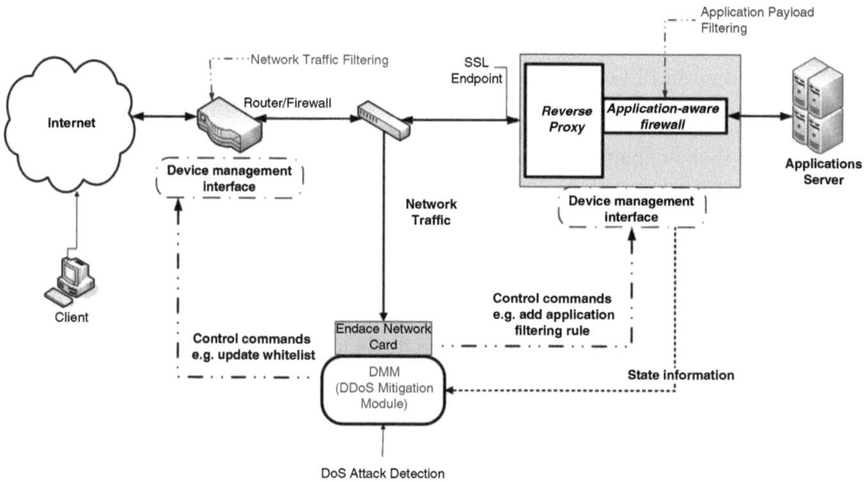

Fig. 7.23 Overview of DoS mitigation solution architecture

and DMM. The resources we want to protect are represented as a set of application servers. A brief description of each of these components is provided below.

Network firewall: this component is the first line of defence against DoS attacks. A network firewall function is mainly to filter out/drop packets based on a set of predefined filtering rules, such as drop packets coming from an IP address which is not in its white-list, drop packets to certain port numbers and so on.

Application-aware firewall: this component performs a deep-packet analysis of network packets to detect any malicious payload targeted at the application-layer services (such as HTTP payload) based on a set of pre-defined rules.

DMM: this component is the 'brain' of the DoS mitigation architecture[9]. It captures all network traffic (which have been allowed into the network by the network firewall) passively and perform DoS attack analysis based on the captured network data. Because of the relatively high overhead involved in analysing the network traffic for DoS attacks, the DMM is *not* positioned in-line. This DMM module is also equipped with a powerful network card, called the Endace DAG7.5G2, which has many functionalities, some of which include the ability to:

- Capture all network traffic with 0% loss
- Perform some network traffic filtering based on some built-in rules
- Split network traffic into separate streams based on configurable characteristics (such as destination machine IP address, etc)

[9]The DMM component is still more of a conceptual idea at this point. It is still a work-in-progress and further research is needed to properly develop this module.

- Allow the captured network traffic to be accessed by software-based modules to do further analysis/filtering
- Be used as an FPGA card
- And many others

Based on these capabilities, we can write several software modules to analyse each network stream (which has been split by the Endance card). The DMM can therefore have many modules, each of the modules serves to analyse certain types of traffic or to detect certain types of DoS attack. For example, we can have a module implementing a *change point detection analysis* to detect any unusual number of new IP addresses in the incoming traffic, while another module can be used to implement the *bloomfilter* analysis to efficiently detect if an Internet protocol (IP) has been 'seen' in the network before, and so on.

When the DMM's analysis results in the detection of DoS attacks, it will then update the network firewall, or the application firewall, or both, with new rules to filter out those packets which are considered to be causing the DoS attacks.

Reverse proxy: a reverse proxy is needed mainly to handle encrypted traffic. Without the reverse proxy, an SSL traffic is likely to 'end' at the application servers themselves. In this scenario, the application-layer firewall and DMM cannot perform any filtering and analysis because the payloads are encrypted. By using a reverse proxy, we can terminate an SSL traffic at the reverse proxy itself, allowing application data to be accessed in plaintext by the application firewall. Furthermore, it is also possible for the application firewall to copy and forward the decrypted traffic to the DMM for further analysis.

7.4.3.2 DoS Mitigation Architecture for Web Services

Figure 7.24 shows how the mitigation techniques discussed in Sect. 7.4.2 fit into the DoS mitigation solution architecture described in Sect. 7.4.3.1.

The XML payload analysis technique is initially designed to be deployed at the application firewall to filter out any malicious payloads. The IP address white-list filtering is to be deployed at the network traffic firewall itself. At the same time, a DMM is installed in the network to capture all traffic which will be analysed by the DMM passively. The long-term goal is to enable a form of machine learning at the DMM such that new types of DoS attack can be detected and have the DMM to update the network firewall and application firewall rules accordingly.

In relation to web services, our main concerns are with the design of the XML payload analysis algorithm that the application firewall should apply. Note that the DMM and application firewalls can detect and apply other types of analysis (not just XML payload analysis). However, for the purpose of this chapter, we are mainly concerned with XML payload analysis only.

Similar situation also applies to IP address white-list. The DMM is responsible, using its analysis techniques, for determining which IP addresses should be added/removed from the whitelist of the network firewall.

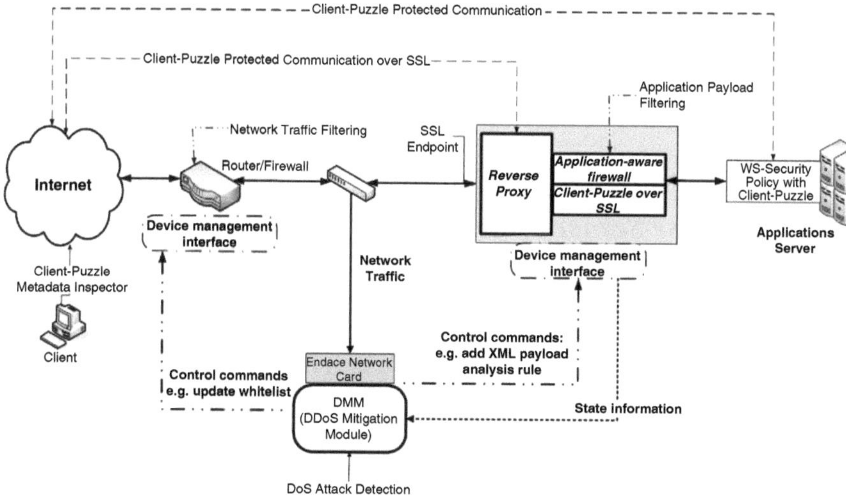

Fig. 7.24 DoS mitigation architecture for web services

The client puzzle solution is to be implemented at the application side itself. In the first usage mode, the client puzzle solution is to be integrated with the underlying transport layer protocol (mainly SSL). If the SSL endpoints are between the client and the application server, then the use of the client puzzle solution in this mode will result in all web services traffic being encrypted. This will cause the DMM and application layer firewall to be unable to perform any meaningful deep-packet analysis. Therefore, as shown in Fig. 7.24, we use a reverse proxy as the SSL termination point on the server's side. In this approach, the reverse proxy can decrypt all the traffic and forward the traffic data to the application firewall and DMM for further analysis. Thus, we do not break either the application firewall or the DMM functionalities.

In the second usage mode, the client puzzle solution is to be integrated with the web services application framework itself as part of the WS-Security Policy definition. In this approach, regardless of the use (or the absence) of transport layer encryption, a client needs to solve the puzzle challenge given by the application server, before the server will even perform any further resource-intensive operations (such as XML signature verification and SOAP payload decryption). Of course, if the SOAP payload is encrypted, the ability of the application firewall and DMM to perform its analysis might be limited. However, encrypted SOAP messages (for web services applications) normally follow a well-defined structure as defined by the various WS-* standards. Thus, a limited form of deep-packet analysis can still be performed (e.g., by examining the SOAP header structure to determine if there are excessive number of signatures or encryptions).

The 'Metadata Inspector' is very much a mechanism to protect DoS on the client side. Therefore, it is only logical to have it installed on the client side.

The architecture shown in Fig. 7.24 and described in this section is still at its draft form. Most of the fine-tuning of the DoS mitigation architecture can only be effectively performed during the implementation stage of the mitigation techniques, which is the main task for the next few milestones.

7.4.4 *Use-Case Scenario*

To give a 'feel' of how the architecture that we have described in Sect. 7.4.3 works in mitigating DoS, we provide several use-case scenarios for illustration purposes. Note that the use-case scenarios detailed below are not exhaustive.

- A group of bots is controlled by an attacker. Each of these bots initially sends legitimate web services requests to the web services applications; thus, their IP addresses are very likely to be included in the network firewall's white-list. At some point, the attacker instructs all the bots to start sending a flood of WSDL requests to the web service server, but only a few requests from each client. The above mitigation strategy can help mitigate a potential DoS situation through several mechanisms:
  - The DMM might notice a sudden spike in the network traffic; thus, a potential DoS situation might be flagged.
  - The DMM could instruct the web application server to start using the client puzzle technique to pace the rate of the requests on the server side.

  Given sufficiently difficult and secure puzzles that are in proportion with the attack load, the server's resources may still be protected from being exhausted.
- The same bots can also be used to launch a WSDL-flooding requests; however this time, instead of asking each bot to send one or two requests, the attacker attempts to get each bot to send hundreds/thousands of WSDL requests. The DoS mitigation strategies detailed in this section can mitigate the effects of such a DoS attack through several mechanisms:
  - The DMM notices a sudden spike in the network traffic; thus, a potential DoS situation is flagged.
  - The DMM could instruct the web application server to start using the client puzzle technique to pace the rate of requests on the server side.
  - The DMM could also detect the fact that repetitive requests for WSDL have been performed by the same clients. Thus, these clients are then removed from the network firewall's white-list.
  - The net effect is that not only those bots controlled by attacker have to solve the puzzles sent by the application server, their WSDL requests might even be dropped by the firewall as their IP address may have already been removed from the white-list.

Given sufficiently difficult and secure puzzles, and given a sufficiently fast reaction time by the DMM to update the white-list, the server's resources may be protected from being exhausted.

- An attacker tries to send a deeply nested XML payload in an attempt to exhaust a server's memory/CPU resources. In this case:
  - The application layer firewall may have already been configured with a signature to detect such a malicious payload and drop the packet
  - The DMM module can also detect such a malicious payload and update the network firewall white-list with a removal of this attacker's IP address.
- An attacker who controls a group of bots commands them to send a few web services requests containing subtly malicious payloads (such as those payloads with only five signatures to verify) which might not be flagged as problematic by the application firewall. In this scenario, the mitigation strategies detailed in this section can be used to mitigate DoS attack through several mechanisms:
  - The web service application can use a blanket approach to require its clients to solve a puzzle before it provides the services. It can do so by integrating the client puzzle solution as part of its WS-Security Policy advertisement. In this case, the server will not even attempt to verify the digital signatures until the client provides the correct answer to the puzzle. Ideally, the level of difficulty of the puzzle should be equivalent to the expected amount of resources that need to be expended to process the request message.
  - The DMM might be able to learn that, although the client payload does not match any of its signatures for malicious XML payload, there is a sudden increase in the server's workload.
  - Consequently, the DMM might apply stricter DoS detection rules which may now cause the client's XML payload to be flagged as malicious.
  - The DMM can now update the network firewall's rule with the removal of those clients who have sent malicious payloads.

  This is a rather difficult attack to be detected by the DMM when the detection rules are quite lenient (which is necessary to reduce the false positive rate). However, the important feature here is the ability of the DMM to intelligently adjust the strictness of its detection parameter based on other variables (such as an increase in the network traffic volume).

7.5 Cryptographic Client Puzzle for Web Services Applications

In this section, we explain our experiences in employing the cryptographic client puzzle technique to mitigate (1) the heavy-cryptographic processing DoS attack (explained in Sect. 7.2.5.1) as well as (2) a more generic flooding attack whereby a server needs to expand a lot of resources to serve a single request (e.g., BPEL stateful processing, or complex mathematical operations).

7.5.1 Cryptographic Client Puzzle

We have integrated the cryptographic client puzzle technique that is also proposed in this monograph in Chap. 6. A succinct description of the client puzzle is provided in this section. Readers who are interested in the details of the implemented cryptographic client puzzle technique (including its security proofs) should refer to Chap. 6.

Client puzzles, also called *proofs of work*, can be used to counter resource-depletion denial-of-service attacks: before a server is willing to perform some computationally expensive operation, it requires that the client commit some of its own resources and solve some moderately hard puzzle. Client puzzles were first proposed by Dwork and Naor [11] to control junk email by having recipients only accept emails if they were accompanied by a correct puzzle solution, and have since been extended to protect cryptographic protocols such as authentication [3, 16] and key exchange [34, 36] protocols, as well as network protocols such as TCP [25] and TLS [10].

The most commonly proposed type of client puzzle is a *hash-based, computation-bound* puzzle, in which a client is required to find a partial preimage in a cryptographic hash function. For example, in the puzzle of difficulty d proposed by Aura et al. [3], the client C and server S supply nonces N_C and N_S, respectively, and the client must find a solution X such that

$$H(N_S, N_C, X) = \underbrace{0\ldots0}_{d} \| Y, \tag{7.1}$$

where H is a cryptographic hash function, such as SHA-1, and the output starts with at least d 0 bits followed by any string Y. If H is a preimage-resistant hash function, then it should take a client approximately 2^{d-1} calls to H to find a valid solution. However, the verification cost for a server is very low, as it only takes a single hash function call to check Eq. 7.1.

Client puzzle can be used in either interactive (with challenge-response) mode or noninteractive (one-way) mode. In the former mode, a client normally contacts a server indicating its intention to consume its service, followed by the server sending the client a challenge to the puzzle of which the client must provide a correct solution before the server agrees to serve the request. This mode is normally used in establishing a connection, such as during an SSL handshake. In the noninteractive mode, no challenge is given by the server. The client simply includes a puzzle solution for a puzzle with difficulty d (see Eq. 7.1). Obviously, in the latter mode, the client generates the necessary nonces by itself. The noninteractive mode is normally used for situation whereby real-time interaction with a server is not the norm, such as sending an email.

There is a notion of clients being able to do precomputation in the interactive mode; thus, to guarantee the freshness of a puzzle solution, the server needs to give a fresh challenge for every client request. In the noninteractive mode, it is assumed that such a precomputation notion is nonexistent, and that freshness is not a concern in the environment in which it is used.

7.5.2 Implementation of Client Puzzle Technique in Web Services Environment

We have integrated the cryptographic client puzzle with a web service developed using the .NET WCF framework. In this section, the approach and the implementation details are briefly explained.

7.5.2.1 Approach

Our main concern is to study the effectiveness of the client puzzle technique implemented in existing web services development frameworks, such as the Microsoft .NET Windows Communication Foundation (WCF) framework, in mitigating known DoS vulnerabilities. Thus, to keep our experiments simple, we have decided to implement the noninteractive client puzzle solution. This is sufficient because a noninteractive mode already allows us to study the amount of resources being consumed at the server's side in verifying clients' solution, and more importantly, it allows us to study if the client puzzle technique allows legitimate client request to be served, while rejecting those malicious clients who simply send random (and incorrect) puzzle solutions.

The implementation and evaluation of the interactive client puzzle will be considered as part of the future work.

The integration of the client puzzle in our .NET framework is performed by making use of the extensible WS-Policy feature. The server advertises in its WSDL document (which also includes WS-Policy statements) the requirement for the client to provide a client puzzle solution, along with the puzzle difficulty level d and the server nonce N_S. To prevent clients from reusing a solution, we also require that a client must include both the value of nonce N_C and a timestamp T (represented as the number of ticks (10^{-7}) since 1 Jan. of year 1AD). In other words, the client must provide a client puzzle solution X, such that Eq. 7.2 is true[10].

$$H(N_S, N_C, T, X) = \underbrace{0 \ldots 0}_{d} \| Y, \tag{7.2}$$

The client upon reading the WSDL will then be aware that a client puzzle solution needs to be included in the SOAP header along with the request.

[10]In our subsequent experiments, we have not included codes to verify the freshness of timestamps however, we assume that such a process is straightforward.

```
<wsdl:definitions ...>
<wsp:Policy wsu:Id="clientPuzzlePolicy">
  <wsp:ExactlyOne>
   <wsp:All>
    <wsp:clientPuzzle a:difficulty="8" xmlns:a="...">
       abcdef
    </wsp:clientPuzzle>
    ...
   </wsp:All>
  </wsp:ExactlyOne>
</wsp:Policy>
<wsdl:types...>...</wsp:types>
...
</wsdl>
```

Fig. 7.25 Custom WS-Policy assertion for client puzzle in WSDL

```
<s:Envelope...>
<s:Header...>
  <ClientPuzzleSolution xmlns="..." ...>
    <timestamp>634243948044717802</timestamp>
    <clientNonce>LMBfqB</clientNonce>
    <puzzleSolution>abcdef634243948044717802LMBfqB9dvdjkdjfsd...</puzzleSolution>
  </ClientPuzzleSolution>
  ......
</s:Header>
<s:Body>...</s:Body>
</s:Envelope>
```

Fig. 7.26 Inclusion of client puzzle solution in a SOAP request

7.5.2.2 Implementation and Experiment Details

We have implemented the client puzzle using the .NET WCF framework. The server (or the DoS victim) runs on a physical desktop with Intel Core 2 Duo 3 Ghz processor and 4 GB RAM. We use a custom WS-Policy assertion to advertise the server's client puzzle requirement. The inclusion and the processing of custom WS-Policy assertion (along with the necessary processing of such custom assertion at both client's and server's side) are supported in the WCF framework through the implementation (and subsequent related configuration) of the `IPolicyExportExtension`, `IPolicyImportExtension`, `IClientMessageInspector`, `IDispatchMessageInspector`, `IEndpointBehaviour` interfaces and the `BehaviourExtensionElement` class. A snapshot of the WSDL advertising the client puzzle requirement and the subsequent SOAP request message being generated are provided in Figs. 7.25 and 7.26, respectively.

7.5.3 Experiments and Results

Given the stated goals of using client puzzle in the beginning of Sect. 7.5, we have designed our victim (the server) to be running a web service application (built on the

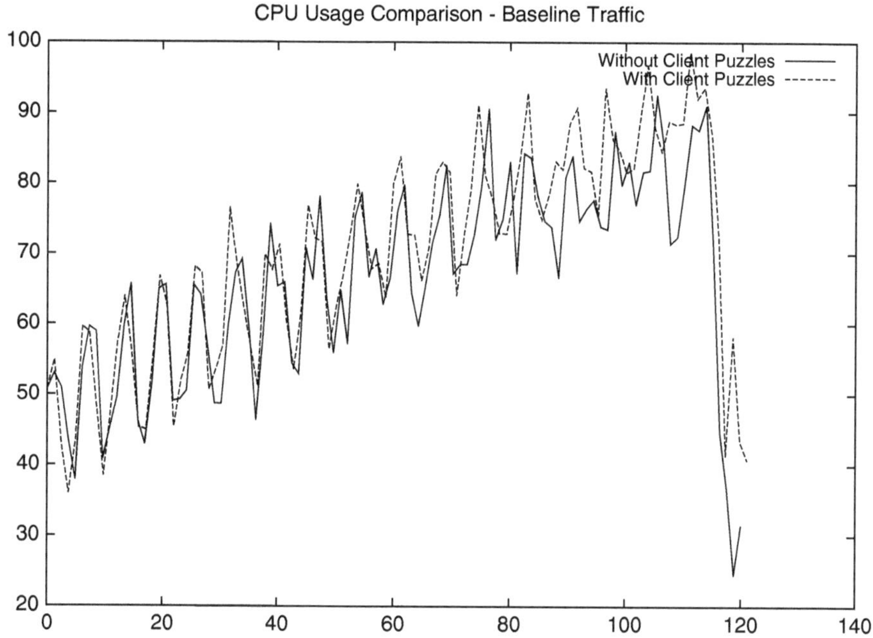

Fig. 7.27 Server CPU usage comparison for server processing 10,000 legitimate requests, with and without client puzzle

WCF framework) whose service requires the server to perform heavy mathematical computation. We have designed three groups of experiments to evaluate different aspects of client puzzle effectiveness. The first group of experiments simply checks if the introduction of client puzzle verification at the server's side causes noticeable increase in the consumption of CPU power. The second group of experiments tests the effectiveness of client puzzle in a generic flooding attack (sending many legitimate-looking requests; see Sect. 7.5.3.2 for details). Finally, the third group of experiments tests the client puzzle effectiveness in mitigating heavy-cryptographic processing attack (explained in Sect. 7.2.5.1).

7.5.3.1 Server Baseline Behaviour with Client Puzzle

To assess the effect of processing client puzzle on the server, we send about 10,000 legitimate requests to the web service with legitimate client puzzle solution, and without any client puzzle. We then compare the CPU usage data of the server in servicing these two group of requests. As we can see from Fig. 7.27, in general, the introduction of the client puzzle does *not* add any significant processing overhead to the server. This proves the notion that the resources required to verify the correctness of a client puzzle solution is minimal.

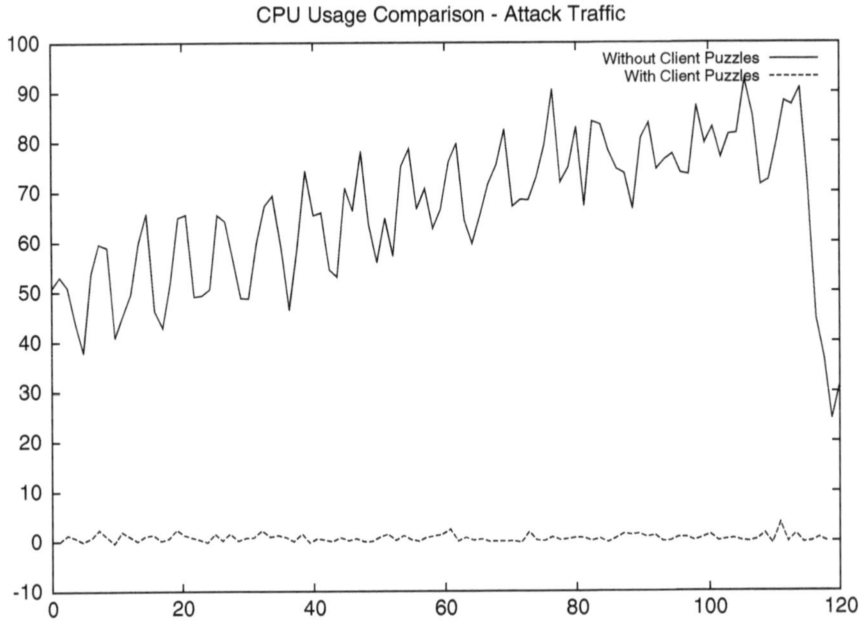

Fig. 7.28 The reduction of CPU consumption on server when client puzzle is implemented

7.5.3.2 Mitigating Generic Flood Attack

In this second experiment, we want to measure the effectiveness of client puzzle solution in filtering out those malicious requests from those legitimate requests. To do so, we conduct several rounds of experiments.

The first one compares the CPU resources expanded on the server in serving 10,000 illegitimate requests with and without the client puzzle implementation. That is, when no client puzzle technique is used, an illegitimate request looks just like any legitimate request (there is nothing in the payload to indicate that it is an illegitimate request). When the client puzzle technique is used, the illegitimate request is manifested by the use of random and incorrect puzzle solution. The results, shown in Fig. 7.28, clearly show the reduction in the CPU resource usage when the client puzzle technique is used.

The result shown in Fig. 7.28 is nevertheless expected because the server does not perform any heavy mathematical operation required once it realises that a client's puzzle solution is incorrect. A more interesting question is how effective is the client puzzle technique in serving legitimate requests when the server is under the stress of receiving thousands of other illegitimate requests. To measure it, we run another round of experiment, this time using two clients. The first client sends 1,000 legitimate requests (indicated by correct client puzzle solution), while the second client sends 10,000 illegitimate requests (indicated by incorrect client puzzle solution). We then compare the server's CPU usage with when the server only

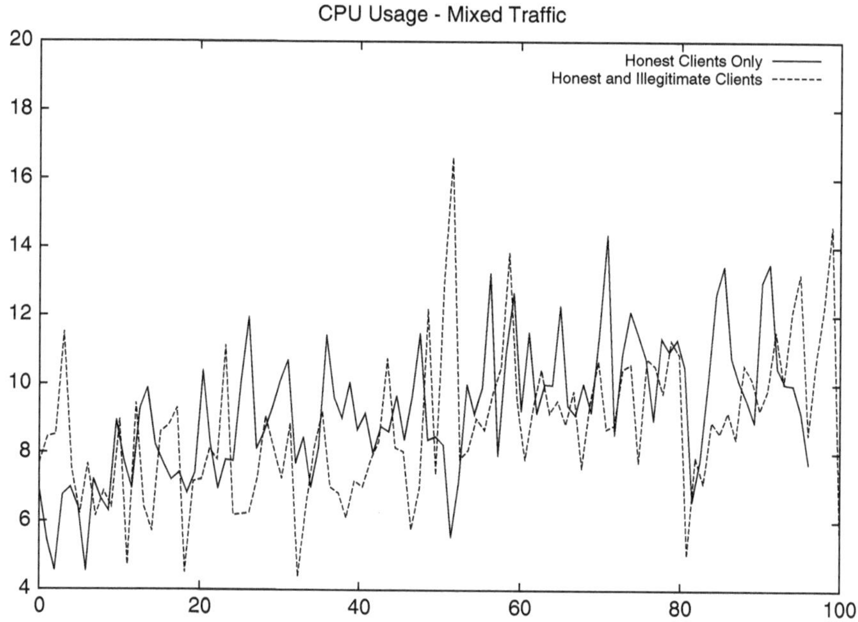

Fig. 7.29 CPU usage remains low when 10,000 illegitimate requests are mixed with 1,000 legitimate requests

Table 7.4 Legitimate requests being served

Scenario	Successful response
1,000 Legitimate requests only	919
Mixed requests (1,000 legitimate, 10,000 illegitimate)	949

serves 1,000 legitimate requests. Figure 7.29 shows that: (1) the CPU usage remains relatively the same between the mixed client scenario (1,000 legitimate and 10,000 illegitimate requests) and the 1,000 legitimate client scenario and (2) the CPU usage in both cases remains very low. These results confirm that the use of client puzzle can protect a server's CPU resources under a generic flooding attack.

Furthermore, Table 7.4 shows that the number of legitimate requests being served successfully remains very high (>90%) even when the server is under attack. This result therefore confirms that the use of client puzzle can protect a web service from serving legitimate client requests when it is under DoS attack.

7.5.3.3 Mitigating Heavy-Cryptographic Processing DoS Attack

In this experiment, we want to assess the effectiveness of client puzzle in mitigating heavy-cryptographic processing of web services. Client puzzles allow a server to

Fig. 7.30 CPU usage of server processing 1,000 honest requests (with and without client puzzle but all with digital signatures)

detect a malicious request by quickly and cheaply verifying that the given solution was incorrect. This avoids wasting resources on expensive cryptographic operations and preserves them for legitimate requests.

We run two rounds of experiments. The first round is simply to obtain the baseline behaviour when 1,000 legitimate requests (with and without client puzzle, but all with digital signatures) are sent to the server. The baseline behaviour is shown in Fig. 7.30. As we expect, the addition of client puzzle processing does not add any noticeable processing overhead to the server.

Having obtained the baseline behaviour, we then run the second round of experiment. This time, we send 10,000 illegitimate requests, all requests with many digital signatures and without client puzzle. We then compare the result with the scenario when 10,000 illegitimate requests with many digital signatures and with client puzzle are processed. The results are shown in Fig. 7.31. This figure shows that (1) the CPU usage for both scenario (with and without client puzzle) is more or less the same, and (2) there is a reduction in the CPU usage consumption as compared to baseline scenario. From these two observations, we conclude that the use of client puzzle in this case does not provide any additional benefits to mitigating heavy-cryptographic processing attack. In fact, given the reduction in the CPU usage

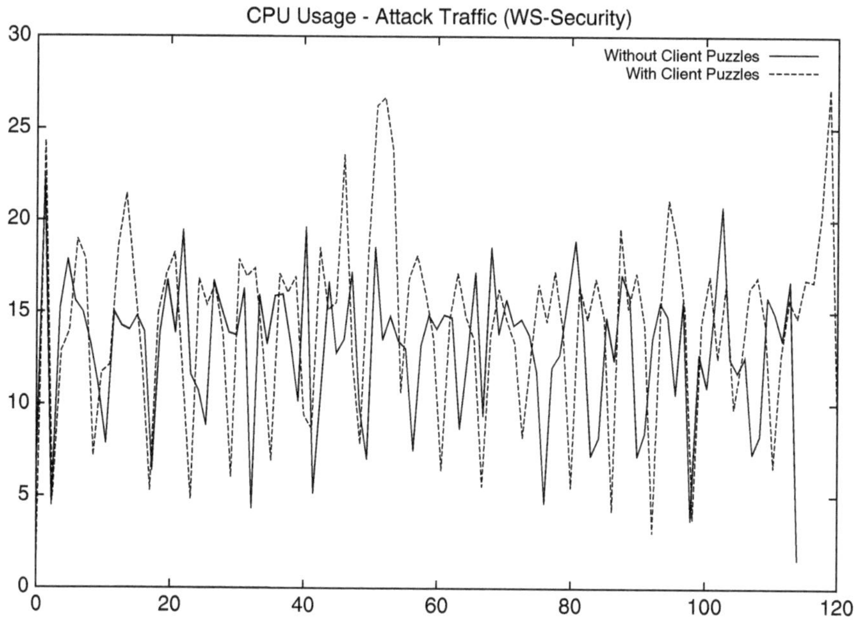

Fig. 7.31 CPU usage processing 10,000 malicious heavy-cryptography requests (with and without client puzzle)

for the scenario when the client puzzle solution is not used, one can deduce that the .NET WCF framework already provides some form of defence against such an attack.

We have not been able to conclusively explain why the client puzzle does not provide any additional protection. One possibility is that the default process of the server is to process the digital signature first before any extension behaviour (in this case, the client puzzle). Thus, assuming that the .NET WCF framework does reject requests with multiple digital signatures (which does not conform to its advertised security policy), this explains the similar CPU-usage pattern between the use and nonuse of client puzzle. However, while we believe that we have configured our .NET server to ensure that the client puzzle is processed first, we still obtain similar results as shown in Fig. 7.31. Further research is required to satisfactorily explain the ineffectiveness of client puzzle in this scenario.

7.6 Conclusions and Future Work

In this chapter, we have explained the vulnerability of web services to DoS attacks. We have also validated some of those theoretical DoS vulnerabilities to show that recent web services platforms are still vulnerable to DoS attacks, although it is more often servers' CPU power that is being attacked, not their memory as often believed. We have also explained the various mitigation techniques that can be employed to address the DoS issues in web services, along with a possible implementation architecture. Furthermore, we have also explained how a noninteractive client puzzle solution can be used to mitigate DoS attacks. Our experiments with client puzzle in web services applications show that (1) the client puzzle does not add any significant processing overhead on the server's side, (2) the client puzzle technique is suitable to mitigate generic flooding attack on resource-intensive web services and (3) the client puzzle technique does not seem to add any significant value in mitigating the heavy-cryptographic processing attack.

We note that there are still many other scenarios in which we can assess the effectiveness of the client puzzle technique. For example, implementing the client puzzle technique on a different platform, such as Java Metro platform, might actually give different results (due to the potentially differing internal default behaviours between the .NET and the Java frameworks).

The noninteractive client puzzle technique which we have implemented can still be extended to provide some safeguards against precomputation by periodically (such as every 1 min) refreshing the server's nonce in the WSDL. Of course, a malicious client can still do some precomputation; however, the precomputed solutions can only be replayed for a maximum period of the interval after which the server challenge is changed. Thus, the window of opportunity for the client to replay its solution can be reduced.

Finally, we still have to implement the interactive client puzzle technique in web services to even reduce the chance of any precomputation by the client, and observe the overhead on the server's resources to study the feasibility of using the more-secure interactive client puzzle.

References

1. Ahmed, E., A. Clark, and G. Mohay. 2008. A novel sliding window based change detection algorithm for asymmetric traffic. In *Proceedings of the IFIP International Conference on Network and Parallel Computing*, 168–175, Oct 2008.
2. Algergawy, A., R. Nayak, and G. Saake. 2009. XML schema element similarity measures: A schema matching context. In *OTM Conferences (2)*, 1246–1253, 2009.
3. Aura, T., P. Nikander, and J. Leiwo. 2000. DoS-resistant authentication with client puzzles. In *Security Protocols Workshop 2000*, 170–181. Cambridge, Apr 2000.
4. Badishi, G., A. Herzberg, I. Keidar, O. Romanov, and A. Yachin. 2008. An empirical study of denial of service mitigation techniques. In *IEEE Symposium on Reliable Distributed Systems. SRDS '08*, 115–124, Oct 2008.
5. Ballinger, K., D. Ehnebuske, C. Ferris, M. Gudgin, C. Liu, M. Nottingham, and P. Yendluri. 2006. Basic profile version 1.1 final material. http://www.ws-i.org/profiles/basicprofile-1.1.html. Accessed 17 Feb 2011.
6. Box, D., D. Ehnebuske, G. Kakivaya, A. Layman, N. Mendelsohn, H. F. Nielsen, S. Thatte, and D. Winer. 2000. Simple object access protocol (soap) 1.1. http://www.w3.org/TR/2000/NOTE-SOAP-20000508/. Accessed 16 Feb 2011.

7. Butek, R. 2005. Which style of WSDL should I use? http://www.ibm.com/developerworks/webservices/library/ws-whichwsdl/. Accessed 17 Feb 2011.
8. Chinnici, R., J.-J. Moreau, A. Ryman, and S. Weerawarana. 2007. Web services description language (WSDL) version 2.0. Part 1 Core language. http://www.w3.org/TR/wsdl20/. Accessed 17 Feb 2011.
9. Christensen, E., F. Curbera, G. Meredith, and S. Weerawarana. 2001. Web services description language (WSDL) 1.1 – W3C note. http://www.w3.org/TR/wsdl.html. Accessed 17 Feb 2011.
10. Dean, D. and A. Stubblefield. 2001. Using Client Puzzles to Protect TLS. In *Proc. 10th USENIX Security Symposium*, 2001.
11. Dwork, C. and M. Naor. 1992. Pricing via processing or combatting junk mail. In *CRYPTO '92 Proceedings of the 12th Annual International Cryptology Conference on Advances in Cryptology*, 139–147, London, 1992. Springer.
12. Eastlake, D., J. Reagle, D. Eastlake, J. Reagle, T. Imamura, B. Dillaway, and E. Simon. 2002. XML encryption syntax and processing. http://www.w3.org/TR/xmlenc-core/. Accessed 16 Feb 2011.
13. Eastlake, D., J. Reagle, and D. Solo. 2002. (Extensible markup language) XML signature syntax and processing. http://tools.ietf.org/html/rfc3275. Accessed 16 Feb 2011.
14. Gudgin, M., M. Hadley, N. Mendelsohn, J.-J. Moreau, H. Nielsen, A. Karmarkar, and Y. Lafon. 2007. SOAP version 1.2. Part 1: Messaging Framework (Second edition). http://www.w3.org/TR/soap12-part1/. Accessed 16 Feb 2011.
15. Gudgin, M., M. Hadley, N. Mendelsohn, J.-J. Moreau, H. Nielsen, A. Karmarkar, and Y. Lafon. 2007. SOAP version 1.2. Part 2: Adjuncts (Second edition) http://www.w3.org/TR/soap12-part2/. Accessed 16 Feb 2011.
16. Juels, A. and J. Brainard. 1999. Client puzzles: A cryptographic defense against connection depletion attacks. In *Proceedings of the Network and Distributed System Security Symposium (NDSS '99)*, 151–165, San Diego, Feb 1999. Internet Society Press, Reston.
17. Kutty, S., T. Tran, R. Nayak, and Y. Li. 2007. Clustering XML documents using closed frequent subtrees: A structural similarity approach. In *INEX*, 183–194, 2007.
18. Lawrence, K., C. Kaler, A. Nadalin, M. Goodner, M. Gudgin, A. Barbir, and H. Granqvist. 2007. 'WS-Trust 1.3,' OASIS Standard ws-trust-200512, Mar 2007. http://docs.oasis-open.org/ws-sx/ws-trust/200512/ws-trust-1.3-os.html. Accessed 31 Aug 2011.
19. Lawrence, K., C. Kaler, A. Nadalin, M. Goodner, M. Gudgin, A. Barbir, and H. Granqvist. 2008. WS-Securitypolicy 1.2 – OASIS standard incorporating proposed errata. http://docs.oasis-open.org/ws-sx/ws-securitypolicy/200702/ws-securitypolicy-1.2-spec-errata-cd-01.pdf. Accessed 17 Feb 2011.
20. Lawrence, K., C. Kaler, A. Nadalin, C. Kaler, R. Monzillo, and P. Hallam-Baker. 2006. Web services security: SOAP message security 1.1 (WS-Security 2004). http://docs.oasis-open.org/wss/v1.1/wss-v1.1-spec-os-SOAPMessageSecurity.pdf. Accessed 16 Feb 2011.
21. Lawrence, K., C. Kaler, A. Nadalin, C. Kaler, R. Monzillo, and P. Hallam-Baker. 2006. Web services security X.509 certificate token profile 1.1. docs.oasis-open.org/wss/v1.1/wss-v1.1-spec-os-x509TokenProfile-01.pdf. Accessed 17 Feb 2011.
22. Lawrence, K., C. Kaler, A. Nadalin, R. Monzillo, and P. Hallam-Baker. 2006. Web services security kerberos token profile 1.1. http://www.oasis-open.org/committees/download.php/16788/wss-v1.1-spec-os-KerberosTokenProfile.pdf. Accessed 17 Feb 2011.
23. Lawrence, K., C. Kaler, A. Nadalin, R. Monzillo, and P. Hallam-Baker. 2006. Web services security: SAML token profile 1.1. http://www.oasis-open.org/committees/download.php/16768/wss-v1.1-spec-os-SAMLTokenProfile.pdf. Accessed 17 Feb 2011.
24. Lawrence, K., C. Kaler, A. Nadalin, R. Monzillo, and P. Hallam-Baker. 2006. Web services security username token profile 1.1. http://docs.oasis-open.org/wss/v1.1/wss-v1.1-spec-os-UsernameTokenProfile.pdf. Accessed 17 Feb 2011.
25. McNevin, T., J.-M. Park, and R. Marchany. 2004. pTCP: A client puzzle protocol for defending against resource exhaustion denial of service attacks. Technical report TR-ECE-04-10, Department of Electrical and Computer Engineering, Virginia Tech, Oct 2004. http://www.ece.vt.edu/parkjm/Research/techReport_pTCP.pdf. Accessed 17 Feb 2011.

26. Mitra, N. and Y. Lafon. 2007. SOAP version 1.2 part 0: Primer (second edition). http://www.w3.org/TR/soap12-part0/. Accessed 16 Feb 2011.
27. Nayak, R. and S. Xu. 2005. XML documents clustering by structures. In *INEX*, 432–442, 2005
28. Nielsen, H. and H. Ruellan. 2004. SOAP 1.2 attachment feature – w3c working group note. http://www.w3.org/TR/soap12-af/. Accessed 17 Feb 2011.
29. Padmanabhuni, S., V. Singh, K. Kumar, and A. Chatterjee. 2006. Preventing service oriented denial of service (PreSODoS): A proposed approach. In *ICWS '06: Proceedings of the IEEE International Conference on Web Services*, 577–584, Washington, DC, 2006. IEEE Computer Society.
30. Reid, J., A. Clark, J. Gonzalez-Nieto, J. Smith, and K. Viswanathan. 2004. Denial of service issues in voice over IP networks. In *First International Conference on E-Business and Telecommunication Networks (ICETE 2004)*, Setubal, Portugal, 25–28 August 2004
31. Rosenberg, J. and D. Remy. 2004. *Securing web services with WS-security: Demystifying WS-security, WS-policy, SAML, XML signature, and XML encryption*. SAMS Publishing.
32. Siddiqui, B. 2002. Developing web services, Part3: SOAP interoperability. http://www.ibm.com/developerworks/webservices/library/ws-intwsdl3.html. Accessed 17 Feb 2011.
33. Singhal, A., T. Winograd, and K. Scarfone. 2007. Guide to secure web services – Recommendations of the national institute of standards and technology. Technical report 800-95. http://csrc.nist.gov/publications/nistpubs/800-95/SP800-95.pdf. Accessed 17 Feb 2011.
34. Smith, J., J. Gonzalez-Nieto, and C. Boyd. 2006. Modelling denial of service attacks on JFK with Meadows's cost-based framework. In R. Buyya, T. Ma, R. Safavi-Naini, C. Steketee, and W. Susilo (eds) ACSW Frontiers 2006, 16–19 January 2006, Australia, Tasmania, Hobart.
35. Snort Project, T. 2011. *SNORT Users Manual*. http://www.snort.org/assets/166/snort_manual.pdf.
36. Stebila, D. and B. Ustaoglu. 2009. Towards denial-of-service-resilient key agreement protocols. In *Proceedings of the 14th Australasian Conference on Information Security and Privacy (ACISP), LNCS*, vol. 5594, 389–406. Springer.
37. Suriadi, S., A. Clark, and D. Schmidt. 2010. Validating denial of service vulnerabilities in Web services. In *4th International Conference on Network and System Security (NSS)*, 175–182, Sept 2010.
38. Tran, T., R. Nayak, and P. Bruza. 2008. Combining structure and content similarities for XML document clustering. In *AusDM*, 219–226, 2008.
39. Vedamuthu, A., D. Orchard, F. Hirsch, M. Hondo, P. Yendluri, T. Boubez, and U. Yalçinalp. 2007. Web Services Policy 1.5 – Attachment. http://www.w3.org/TR/ws-policy-attach/. Accessed 17 Feb 2011.
40. Vedamuthu, A., D. Orchard, F. Hirsch, M. Hondo, P. Yendluri, T. Boubez, and U. Yalçinalp. 2007. Web Services Policy 1.5 – Framework. http://www.w3.org/TR/ws-policy/. Accessed 16 Feb 2011.

Chapter 8
DoS Vulnerabilities in IPv6

J. Smith, E. Ahmed, C. Chellappan, S.P. Meenakshi, S.V. Raghavan, S. Suriadi, and A.B. Tickle

8.1 Introduction

Central to the functioning of the Internet itself as well as most corporate and organisational intranets is the TCP/IP suite of protocols. Within the TCP/IP suite, the transmission control protocol (TCP) offers a robust delivery mechanism for all kinds of data across a network of arbitrary complexity. The other key protocol component, the Internet protocol (IP), primarily manages the routing of messages (aka packets or datagrams) between communicating entities. The Internet protocol (IP) also deals with issues related to network and computer addresses, that is so-called IP addresses. The current version of the Internet protocol (IP) is IPv4. As has been discussed in the previous chapters, the vulnerabilities of IPv4 have been exploited in denial-of-service (DoS) attacks. IPv4 also has a number of design limitations of which the impending exhaustion of available IPv4 addresses is one of the more critical. Development of IPv6, the designated successor to IPv4, has been underway since 1998. IPv6 attempts to address some of the security limitations of IPv4 but, importantly, also solves the address shortage problem by using 128-bit addresses compared to the 32-bit addresses adopted in IPv4. This creates a potential address space within IPv6 that is more than 20 orders of magnitude larger than IPv4's address space.

J. Smith (✉) • E. Ahmed • S. Suriadi • A.B. Tickle
Information Security Institute, Queensland University of Technology, Brisbane, Australia
e-mail: j4.smith@qut.edu.au; e.ahmed@qut.edu.au; s.suriadi@qut.edu.au; ab.tickle@qut.edu.au

C. Chellappan
College of Engineering Guindy, Anna University, Chennai, India
e-mail: drcc@annauniv.edu

S.P. Meenakshi • S.V. Raghavan
Department of Computer Science and Engineering, Indian Institute of Technology Madras, Chennai, India
e-mail: spmeena@cse.iitm.ac.in; svr@cs.iitm.ernet.in

S.V. Raghavan and E. Dawson (eds.), *An Investigation into the Detection and Mitigation of Denial of Service (DoS) Attacks: Critical Information Infrastructure Protection*, DOI 10.1007/978-81-322-0277-6_8,

Apart from the prospect of finally resolving the pressing issue of depletion of available IPv4 network addresses, the change to IPv6 offers a number of other benefits. These include, for example, the simplified assignment of addresses which significantly reduces the manual tasks required to administer a large group of devices. IPv6 also directly supports a number of significant security enhancements in the areas of authentication and encryption and also affords the prospect of strengthening the resilience of core network services against being compromised.

However, it is common for any new technology to come with its own inherent set of new vulnerabilities. For IPv6, this is true both in its ultimate deployment as well as during the period of transition from IPv4. The purpose of this chapter is to explore these vulnerabilities in more detail and, in particular, those vulnerabilities that could be exploited in a denial-of-service (DoS) attack. By way of example, as indicated above, one of the drivers for the adoption of Internet protocol version 6 (IPv6) is a significant increase in address space. A key assumption has been that the sparseness of addresses within the total IPv6 address space would inhibit the propagation of worms which constitute one of the main vectors of Denial of Service (DoS) attacks. However, experiments have been conducted that demonstrate this may not be the case [44]. In fact, worms that are programmed to utilise neighbour cache information and multicast addresses, two key features of IPv6, may identify propagation targets within IPv6 more quickly than their Internet protocol version 4 (IPv4) counterparts. A discussion of this and other issues relating to IPv6 vulnerabilities and their potential for exploitation in DoS attacks is covered in Sect. 8.2.

The ubiquity of the TCP/IP suite of protocols and their widespread availability in commercial-off-the-shelf (COTS) products means that the impact of the exploitation of vulnerabilities in IPv6 has the potential to extend DoS attacks well beyond those of the Internet. One such domain, and one that is the subject of discussion in this chapter, involves computer- and network-based systems that are used to monitor and control critical infrastructure, particularly in utilities such as energy and water. In such sectors, a DoS attack can have potentially catastrophic consequences. Section 8.3 discusses this issue and examines the efficacy of tools to model and review risks emanating from the propagation of identifiable DoS vulnerabilities in IPv6 deployment in such monitoring and control systems that would form the basis for a DoS attack.

8.2 An Overview of IPv6 DoS Vulnerabilities and Attacks

Given the importance of IPv6 to the future of the Internet, IPv6 vulnerabilities and attacks have been the subject of a number of detailed and comprehensive studies (see, for example, [6, 8, 9, 16, 17, 25, 29, 34]). The ensuing discussion on IPv6 DoS vulnerabilities and attacks essentially summarises some of the key findings of those studies that are of immediate relevance to detecting and mitigating

DoS attacks. Of specific interest in this discussion are those attacks that exploit vulnerabilities in:

- Internet control message protocol version 6 (ICMPv6) (the signalling and control protocol employed in IPv6 networks)
- IPv6 Mobility Extensions
- IPv6 Extension Headers
- IPv6 Transition and Coexistence

Prior to commencing the discussion, it should be noted that, according to RFC 4294 (IPv6 Node Requirements), a functioning IPv6 implementation will need to be compliant with over 50 other request for comments (RFCs) [25, Section 12]. Consequently, the exact behaviour of any given IPv6 node will be determined by the version of the standards it implements, especially the quality of the implementation and the interpretations made by the implementer where the standards were unclear or ambiguous. Moreover, determining the level of exposure faced by any specific node to the vulnerabilities discussed below will primarily need to be based on a case-by-case assessment.

8.2.1 ICMPv6

The ICMPv6 [6] is the signalling and control protocol employed in IPv6 networks. In addition to supporting the set of management and control functions found in IPv4, ICMPv6 also offers:

- Address autoconfiguration
- Neighbour discovery
- Path maximum transmission unit (MTU) discovery
- Multicast group management
- Informational messaging
- Error messaging

As with ICMP in IPv4, ICMPv6 datagrams carry information used for control and signalling purposes. Because ICMPv6 datagrams can alter the behaviour of IPv6 nodes, one of the key roles of Internet protocol security (IPsec) is to prevent or at least mitigate attacks via spoofed and malicious ICMPv6 datagrams. Nevertheless DoS attacks that exploit this vector include:

8.2.1.1 Error Message Misuse

The ICMPv6 specification seeks to limit the way in which IPv6 nodes can be used as traffic generators for denial-of-service attacks. In particular [6, Section 2.4] requires that ICMPv6 error messages not be generated as a result of receiving:

- An ICMPv6 error message
- An ICMPv6 redirect

- A packet destined to an IPv6 multicast address, except for packet too big and parameter problem messages
- A packet sent as a link-layer multicast or broadcast
- A packet whose source address does not uniquely identify a single node

While such message processing requirements reduce the mechanisms by which IPv6 nodes can be turned into traffic generators, they do not eliminate them. In particular the exception for processing packets destined to a multicast address that permit parameter problem messages to be generated may be exploited (via the inclusion of unknown destination options) to instantiate a denial-of-service attack [6, Section 5.2].

8.2.1.2 ICMPv6 Attacks Against TCP

As with IPv4, there are interdependencies between ICMPv6 and upper-layer connection-oriented protocols such as TCP which can be exploited. For example it is possible for an attacker, through the generation of certain ICMPv6 error messages, to blindly reset a TCP connection [13, 27].[1] Significantly, this attack will work even where TCP layer authentication mechanisms are employed. Exacerbating the impact of this attack is the fact that some host implementations will not only abort the specific connection referred to by the attacking ICMPv6 datagram, but also all other connections between the identified peers [13, Section 5.1].

8.2.1.3 Informational Message Misuse

A significant problem in IPv4 is that nodes respond to directed broadcast echo requests. In this attack a single ping packet sent to a broadcast address results in a flood of traffic being sent to a victim system (the spoofed source of the echo request). While broadcast addresses have been deprecated in IPv6, certain multicast addresses (e.g. all nodes multicast address) may be used in a similar fashion. Freely available IPv6 attack tools make use of this vulnerability to amplify traffic in denial-of-service attacks [40].

8.2.1.4 Node Information Query

Two new message types are introduced in ICMPv6 to assist in discovering information about nodes in an IPv6 network. IPv6 node information queries and responses

[1] The message generated needs to be considered a *hard error* and includes ICMPv6 error messages of type 1 (destination unreachable) codes 1 (communication with host administratively prohibited) and 4 (port unreachable).

are indicated by message type numbers 139 and 140, respectively [8]. These queries can be used to ascertain:

- The nodes fully qualified domain name
- The nodes IPv6 addresses (global, site-local, link-local, or all unicast addresses)
- The nodes IPv4 addresses

However support for node information queries may expose sensitive internal addressing information and should be controlled. A detailed discussion of security considerations related to the node information query capabilities of ICMPv6 is contained in [8, Section 8].

8.2.1.5 Neighbour Discovery

RFC 4861 [26] specifies the requirements for IPv6 nodes to implement the neighbour discovery protocol. This protocol introduces additional ICMPv6 types that support the solicitation and advertisement of node and router addresses on a link that are necessary to support stateless address autoconfiguration [41].

The neighbour discovery protocols introduce three main threats [26, Section 11]:

1. Denial-of-service attacks
2. Address spoofing
3. Router spoofing

A rigourous analysis of trust models and threats related to neighbour discovery, which recategorise the threats, is presented in RFC 3756 [29]. The threats identified include:

1. *Redirect attacks*: in which a malicious node redirects packets away from the intended recipient to another node on the link. Redirect attacks can impact availability, confidentiality and integrity of data if that data is cryptographically unprotected.
2. *Denial-of-service attacks*: in which a targeted node is prevented from communicating with any other nodes.
3. *Flooding denial-of-service attacks*: in which a malicious node redirects other hosts' traffic to a target victim node, flooding that target with the traffic.

The analysis in RFC 3756 [29, Section 4] further categorises attacks based on whether they involve routers or routing infrastructure and whether they employ replay techniques.

For example RFC3756 [29, Section 4] describes attacks that exploit the unauthenticated nature of neighbour discovery (ND), neighbour unreachability detection (NUD) and duplicate address detection (DAD) messages utilised in IPv6 address autoconfiguration [41]. As such, attacks in this category do not require any interaction with routers or routing infrastructure. Hence vulnerability to these attacks cannot be mitigated by improved configuration of routers or filtering strategies implemented by routers.

As mentioned previously, the ND protocol in IPv6 is used by IPv6 nodes to discover peers, determine and cache link layer addressing information for each other, to discover routers and to monitor the reachability of peer routers and nodes. This information is cached by each node in a set of per interface data structures [26, Section5.1]. Entries in these caches can then be maliciously manipulated in a number of ways. For example neighbour solicitation NS and neighbour advertisement (NA) messages implement address resolution protocol (ARP) like functionality in an IPv6 network. That is the information contained in solicitations and advertisements is used to create and maintain the mapping between IP layer and link-layer addresses. Furthermore, neighbour solicitations (NSs) are also used for reachability detection (via the NUD procedure), and DAD during stateless address autoconfiguration.

Specifically, nodes seeing a NS or NA will update the neighbour cache entry for the source IP address of the message, with the source link layer address option (for NSs) and the target link layer address option (for router advertisements (RAs)). Packets will then be redirected to this link layer address, until the reachability timer expires and the NUD algorithm is activated. To maintain the remapping of IP and link layer address in the neighbour cache, the attacker merely has to respond to NUD messages appropriately.

When such spoofing is used to redirect traffic to another host, this may have confidentiality and integrity implications. Where this spoofing is used to 'black hole' packets, it is an attack on availability.

Another example of a non-router-related attack involves duplicate address detection (DAD) denial-of-service attack. The DAD procedure is executed whenever a node is attempting to establish an address on a link using stateless address autoconfiguration [41]. A malicious node on the link is able to interfere with DAD by either (1) pretending to be executing DAD for the same address by spoofing appropriately formed NSs or (2) replying with NAs that the requested address is already in use on the link. Either of these attacks will prevent the target node from establishing an address on the link and result in denial of service for that node.

A second set of examples discussed in RFC3756 involve router-related attacks [29, Section 4.2]. Router and routing-related attacks necessarily involve the RA and router solicitation (RS) messages used in ND. For example in an attack involving a malicious default router, an attacker can masquerade as the default router for the local link by multicasting spoofed RA messages or unicasting RA messages in response to RSs from nodes entering the network. In order to ensure that the malicious node is selected as the default router, the attacker may additionally spoof RA messages from the legitimate router and specify a lifetime of zero, thereby expiring the entry from the target's default router list. When used as the default router the attacker can access, modify, redirect or drop packets. Such an attack, therefore, can affect confidentiality, integrity and availability.

RFC 3756 [29, Section 4.2.2] also states that the removal of all entries from the default routers list will cause nodes to assume that all address prefixes are on-link and attempt to connect via ND address resolution. By spoofing a NA, an attacker could intercept traffic destined for any node. It should be noted that the current ND

RFC [26] changes the behaviour when the default routers list is empty and this attack would be ineffective against hosts implementing RFC 4861. The attack is listed here as there may be legacy IP stacks in use that remain vulnerable to this attack.

One other example of a router-related attack involves redirect message spoofing. By spoofing a first-hop router's layer two address an attacker is able to issue bogus ICMPv6 redirect messages to victims. This redirect message will update the victim node's destination cache, causing all packets to the targeted address being (re)directed to an address chosen by the attacker. The effects of this attack will persist as long as the attacker continues to respond to NUD messages.

RFC 3756 [29, Section 4.2] also describes a router-related attack involving a bogus on-link prefix. By spoofing an appropriately crafted RA an attacker can convince nodes that an arbitrary length network prefix is on-link (when in fact it is not). Whenever a sending host believes a prefix to be on-link, it will not attempt to send packets destined for that prefix to the router. Instead it will attempt address resolution on-link. Recall that the lifetime of an entry in the Prefix list can be specified as infinite, in which case the entry will only be removed when the node is restarted, or a RA with a lifetime of zero is received for the same prefix.

Another router-related attack involves bogus address configuration prefix. A bogus address prefix can be placed in a fabricated RA and multicast by the attacker. Any nodes performing stateless address autoconfiguration will use this information to form an address. Such an address will be of no use in communicating with off-link peers, resulting in denial of service for the victim node.

One final example of a router-related attack described in RFC 3756 [29, Section 4.2] involves parameter spoofing. In addition to providing address prefix information, RAs are also used to carry additional configuration parameters to nodes undergoing stateless address autoconfiguration. The manipulation of these additional parameters can impact on the nodes ability to utilise the network. One attack involves the construction of a RA that contains a very low *current hop limit.* Using this parameter the victim node will be unable to communicate with off-link hosts as their packets are dropped (due to time to live (TTL) expiration).

8.2.2 Mobility Extensions

Mobility, the seamless continuation of application layer communications across multiple points of attachment to a network, is identified as an essential feature for IP networks. Mobility support is an optional add-on feature for IPv4 networks [31] but an implicit feature for IPv6 networks [20]. An important consideration when managing vulnerabilities in IPv6 networks is that vulnerabilities related to mobility functions must be addressed, even if the deployment is not exercising that functionality.

Given the predominance of information systems attached to IP networks and the fact that IP networks in general, and the Internet in particular, interconnect networks that adopt differing layer 2 technologies, the Internet engineering task force (IETF) began investigating network layer mobility solutions that would permit mobility, independent of the specific data link layer being utilised. The IETF 'IP Routing for Wireless/Mobile Hosts' working group was formed in 1992 and tasked with investigating network layer solutions to the IP mobility problem.

Whilst the IETF had successfully addressed many of the requirements driving the design of mobile Internet protocol (MIP), it appeared that the use of binding updates for route optimisation had introduced a number of new security threats. These threats were considered so serious that they would have to be addressed before the architecture could be accepted for standardisation by the IETF. Aura and Roe [2] note that the threats posed by unsecured binding updates caused the IETF to halt the standardisation process as it was believed they could threaten the stability of the entire Internet. Detailed analyses of these threats are presented by Aura and Roe [2], Nikander et al. [28], Kempf et al. [21] and Zhang et al. [43].

8.2.3 Extension Headers

The IPv6 specification defines the notion of an extension header [11, Section 4] which can be used to convey additional information to intermediate or destination nodes. Extension headers allow the basic header to be maintained at a fixed size (and therefore incur a fixed process cost at each router).

The following extension headers are identified as being required in a full IPv6 implementation [11, p. 7]: hop-by-hop options, routing, fragment, destination options, authentication [12, 22] and encapsulating security payload [12, 22].

The basic routing header (type 0) influences the trajectory of packets through a network by specifying waypoint routers that the packet must pass. A significant vulnerability related to type 0 routing header extensions was identified in 2001 [33]. This vulnerability was demonstrated in 2007 to be easily exploitable. It permits an attacker to amplify traffic by a factor of 88 and potentially bypass network perimeter access control lists (ACLs) [3]. This vulnerability has also been verified by Kim et al. [23]. In light of this, the IETF deprecated the use of type 0 routing headers in RFC 5095 [1]. It may be some time before all deployed IPv6 devices are using an IPv6 stack compliant with the latest version of the IETF specification however. A summary of support for routing header 0 was conducted in 2007 and is presented in Table 8.1.

It must be noted that the type 0 routing header related to this vulnerability is distinct from the routing headers used to support mobile IP. Routing headers used in support of mobile IP are type 2 headers not known to be subject to such vulnerabilities.

A router alert option [30] is defined as a hop-by-hop extension that allows routers that support the option to further interrogate an IPv6 packet before forwarding. Such a capability is deemed as necessary to support quality of service initiatives. It is possible that a flood of packets with this option set will increase the load on routers processing the option, thereby degrading the performance of the router.

Table 8.1 Type 0 Routing Header Support as of December 2007 [3]

OS	Host	Router	Configurable?
Linux 2.6	dropped	processed	no
FreeBSD 6.2	processed	processed	no
NetBSD 3.1	processed	processed	no
OpenBSD 4.0	processed	processed	no
Mac OS X	processed	processed	no
Cisco IOS	n/a	processed	yes
Cisco PIX	n/a	dropped	n/a
Juniper RTR	n/a	processed	no
Netscreen FW	n/a	dropped	n/a
Windows XP SP2	dropped	n/a	n/a
Windows Vista	dropped	n/a	n/a

Other examples that exploit header extensibility in IPv6 can be found in [32].

8.2.4 *IPv6 Transition and Coexistence*

The migration from IPv4 to IPv6 will take decades and may never be complete. As such the IETF has standardised numerous ways to facilitate transition and coexistence of IPv4/IPv6. Vulnerabilities associated with transition and coexistence will stem from issues related to the IPv6 protocols, issues related to the transition mechanism or issues related to IPv6 deployment [9].

A detailed description of transition mechanisms and their implementation is provided in the 6net consortium's IPv6 deployment guide [39, Chapter 5]. The IETF have also released numerous informational and standards track RFCs that identify and address security concerns related to transition and coexistence. A summary listing of these includes:

- RFC 3964, Security Considerations for 6to4 [34]
- RFC 4380, Teredo: Tunneling IPv6 over UDP through NATs [17, Section 7]
- RFC 4942, IPv6 Transition/Coexistence Security Considerations [9]
- Draft, Security Concerns with IP Tunneling [16]

These studies show that a major vulnerability related to IPv4/IPv6 transition and coexistence is the potential for inconsistent application of security policy and control measure across all avenues of access to information and information systems. For example the general security model for IPv4 networks of a well perimeterised subnetwork with ingress and egress mediated by a network access control device (firewall) does not apply directly, or completely, in an IPv6 environment. Moreover the application of policy at a network perimeter is made more complex by addressing issues (privacy extensions for example) and extensible header formats that may require firewalls and intrusion detection systems to store large amounts of state and perform significant packet processing to extract parameters (such as transport layer port information) before policy can be meaningfully applied. Furthermore, transition mechanism that employs tunnelling makes the application of policy difficult.

In addition the configuration interfaces provided by vendors may inadvertently expose services and device interfaces. Sometimes the default IPv6 configuration is insecure. For example secure shell (SSH) access may be provisioned for remote management of a router, yet separate commands are required to disable telnet access to the device via IPv6 and enforce the use of a secure protocol alternative [9, Section 4.8].

Controls available in IPv4 networks may also not have analogues in the IPv6 world. While primitive filtering may be possible in IPv6 firewalling products, they do not usually offer the same range of features as their more mature IPv4 counterparts. A 2007 survey conducted by the Internet Corporation for Assigned Names and Numbers (ICANN) Security and Stability Advisory Committee (SSAC) concluded that IPv6 was not broadly supported by commercial firewalls and that '...the survey results do suggest that an organization that adopts IPv6 today may not be able duplicate IPv4 security feature and policy support.' [35, p. 20]. While this situation is expected to improve as IPv6 sees greater deployment, those deploying IPv6 infrastructure must ensure that they are able to apply a policy and its required controls consistently, regardless of the network layer used to access resources.

Dual stack and tunnelling-based transition mechanisms are emerging as the dominant approach for IPv6 transition and coexistence. Where a dual stack migration is employed, the issues of IPv6 specific vulnerabilities and security policy control consistency must be addressed. The ensuing discussion focuses on vulnerabilities and challenges presented by the use of tunnelling-based mechanisms for the transition from IPv4 to IPv6 and the use of 6to4 dual-state mechanisms.

8.2.4.1 General Tunnelling Issues

RFC 4942 articulates three generic dangers associated with tunnelled traffic: (1) it may be easier to bypass ingress/egress filtering checks; (2) sending attack traffic via a tunnel interface allows an attacker to inject traffic from remote locations that has a TTL of 255 and this may be easier than gaining physical access to a local link and (3) automatic tunnelling mechanisms require end-points to accept and decapsulate traffic from any location on the Internet.

Where IPv6 is tunnelled in IPv4, it may violate the security assumptions of the IPv4 network security design. For example, an IPv4 host behind a firewall and network address translation (NAT) may be considered to be unaddressable to the Internet at large. Through the establishment of an IPv6 connection, via a User Datagram Protocol (UDP)-based Teredo tunnel, however, the host will be directly accessible to the IPv6 Internet. Under such circumstances, the host may not be deployed with the requisite controls (host-based firewall, etc.) to permit its secure operation.

The security implications of Teredo tunnels are analysed in [17, Section 7] as well as [15]. These analyses discuss vulnerabilities that are exposed owing to the fact that Teredo facilitates ingress filtering bypass and makes hosts behind NAT gateways accessible to the IPv6 Internet.

8.2.4.2 6to4 Denial of Service Issues

An alternative vector of attack and one that could potentially arise in the IPv4-IPv6 transition process is to exploit the 6to4 tunnelling process. An attacker can spoof the source address on the inner IPv6 packet to a victim's address (e.g. a 6to4 relay router). Without proper security checks, the attacker's IPv4 address (which is contained in the outer IPv4 packet) is discarded when the outer IPv4 header is de-capsulated. The net effect is:

- To make the attackers' actual IP address untraceable
- To 'reflect' reply packets to the victim's IP address thereby creating a (distributed) denial-of-service attack as described in RFC 3964 [34]

8.2.5 Summary of IPv6 Vulnerabilities and Attacks

The IPv6 suite of protocols introduces enhanced features and functionality, including address autoconfiguration, mobility support and mechanisms for transition from existing IPv4 networks. The preceding discussion has identified and detailed the vulnerabilities that are present in the current IPv6 protocol specifications and which can be exploited as part of a DoS attack. The discussion has focused on vulnerabilities as they relate to:

- The protocol in general
- ICMPv6
  - Informational messages
  - Error messages
  - Neighbour discovery
- Mobility extensions
- Vulnerabilities associated with transition mechanisms

The secure deployment and operation of an IPv6 infrastructure that is resilient to DoS attacks will require that the vulnerabilities identified in the preceding discussion are adequately anticipated and managed.

8.3 DoS Threats in Control Systems Using IPv6

8.3.1 Overview

As indicated previously, the ubiquity of the TCP/IP suite of protocols and its widespread availability in commercial-off-the-shelf (COTS)computer and network-based products means that the TCP/IP protocol can be found in a diverse range of application domains. Consequently the migration from IPv4 to IPv6 will affect applications well beyond the Internet as will the impact of the exploitation of DoS vulnerabilities in the IPv6 which has been detailed in the preceding discussion.

One such application domain, and the one that provides the context for the ensuing discussion on DoS vulnerabilities in IPv6, is the monitoring and control of industrial and engineering processes. Such processes can be found in sectors such as water distribution, waste water management, electricity generation and distribution, telecommunications and oil and gas refining and transportation. A characteristic feature of these applications is the demand for very high if not continuous availability and the potential societal impacts of any significant periods of service interruption to critical infrastructure.

In this application domain, the set of networked components to be interconnected comprises devices such as remote terminal units (RTUs) and programmable logic controllers (PLCs). The TCP/IP suite of protocols is firmly established in devices used in monitoring and control systems because TCP/IP satisfies the inherent requirement to interconnect such devices reliably and cost-effectively. Consequently, as IPv6 capability and support becomes more widespread, IPv6-enabled network devices will also become more commonplace in this application domain. Hence the security issues arising from the DoS vulnerabilities of IPv6 must also be addressed in this context as the exploitation and propagation of IPv6 DoS vulnerabilities may have a serious domino-effect: the denial of *critical services* provided by these control systems.

Given the scale and complexity of current monitoring and control systems and their key role in delivering basic services, in order to minimise the risk of service disruption, there is a need for organisations using such systems to be able to model and review the risks emanating from the propagation of identifiable vulnerabilities in IPv6 prior to actual operational deployment. Currently, there is a lack of robust and reliable tools to perform this task. The main goal of this section is therefore to propose a methodology to (1) identify potential threats (and their propagation in a control system) as a result of exploiting IPv6 DoS vulnerability, and (2) use the threat information to analyse the *risk* of a control system getting into the undesirable states of denying critical services to users.

For the purposes of this discussion, the control system selected is the Supervisory Control and Data Acquisition (SCADA) system. SCADA is one of the dominant members of the set of systems used in monitoring and control and IPv6-enabled devices are available in SCADA-based systems [42].

As discussed in [37], the first step is to identify and document threats to SCADA systems as a result of exploiting known IPv6 DoS vulnerabilities. There are several techniques that could be used to achieve this end including:

1. Utilise the availability of IPv6 in current COTS (commercial-off-the-shelf) network products to deploy a prototype SCADA over a working IPv6 system thereby allowing threats resulting from IPv6 vulnerabilities to be identified directly
2. Use formal analysis techniques to model a SCADA over IPv6 system and identify the vulnerabilities
3. Use informal analysis of how IPv6 vulnerabilities can lead to threats to SCADA over IPv6 by studying the existing literature e.g. [36, 42] and drawing upon past experience

Suriadi [37] analyses the merits of each of these options. For example, the first approach, whilst being the most authoritative, is also the most problematic given the time and cost of synthesising a credible real-world deployment of SCADA over IPv6. Conversely an informal analysis (Option 3) is the cheapest and least time consuming of the three options. It is also the approach that has been used in many of existing publications of security analysis of IPv6 networks such as RFC 3756 [29] and RFC 3964 [34]. However, informal analysis is not exhaustive. Also the large number of variables potentially renders the approach incomplete and prone to errors. As discussed in [37], applying formal methods (Option 2) to the task offers the prospect of mathematical precision, completeness and accuracy without incurring the high costs associated with an actual operational deployment. However, applying formal methods to a complete SCADA system is potentially a large and exacting undertaking. On this basis, the approach discussed below seeks to find an eclectic combination of informal and formal analysis.

The formal analysis tools used are Coloured Petri Nets (CPNs) [18] along with the supported simulation and state space analysis to study the effects of known IPv6 DoS vulnerabilities in the SCADA system. The justification for using Coloured Petri Nets (CPNs) to model and analyse the SCADA system is that CPNs are an established approach to modelling distributed systems. Moreover CPNs have been used successfully in a number of similar situations including, for example, a vulnerability assessment of a SCADA system [38], a SCADA substation operation [19] and hazardous liquid loading SCADA system [14]. In the remainder of this section, we demonstrate how CPN can be used to detect and identify threats to a SCADA system manifested from IPv6 DoS vulnerabilities. Once this process has been completed, the output is a set of identified and documented threats to a SCADA system based on known IPv6 DoS vulnerabilities.

However, at least from a management point of view, the ultimate goal of this exercise is to be able to proportionately allocate resources to mitigate the identified threats according to their risks. Hence there is a requirement for a mechanism to study how 'risky' the identified threats are and to perform an analysis of the corresponding risks. In this section, we seek to demonstrate how we can study the risk of a control system going into those undesirable states whereby it may start denying services to legitimate clients.

For the purposes of this exercise, the tool selected for the risk analysis phase is the ISM Risk Modelling tool [4, 10]. This tool was developed based on the information security management and modelling approach proposed by Longley et al. and has been developed over more than 15 years [4, 5, 24]. It can be used for several purposes, including:

- As a tool to aid threat documentation
- As a risk modelling tool: by reading the threat documentation that has been fed into the ISM Tool
- As a tool to analyse the effectiveness of threat countermeasures

Depending on the system being studied, organisations may already have such threat documentation, or threats can be identified and documented using the approach explained previously. The ISM Tool allows these threats to be properly documented in a form that can be readily used for risk analysis. The purpose of the ensuing discussion is to provide an example of how threat documentation and threat propagation are documented using ISM Tool.

Once threats are documented into the ISM Tool, they can be automatically scanned to generate a risk simulation by creating a threat network diagram [4]. These diagrams show both the risk measurement of a threat and the propagation of threat. In this context, the 'risk measurement' is derived based on the consequences of a documented threat (i.e. catastrophic, major, moderate, minor, or insignificant) as well as the probability of the threat happening [10]. In the context of this discussion the aim is to demonstrate how we can use the ISM Tool to derive a risk simulation of a SCADA system (which runs over IPv6 network) reaching the undesirable states whereby it may start *denying critical services* to users.

The ISM tool also captures the set of mitigation techniques that have been (or planned to be) applied. Similar to threat documentation, such information may already be well-documented in an organisation, or a further study (using methods described previously) is needed to verify the effectiveness of a countermeasure technique. An analysis of the effectiveness of countermeasure techniques to mitigate IPv6 threats to SCADA systems is not part of this immediate discussion but offers an avenue for future work.

8.3.2 A Subset of IPv6 DoS Vulnerabilities to SCADA Systems to be Modelled

As discussed previously, IPv6 DoS vulnerabilities can be broadly categorised into two main groups. The first group comprises of DoS vulnerabilities that are intrinsic to the IPv6 itself. The second group of DoS vulnerabilities arises from the IPv4

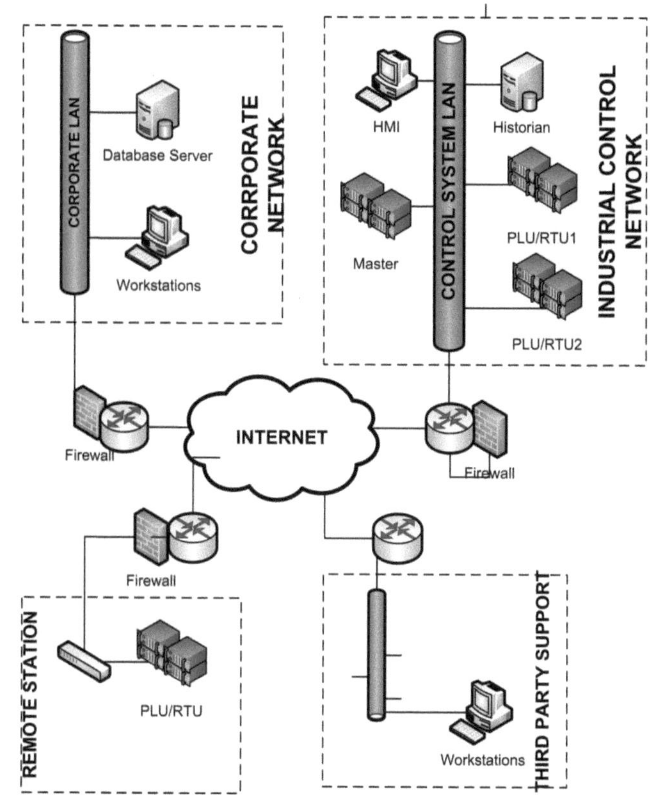

Fig. 8.1 A simple SCADA system comprising one Master controller and two RTUs

to IPv6 transition where the two protocols stacks must co-exist (RFC 4942 [9]). For the purposes of demonstrating the efficacy of the techniques under discussion, Suriadi et al. [37] selected the following subset as being some of the more significant IPv6 DoS vulnerabilities to SCADA systems:

- The Neighbour Advertisement (NA)/Neighbour Solicitation (NS) (NA/NS) spoofing vulnerability (RFC 3756 [29])
- The remote injection of ND messages vulnerability (RFC 3964 [34])
- The reflection of traffic to 6to4 relay vulnerability (RFC 3964 [34])

For the purposes of this discussion only the first of these vulnerabilities, i.e. Neighbour Advertisement (NA)/Neighbour Solicitation (NS) (NA/NS) spoofing vulnerability is covered. The details of this NA/NS spoofing vulnerability have been discussed earlier in Sect. 8.2.1.

As in [37], the simple SCADA system shown in Fig. 8.1 comprising of one Master controller and several Remote Terminal Units RTUs (scattered throughout various networks) can be used to illustrate the basic ideas. In this simplified model it is assumed that the master RTU collects information from various sensors and RTUs from both industrial control network as well as from remote stations, and the master RTU sends control commands to these devices. Additionally, control commands can also be sent from human SCADA operators and third-party support operators who may be located in the corporate network or remotely. It should be noted that an actual operational environment would provide a richer level of contextual information than that of Fig. 8.1. Moreover it is important to note that contextual information is absolutely indispensable to delivering a sensible and coherent risk analysis.

Within IPv6, the role of the Neighbour Discovery (ND) mechanism is to create and maintain the mapping between the IP-layer address and the link-layer address. This is achieved through a process of Neighbour Solicitation (NS) and Neighbour Advertisement (NA). However this same process offers a vector through which an attacker is able to compromise system integrity by redirecting traffic away from the intended recipient. This attack could, for example, be initiated by injecting spoofed NA and/or NS messages either remotely or locally, thereby corrupting the link between IP address and the corresponding link-layer address (see RFC 3756 and RFC 3964 for details [29, 34]). Applying this scenario to Fig. 8.1, the (legitimate) Master controller node could, for example, be given the link layer address for the Remote Terminal Unit RTU2 as the address for the Remote Terminal Unit RTU1 by one of its neighbours. Consequently control messages intended for the Remote Terminal Unit RTU1 could instead be (incorrectly) delivered to the Remote Terminal Unit RTU2. The net result of RTU2 executing the control messages intended for RTU1 could be *system instability and potential failure – hence in a DoS-state.*

While such an attack is similar to the ARP spoofing attack within IPv4, there are few known mitigation techniques for the NA spoofing attack [7] particularly since, in IPv6, such an attack can be launched remotely (see RFC 3964 [34] for details).

As will be discussed, the key purpose of the work undertaken is to demonstrate how to detect threats and analyse their risks should the vulnerabilities described above be exploited in a control system environment.

8.3.3 Modelling the NA/NS Spoofing Vulnerability with Coloured Petri Nets (CPN)

The intent here is to establish a proof-of-concept formal model of a SCADA system to demonstrate the practicality of using the formal method approach to detect threats to SCADA systems. Once the correctness of the Coloured Petri Nets (CPN) model is established, the next step is to model an attack which exploits the IPv6 vulnerability to spoofing of NA message. It is then possible to study the behaviour of the SCADA system under attacks through simulation as well as through formal model checking analysis (such as state space analysis). Finally, the intent is to show how it is possible to use formal analysis to understand the consequence of a threat. Such an understanding is crucial for the purpose of risk analysis.

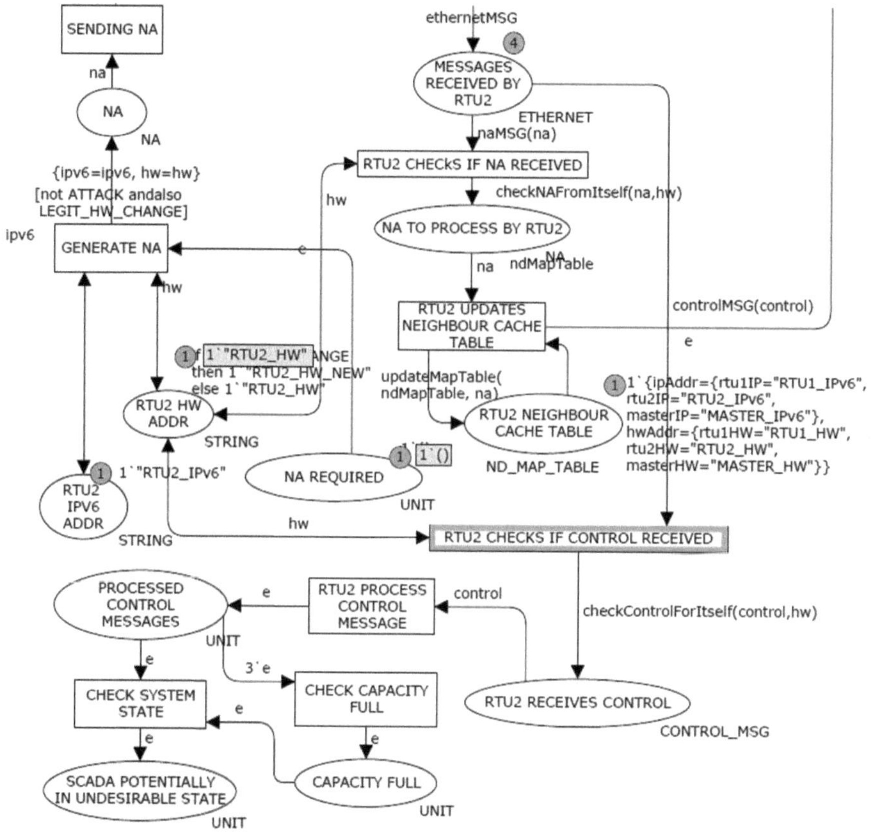

Fig. 8.2 A simplified CPN model of the transmission of a Neighbour Advertisement (NA) message by RTU RTU2

The top right quadrant of the simplified SCADA network shown in Fig. 8.1 is labelled as the 'Industrial Control Network'. Figure 8.2 shows the Coloured Petri Net (CPN) model of the sending of a Neighbour Advertisement (NA) message by the Remote Terminal Unit RTU2 within this portion of the SCADA network. The lower part of Fig. 8.2 shows the receiving of an NA message (and the updating of the Neighbour Cache Table) at each node, and the operations that RTU2 executes when a control message is received.

It should be noted that, for the purposes of this illustration, Fig. 8.2 shows an abstraction of the Neighbour Advertisement (NA) packet and the control message packet. For example the model in Fig. 8.2 does not show all of the detail involved in the processing of a received control message by the Remote Terminal Unit RTU2. Instead the focus is on showing how the Remote Terminal Unit RTU2 can reach an *unstable or undesirable state*. Again, to simplify the discussion, in the model it is assumed that RTU2 has almost reached its 'maximum capacity' (in a real world system, imagine RTU2 as a controller of an actuator, such as a pump, that pumps oil into an almost-full tank).

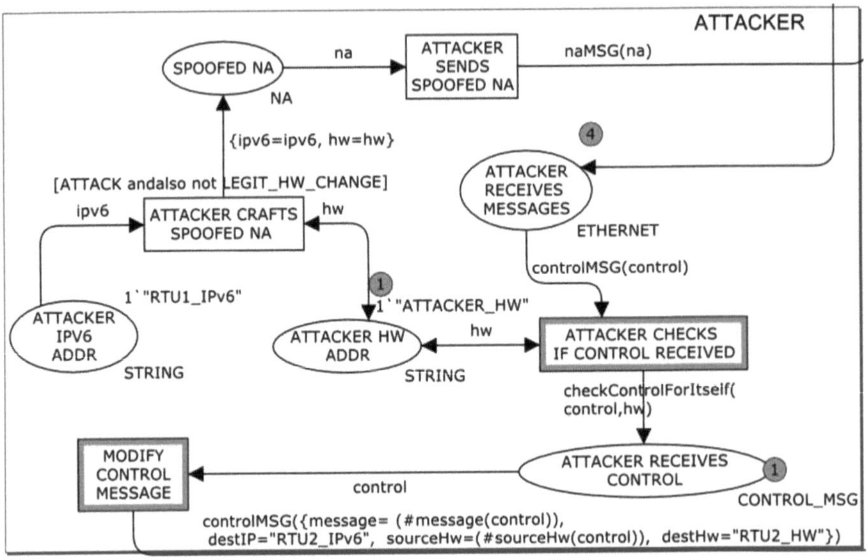

Fig. 8.3 A snippet of a CPN Model depicting an attacker exploiting IPv6 spoofed NA vulnerability

Using the above analogy, RTU2 is modelled such that it can only handle additional 3 'pump' control messages before the tank is full. Overall, RTU2 will send an NA message to be processed by both RTU1 and Master controller. After a successful processing of the NA messages by all nodes, the Master controller sends 5 control messages to RTU1.

Verification that the modelled SCADA system behaves correctly in the generation and processing of both NA and control messages is achieved through the simulation of the CPN model.

The next step is to model the exploitation of the spoofed NA vulnerability. This is performed by adding an attacker model as shown in Fig. 8.3. The attacker injects a spoofed NA message claiming that the corresponding hardware address of an IPv6 address (which actually belongs to RTU1) is the attacker's hardware address. The attacker then modifies any control messages it receives so that the messages are directed to the low-capacity RTU2 with the aim of causing instability in the system controlled by RTU2.

Leaving the rest of the model intact, a simulation is then run to study what happens to the SCADA system when the attacker injects the spoofed NA message followed by the Master controller sending a series of 5 control messages to RTU1. From the simulation of the model, it is noted that the control message is mistakenly delivered to RTU2 who then executes the control messages. As a result, *the SCADA system may reach a potentially undesirable state which may translate to it going to failure or start denying services to users* (see Tables 8.2 and 8.3 for summary of our threats analysis and their propagation).

Table 8.2 Summary of threats

Threat name	Threat entity	Consequences
Incorrect link address	– RTU1/RTU2/Master/HMI/Historian at Industry Control Network (ICN) – RTU at Remote Station	Minor
Internal control messages communication errors	– Communication between Master controller and RTU1/RTU2 at ICN	Moderate
HMI provides inaccurate feedback	– HMI at ICN	Moderate
Incorrect control message delivered	– All entities at ICN	Moderate
SCADA reaches undesirable state(s)	– SCADA Site Network	Major

Table 8.3 Summary of threat propagation

Incident threat entity	Target threat entity
Incorrect link layer address	Internal messages communication errors
Internal messages communication errors	HMI provides inaccurate feedback
HMI provides inaccurate feedback	Incorrect control messages delivered
Incorrect control messages delivered	SCADA system reaches undesirable state (e.g. failure/denying services)

8.3.4 Risk Analysis Using ISM Tool

The preceding discussion has described how it is possible to identify threats to a SCADA system based on known IPv6 DoS vulnerabilities which may lead to the SCADA system itself being in an undesirable state (such as failure). The next phase is to proportionately allocate resources to mitigate the identified threats according to their risks using the ISM Risk Modelling tool [4, 10]. In particular the aim is to:

1. Document detected threats and threats propagation
2. Generate a simulation of risk which shows the risk level of documented threats

The risk simulation can be performed in two styles: 'forward' and 'backward'. In the 'forward' style, a user chooses the starting threats that are assumed to have happened, and then either (a) uses the tool to show the ensuing chains of threats (as well as the risk level for the ensuing threats), or (b) also chooses the resultant threats so that ISM Tool can show how the starting threats may lead to the chosen resultant threats as well as the risk level associated with the starting threats, the propagated threats, and the causal threats.

In the 'backward' style, a user simply chooses the resultant threats (such as a DoS threat to the SCADA system). From there, the ISM Tool will do a 'backward' analysis to show all the threats that may lead to the chosen resultant threat [4], their propagation and their risk levels.

To document threats, some key information is required, including (1) the entities of the system (such as platforms, hardware, softwares, networks, information assets and so on) which may become the victims of some threats, (2) the relationships between those entities (e.g. a master controller and an RTU may have a communication relationship) and (3) the detected threats and their propagations.

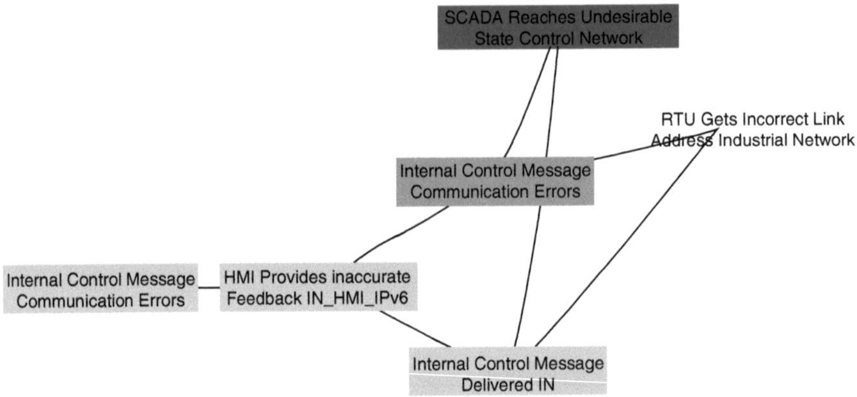

Fig. 8.4 Backward risk analysis diagram

Table 8.4 Backward risk analysis report

Threat name	Risk
RTU Gets incorrect link address	Low
Internal control message communication errors	High
HMI provides inaccurate feedback	Moderate
Incorrect control message delivered	Moderate
SCADA reaches undesirable state	Extreme

The first two pieces of information (entities involved and their relationships) can be easily documented based on the simple SCADA system shown in Fig. 8.1. Furthermore, information about the detected threats and their propagations is obtained from the results of both the formal and informal analysis of IPv6 threat detection explained in Sect. 8.3.3. These results are summarised in Tables 8.2 and 8.3.

Once the threats are documented, it is simple to obtain the risk simulation. Figure 8.4 shows a threat network generated using a 'backward' risk simulation whereby a causal threat (in this case the 'SCADA Reaches Undesirable State' threat) is chosen, and we let the ISM Tool to calculate all the other threats that may lead to the chosen resultant threat. The result of the backward risk analysis is shown in Fig. 8.4. The colour of each of the node in Fig. 8.4 is used to convey the risk level of each of the threat. The ISM Tool can also generate a risk measurement report (see Table 8.4) which summarises the resulting threats (including the initiative threats, the intermediary threats and the resulting threats) and their risk level.

From Fig. 8.4 and Table 8.4, it is possible to conclude that there is an extreme risk of a seemingly minor threat (e.g. a node obtains incorrect link layer address) leading to a major threat whereby a SCADA system becomes unstable (and may potentially lead to a system failure or the system behaving in an unpredictable manner, thus potentially denying proper services to users). A 'forward' style analysis (not discussed in this section) can also be performed to determine the risks of certain initiating threats which are assumed to have happened.

8.4 Conclusions

The transition of the TCP/IP protocol suite from the current IPv4 to IPv6 will open a new era in computer and data communications by facilitating the interconnection of a plethora of new networked devices. However, as shown in the preceding discussion, the transition is not risk-free. DoS vulnerabilities in IPv6 itself as well as in the composite IPv4/IPv6 protocol environments during the transition process can and will be exploited to create a vector for launching DoS attacks that could extend beyond the Internet to critical infrastructure such as utilities supplying basic services such as energy and water. The preceding discussion has examined the origin and nature of such attacks as well as how practical solutions can be developed to mitigate the DoS threats in control systems.

References

1. Abley, J., P. Savola, and G. Neville-Neil. 2007. Deprecation of type 0 routing headers in IPv6. http://tools.ietf.org/html/rfc5095. Accessed 24 Feb 2011.
2. Aura, T., and M. Roe. 2006. Designing the mobile IPv6 security protocol. *Annales des Télécommunications* 61(3–4): 332–356.
3. Biondi, P., and A. Ebalard. 2007. IPv6 routing header security. http://www.secdev.org/conf/IPv6_RH_security-csw07.pdf. Accessed 24 Feb 2011.
4. Branagan, M., R. Dawson, and D. Longley. 2006. Security risk analysis for complex systems. In *Proceedings of the Information Security for South Africa 2006 from Insight to Foresight Conference*, Pretoria, South Africa, 2006.
5. Caelli, W.J., D. Longley, and A.B. Tickle. 1992. A methodology for describing information and physical security architectures. In *Eighth International Conference on Information Security (SEC)*, 277–296, Singapore, 1992. IFIP Transactions.
6. Conta, A., S. Deering, and M. Gupta. 2006. Internet control message protocol (ICMPv6) for the internet protocol version 6 (IPv6) Specification. http://tools.ietf.org/html/rfc4443. Accessed 24 Feb 2011.
7. Convery, S. and D. Miller. 2004. IPv6 and IPv4 threat comparison and best-practice evaluation. http://www.cisco.com/web/about/security/security_services/ciag/documents/v6-v4-threats.pdf. Accessed 24 Feb 2011. http://www.6journal.org/archive/00000180/01/v6-v4-threats.pdf
8. Crawford, M. and B. Haberman. 2006. IPv6 node information queries. http://tools.ietf.org/html/rfc4620. Accessed 24 Feb 2011.
9. Davies, E., S. Krishnan, and P. Savola. 2007. IPv6 transition/co-existence security considerations. http://tools.ietf.org/html/rfc4942. Accessed 24 Feb 2011.
10. Dawson, R.E. 2008. Secure communications for critical infrastructure control systems. Master's thesis, Queensland University of Technology, Brisbane.
11. Deering, S. and R. Hinden. 1998. Internet protocol, version 6 (IPv6) specification. http://tools.ietf.org/html/rfc2460. Accessed 24 Feb 2011. Updated by RFC 5095.
12. Eastlake 3rd, D. 2005. Cryptographic algorithm implementation requirements for encapsulating security payload (ESP) and authentication header (AH). http://tools.ietf.org/html/rfc4305. Accessed 24 Feb 2011. Obsoleted by RFC 4835.
13. Gont, F. 2008. ICMP attacks against TCP. Draft, IETF. http://tools.ietf.org/id/draft-ietf-tcpm-icmp-attacks-04.txt. Accessed 24 Feb 2011.

14. Henry, M.H., R.M. Layer, K.Z. Snow, and D.R. Zaret. 2009. Evaluating the risk of cyber attacks on SCADA systems via Petri net analysis with application to hazardous liquid loading operations. In *IEEE Conference on Technologies for Homeland Security, 2009. HST '09*, 607–614, 2009.
15. Hoagland, J. 2006. The Teredo protocol: Tunneling past network security and other security implications. Technical report, Symantec. http://www.symantec.com/avcenter/reference/Teredo_Security.pdf. Accessed 24 Feb 2011.
16. Hoagland, J., S. Krishnan, and D. Thaler. 2008. Security concerns with IP tunneling. "internet-draft", "Internet Engineering Task Force". http://www.ietf.org/internet-drafts/draft-ietf-v6ops-tunnel-security-concerns-01.txt. Accessed 24 Feb 2011.
17. Huitema, C. 2006. Teredo: Tunneling IPv6 over UDP through network address translations (NATs). http://tools.ietf.org/html/rfc4380. Accessed 24 Feb 2011.
18. Jensen, K., L.M. Kristensen, and L. Wells. 2007. Coloured Petri Nets and CPN Tools for modelling and validation of concurrent systems. *STTT* 9(3–4): 213–254.
19. Jingbo, H. and M. Longhua. 2006. Fault diagnosis of substation based on Petri nets technology. In *International Conference on Power System Technology, 2006. PowerCon 2006*, 1–5, 2006.
20. Johnson, D., C. Perkins, and J. Arkko. 2004. Mobility support in IPv6. http://tools.ietf.org/html/rfc3775. Accessed 24 Feb 2011.
21. Kempf, J., J. Arkko, and P. Nikander. 2004. Mobile IPv6 security. *Wireless Personal Communications* 29(3–4): 389–414.
22. Kent, S. 2005. IP authentication header. http://tools.ietf.org/html/rfc4302. Accessed 24 Feb 2011.
23. Kim, J.-W., H.-H. Cho, G.-J. Mun, J.-H. Seo, B.-N. Noh, and Y.-M. Kim. 2007. Experiments and countermeasures of security vulnerabilities on next generation network. *Future Generation Communication and Networking* 2: 559–564.
24. Kwok, L.F., and D. Longley. 1999. Information security management and modeling. *Information Management and Computer Security* 7: 30–39.
25. Loughney, J. 2006. IPv6 node requirements. http://tools.ietf.org/html/rfc4294. Accessed 24 Feb 2011. Updated by RFC 5095.
26. Narten, T., E. Nordmark, W. Simpson, and H. Soliman. 2007. Neighbor discovery for IP version 6 (IPv6). http://tools.ietf.org/html/rfc4861. Accessed 24 Feb 2011.
27. National Infrastructure Co-ordination Centre. 2005. Vulnerability issues in ICMP packets with TCP payloads. Vulnerability Advisory 532967/NISCC/ICMP, National Infrastructure Co-ordination Centre.
28. Nikander, P., J. Arkko, T. Aura, G. Montenegro, and E. Nordmark. 2005. Mobile IP version 6 route optimization security design background. http://tools.ietf.org/html/rfc4225. Accessed 24 Feb 2011.
29. Nikander, P., J. Kempf, and E. Nordmark. 2004. IPv6 neighbor discovery (ND) trust models and threats. http://tools.ietf.org/html/rfc3756. Accessed 24 Feb 2011.
30. Partridge, C., and A. Jackson. 1999. IPv6 router alert option. http://tools.ietf.org/html/rfc2711. Accessed 24 Feb 2011.
31. Perkins, C. 2002. IP mobility support for IPv4. http://tools.ietf.org/html/rfc3344. Accessed 24 Feb 2011. Updated by RFC 4721.
32. Potraj, C. 2007. Firewall design considerations for IPv6. Report I733-04IR-2007, National Security Agency.
33. Savola, P. 2001. Security of IPv6 routing header and home address options. Technical report, IETF. http://tools.ietf.org/html/draft-savola-ipv6-rh-ha-security-00. Accessed 25 Aug 2011.
34. Savola, P., and C. Patel. 2004. Security considerations for 6to4. http://tools.ietf.org/html/rfc3964. Accessed 24 Feb 2011.
35. ICANN Security and Stability Advisory Committee. 2007. Survey of IPv6 support in commercial firewalls. Technical Report SAC 021, Internet Corporation for Assigned Names and Numbers (ICANN). http://www.icann.org/en/committees/security/sac021.pdf. Accessed 25 Aug 2011.

36. Stouffer, K., J. Falco, and K. Kent. 2006. *Guide to supervisory control and data acquisition (SCADA) and industrial control systems security.* NIST, USA, initial public draft edition. http://www.goes-r.gov/procurement/ground_documents/Applicable%20Documents/NIST%20Draft-SP800-82.pdf. Accessed 25 Aug 2011.
37. Suriadi, S., A. Tickle, E. Ahmed, J. Smith, and H. Morarji. 2010. Risk modelling the transition of scada system to ipv6. In *What Kind of Information Society? Governance, Virtuality, Surveillance, Sustainability, Resilience*, eds. J. Berleur, M. Hercheui, and L. Hilty, *IFIP advances in information and communication technology*, vol. 328, 384–395. Boston: Springer. 10.1007/978-3-642-15479-9_36.
38. Ten, C.-W., C.-C. Liu, and M. Govindarasu. 2007. Vulnerability assessment of cybersecurity for SCADA systems using attack trees. In *Power Engineering Society General Meeting, 2007. IEEE*, 1–8.
39. The 6net Consortium. 2005. An IPv6 deployment guide. The 6net Consortium. http://www.6net.org/book/deployment-guide.pdf. Accessed 25 Aug 2011.
40. The Hackers Choice. 2006. THC IPv6 attack toolkit. http://freeworld.thc.org/thc-ipv6/. Accessed 24 Feb 2011.
41. Thomson, S., T. Narten, and T. Jinmei. 2007. IPv6 stateless address autoconfiguration. http://tools.ietf.org/html/rfc4862. Accessed 24 Feb 2011.
42. Van Leeuwen, B. 2007. Impacts of ipv6 on infrastructure control systems. *SANDIA REPORT* (SAND2007-0383P), Sept 2007.
43. Zhang, Y.-M., Z.-W. Yu, and H.-H. Cao. 2008. Insider attacks study against mobile IPv6 protocol. In *4th International Conference on Wireless Communications, Networking and Mobile Computing, 2008. WiCOM '08*, 1–4 October 2008.
44. Zheng, Q., T. Liu, X. Guan, Y. Qu, and N. Wang. 2007. A new worm exploiting IPv4-IPv6 dual-stack networks. In *WORM '07: Proceedings of the 2007 ACM workshop on Recurring malcode*, 9–15, New York, 2007. ACM.

Appendix

A.1 Australia

Some Australian organisations involved and/or having an interest in policy, legal and related matters relevant to the DoS/DDoS and cyber security threat to national security, national critical infrastructure (NCI) and, more particularly, national information infrastructure protection (NIIP) include:

ACMA: Australian Communications and Media Authority

'... statutory authority within the federal government portfolio of Broadband, Communications and the Digital Economy. The ACMA is responsible for the regulation of: broadcasting, the internet, radio communications and telecommunications.'

Established 1 July 2005. http://www.acma.gov.au

AGD: Department of the Attorney General

'The Australian Government Attorney-General's Department serves the people of Australia by providing essential expert support to the Government in the maintenance and improvement of Australia's system of law and justice and its national security and emergency management systems. The Department is the central policy and coordinating element of the Attorney-General's portfolio for which the Attorney-General and Minister for Home Affairs are responsible. The mission of the Attorney-General's Department is achieving a just and secure society.'

http://www.ag.gov.au

S.V. Raghavan and E. Dawson (eds.), *An Investigation into the Detection and Mitigation of Denial of Service (DoS) Attacks: Critical Information Infrastructure Protection*,
DOI 10.1007/978-81-322-0277-6,

ANU-SDSC: Australian National University, Strategic & Defence Studies Centre (SDSC)

SDSC is Australia's oldest-established centre for the study of strategic, defence and wider security issues. We aim to apply the rigour and discipline of good scholarship to the key public policy questions concerning Australia's security and broader regional and global security questions.

http://ips.cap.anu.edu.au/sdsc

ASPI: Australian Strategic Policy Institute

ASPI is an independent, non-partisan policy institute. It has been set up by the (Australian) government to provide fresh ideas on Australia's defence and strategic policy choices. Established in 2001 by the former Liberal/National Party Government of Australia under Prime Minister Howard to provide independent advice to the government, it is an independent company limited by guarantee with a board of directors. Since 2004 ASPI has sought revenue from sources outside the Australian Federal Government.

http://www.aspi.org.au

CERT-Australia: Computer Emergency Response Team – Australia

'...a new national computer emergency response team' ...ensures that all Australians and Australian businesses have access to information on cyber threats and vulnerabilities. '...initial point of contact for cyber security incidents impacting upon Australian networks. It is the point of contact for Australia's international cyber security counterparts and has a coordination role in the event of a serious cyber incident.'

Became operational on 28 January 2010. Australian Government Computer Emergency Readiness Team (GovCERT.au) within the Attorney-General's Department, it is now part of CERT Australia. It is managed by the Australian Government's Attorney-General's Department.

http://www.cert.gov.au

CSOC: Cyber Security Operations Centre, Defence Signals Directorate (DSD)

DSD capability that serves all government agencies. Has two main roles: provides government with a comprehensive understanding of cyber threats against Australian interests and coordinates operational responses to cyber events of national importance across government and critical infrastructure.

Helps CERT Australia in the Department of the Attorney General in relation to critical infrastructure protection, including where this is owned by the private sector.

http://www.dsd.gov.au/infosec/csoc.html

DBCDE: Department of Broadband, Communications and the Digital Economy

Within the Department the following sections exist: Cyber Security & Asia-Pacific Engagement and Cyber-Safety & Trade.

As of January 2011, the relevant Minister responsible is Senator Stephen Conroy.

http://www.dbcde.gov.au

DSD: Defence Signals Directorate

Started as Defence Signals Bureau, 12 Nov 1947 '...role to exploit foreign communications and be responsible for communications security in the armed services and government departments ... renamed Defence Signals Branch in October 1949 ... again renamed Defence Signals Division in Jan. 1964. ... intelligence role formally acknowledged in 1977 and renamed the Defence Signals Directorate and made directly responsible to the Secretary of the Department of Defence.'

'From early 1942, Australian, American and British personnel worked together in Central Bureau, Melbourne, attached to General MacArthur's headquarters, while the joint Australian-American Fleet Radio Unit Melbourne (FRUMEL) supported the US Navy's 7th Fleet. Late in 1942, Central Bureau moved to Brisbane, following MacArthur's headquarters, and elements of the bureau deployed with him to the Philippines later in the war.'

http://www.dsd.gov.au

KOKODA: The Kokoda Foundation

'Established as an independent, not-for-profit think tank to research, and foster innovative thinking on Australia's future security challenges.'

http://www.kokodafoundation.org

OCC: Online and Communications Ministerial Council

'...peak ministerial forum across governments to consider and reach agreement on strategic approaches on information and communications issues of a national importance.'

'...chaired by the Australian Government Minister for Broadband, Communications and the Digital Economy. Membership includes senior ministers from state and territory governments and the President of the Australian Local Government Association (ALGA). The Australian Government Minister for Finance and Deregulation is also a member and chairs agenda items relating to e-government. The Council meets annually to discuss policy issues relating to the information economy with a focus on online and communications issues.'

http://www.occ.gov.au

PMC: PMC-NSST Department of the Prime Minister and Cabinet

'... principal function is to provide high-quality policy advice to the Prime Minister and the Cabinet on matters that are at the forefront of public and government administration, including domestic and international affairs and, in particular, the implications of proposals for Commonwealth-State relations.'

Relevant groups include: National Security and International Policy Group, National Security Science and Technology (NSST) Branch within the Defence, Intelligence and Research Coordination Division.

Relevant positions include: National Security Advisor and Deputy National Security Advisor, National Security Chief Information Officer (NSCIO).

http://www.dpmc.gov.au/national_security/index.cfm

QUT-ISI: Information Security Institute – Queensland University of Technology

http://www.isi.qut.edu.au

RAAF-462

'No. 462 Squadron seeks to exploit and protect against exploitation of the information domain and supports operational commanders in providing a secure information environment to support air operations.'

'No. 462 Squadron was reformed in April 2005 as a non-flying squadron within the Information Warfare Wing of the RAAF's Aerospace Operational Support Group. The Squadron's role is to 'protect the Air Force's capability through the conduct of information operations.'

http://www.airforce.gov.au/bases/edinburgh.aspx
http://en.wikipedia.org/wiki/No._462_Squadron_RAAF
http://www.defence.gov.au/news/raafnews/editions/4901/topstories/story9.htm

UNISA-DASI: University of South Australia, Defence and Systems Institute

'...provider of research and education in systems engineering and information assurance for Defence, industry and government agencies.'

http://www.dasi.unisa.edu.au

A.2 India

Some examples of Indian organisations involved and/or having an interest in policy, legal and related matters relevant to the DoS/DDoS and cyber security threat to national security, national critical infrastructure (NCI) and, more particularly, national information infrastructure protection (NIIP) include:

CBI: Central Bureau of Investigation

'Central Bureau of Investigation traces its origin to the Special Police Establishment (SPE) which was set up in 1941 by the Government of India.'

http://cbi.nic.in

CHCIT: Cyber and Hi-Tech Crime Investigation & Training (CHCIT) Centre

Education and training initiative between the Central Bureau of Investigation (CBI) and NASSCOM – DCSI.

http://www.nasscom.in/Nasscom/templates/NormalPage.aspx?id=60319

CYBERLAW INDIA

Website: Cyberlaw India is one of the pioneering names in Indian Cyberlaw.

Claim: Cyberlaw India is a unique platform for collection and sharing of ideas and information, relating to the growth of Cyberlaw in India.

http://www.cyberlaws.net/cyberindia/index1.htm

DRDO: Defense Research and Development Organization

Responsible for development of military technologies.

'DRDO while striving to meet the cutting-edge weapons technology provides ample spinoff benefits to the society at large, thereby contributing to nation building.'

http://www.drdo.gov.in/drdo/English/index.jsp?pg=homebody.jsp

DSCI: Data Security Council of India

'DSCI started functioning as an independent company with its own Board, and a small core team comprising technical experts, guided by the Steering Committee in August 2008.' Entered into MoU with CERT-IN in July 2010.

http://www.dsci.in

NASSCOM: National Association of Software & Services Companies

http://www.nasscom.in
http://www.nasscom.in/Nasscom/templates/LandingPage.aspx?id=5894

SETS: Society for Electronic Transactions and Security

'A premier Information Security R&D organisation; Hub of Information Security knowledge; Centre of excellence in Cryptology and Network Security; Initiated by the office of the Principal Scientific Adviser (PSA) to Govt. of India; Unique Public-Private Partnership (PPP) base institution; Department of Scientific and Industrial Research(DSIR) recognised R&D organisation.'

http://www.setsindia.org

A.3 Other Institutions

Examples of some other organisations involved and/or having an interest in policy, legal and related matters relevant to the DoS/DDoS and cyber security threat to national security, national critical infrastructure (NCI) and, more particularly, national information infrastructure protection (NIIP) include:

ENISA: European Network and Information Security Agency

'ENISA is helping the European Commission, the Member States and the business community to address, respond and especially to prevent Network and Information Security problems. ENISA is as a body of expertise set up by the EU to carry out very specific technical, scientific tasks in the field of Information Security. This work is only performed within the 'Community domain' ('first pillar' and internal market of the EU): as a 'European Community Agency'.

http://www.enisa.europa.eu

APEC: Asia-Pacific Economic Cooperation

Members are (as on Jan 2011): Australia; Brunei Darussalam; Canada; Chile; People's Republic of China; Hong Kong, China; Indonesia; Japan; Republic of Korea; Malaysia; Mexico; New Zealand; Papua New Guinea; Peru; The Republic of the Philippines; The Russian Federation; Singapore; Chinese Taipei; Thailand; United States of America; Viet Nam.

http://www.apec.org

APEC-TEL: APEC Telecommunications and Information Working Group

'The APEC Telecommunications and Information Working Group (TEL) aims to improve telecommunications and information infrastructure in the Asia-Pacific region by developing and implementing appropriate telecommunications and information policies, including relevant human resource and development cooperation strategies.'

http://www.apectelwg.org

CERIAS: Center for Education and Research in Information Assurance and Security – Purdue University, USA

'The Center for Education and Research in Information Assurance and Security (CERIAS) is currently viewed as one of the world's leading centers for research and education in areas of information security that are crucial to the protection of critical computing and communication infrastructure. CERIAS is unique among such national centers in its multidisciplinary approach to the problems ... '

http://www.cerias.purdue.edu

CSIS: Center for Strategic and International Studies

'A bipartisan, nonprofit organisation headquartered in Washington, DC, CSIS conducts research and analysis and develops policy initiatives that look into the future and anticipate change.... Understanding the many issues related to building a comprehensive cyber security strategy that recognises the threat as one of the major national security problems facing the United States while respecting American values related to privacy and civil liberties.'

http://csis.org/category/topics/technology/cybersecurity

CSPRI: Cyber Security Policy and Research Institute, George Washington University, Washington, DC, USA

'...to promote technical research and policy analysis of problems that have a significant computer security and information assurance component. CSPRI's mission is to encourage, promote, facilitate, and execute interdisciplinary research in these areas, including the exploration of the norms, governance issues, and operating systems of cyberspace.'

http://www.cspri.seas.gwu.edu

CYLAB-CMU: CyLab – Carnegie Mellon University, USA

'CyLab was founded in 2003 and is one of the largest university-based cyber security research and education centers in the U.S.'

http://www.cylab.cmu.edu

FIRST: Forum of Incident Response and Security Teams

'FIRST is the premier organization and recognized global leader in incident response...FIRST brings together a variety of computer security incident response teams from government, commercial, and educational organizations.'

http://www.first.org

IANA: Internet Assigned Numbers Authority

'The Internet Assigned Numbers Authority (IANA) is responsible for the global coordination of the DNS Root, IP addressing, and other Internet protocol resources.'

IANA is responsible for some of the key elements that keep the Internet running smoothly, e.g. domain names, number resources and protocol assignment.

http://www.iana.org

ICANN: Internet Corporation for Assigned Names and Numbers

'ICANN was formed in 1998. It is a not-for-profit public-benefit corporation with participants from all over the world dedicated to keeping the Internet secure, stable and interoperable. It promotes competition and develops policy on the Internet's unique identifiers.'

ICANN does not control content on the Internet or access to it.

http://www.icann.org

OECD: Organisation for Economic Cooperation and Development

'The mission of the Organisation for Economic Co-operation and Development (OECD) is to promote policies that will improve the economic and social well-being of people around the world. The OECD provides a forum in which governments can work together to share experiences and seek solutions to common problems.'

50 years old as in 2011. Australia and India are both members.

http://www.oecd.org

RAND-NSRD: Rand Corporation – National Security Research Division, USA

'The RAND Corporation is a nonprofit institution that helps improve policy and decision making through research and analysis.'

'The RAND National Security Research Division (NSRD) conducts research and analysis for all national security sponsors other than the U.S. Air Force and the Army.'

http://www.rand.org/nsrd.html

US-CCU: The U.S. Cyber Consequences Unit

'The U.S. Cyber Consequences Unit (US-CCU) is an independent, non-profit (501c3) research institute. It provides assessments of the strategic and economic consequences of possible cyber-attacks and cyber-assisted physical attacks. It also investigates the likelihood of such attacks and examines the cost-effectiveness of possible counter-measures...The reports and briefings the US-CCU produces are supplied directly to the government, to entire critical infrastructure industries, and to the public ...'

Now associated with the Fletcher School of Law and Diplomacy at Tufts University. Scott Borg is its Director and Chief Economist. He founded the US-CCU at the request of senior government officials, who wanted an independent, economically oriented source of cyber-security research.

http://www.usccu.us

Index

S.V. Raghavan and E. Dawson (eds.), *An Investigation into the Detection and Mitigation of Denial of Service (DoS) Attacks: Critical Information Infrastructure Protection*, DOI 10.1007/978-81-322-0277-6, © Springer India Pvt. Ltd. 2011

MIX
Papier aus verantwortungsvollen Quellen
Paper from responsible sources
FSC® C105338

If you have any concerns about our products,
you can contact us on
ProductSafety@springernature.com

In case Publisher is established outside the EU,
the EU authorized representative is:
Springer Nature Customer Service Center GmbH
Europaplatz 3, 69115 Heidelberg, Germany

Printed by Libri Plureos GmbH
in Hamburg, Germany